THEORY OF GENERAL ECONOMIC EQUILIBRIUM

THEORY OF GENERAL ECONOMIC EQUILIBRIUM

Hans Keiding
University of Copenhagen, Denmark

NEW JERSEY · LONDON · SINGAPORE · BEIJING · SHANGHAI · HONG KONG · TAIPEI · CHENNAI · TOKYO

Published by

World Scientific Publishing Co. Pte. Ltd.
5 Toh Tuck Link, Singapore 596224
USA office: 27 Warren Street, Suite 401-402, Hackensack, NJ 07601
UK office: 57 Shelton Street, Covent Garden, London WC2H 9HE

Library of Congress Cataloging-in-Publication Data
Names: Keiding, Hans, 1945–   author.
Title: Theory of general economic equilibrium / Hans Keiding, University of
   Copenhagen, Denmark.
Description: Singapore ; Hackensack, NJ : World Scientific Publishing Co.
   Pte. Ltd., [2020] | Includes bibliographical references and index.
Identifiers: LCCN 2020006477 | ISBN 9789811214387 (hardcover) |
   ISBN 9789811214394 (ebook)
Subjects: LCSH: Equilibrium (Economics) | Macroeconomics.
Classification: LCC HB145 .K45 2020 | DDC 339.5--dc23
LC record available at https://lccn.loc.gov/2020006477

British Library Cataloguing-in-Publication Data
A catalogue record for this book is available from the British Library.

Copyright © 2020 by World Scientific Publishing Co. Pte. Ltd.

All rights reserved. This book, or parts thereof, may not be reproduced in any form or by any means, electronic or mechanical, including photocopying, recording or any information storage and retrieval system now known or to be invented, without written permission from the publisher.

For photocopying of material in this volume, please pay a copying fee through the Copyright Clearance Center, Inc., 222 Rosewood Drive, Danvers, MA 01923, USA. In this case permission to photocopy is not required from the publisher.

For any available supplementary material, please visit
https://www.worldscientific.com/worldscibooks/10.1142/11668#t=suppl

Desk Editors: Balasubramanian/Shreya Gopi

Typeset by Stallion Press
Email: enquiries@stallionpress.com

Preface

The economic theory of general equilibrium has its origin in the interests of the neoclassical economists from about 1870 and onwards with the determination of market prices from demand and supply. Since demand and supply for any single commodity depends on what happens in the other markets, it was necessary to consider simultaneous equilibria in all the markets. The welfare theoretical aspects of market equilibria were developed in the 1930s, but the real breakthrough came in the 1950s with the use of an updated mathematical toolbox. The theory had its golden days in the sixties and seventies. After this, the number of remarkable new insights became fewer, researchers felt less attracted to the field, and it slowly began to be filtered out of the curriculum in many universities. However, general equilibrium theory remains very much alive and, much if not most, of contemporary economic science draws upon it in one way or another.

The mathematical appearance of the theory has attracted criticism from the less mathematically oriented part of the procession, which over time has accused general equilibrium theory of many sins, from outright logical failures over misconception of economics, and in the later stages of mainly being a failure with no definite and precise explanation of what happens in markets. Thus, the Sonnenschein–Debreu result that all functions can be aggregate excess demand functions of a suitably defined set of consumers (to be described in Chapter 6, Section 3) is taken as stating that the theory is empty since it cannot single out a particular class of excess demand functions as compatible with equilibrium in the market (cf. e.g. Ackerman, 2002). In our context, the result is less devastating, showing as it does that the stability of dynamical approaches toward equilibrium in a market does not come about by itself but must be based on other features. Indeed, much of the general equilibrium theory does exactly this, it points out that nothing miraculous occurs in the market, any particular phenomenon must be explained by particular circumstances rather than being just a general outcome of "market forces".

This controversy may have roots in the differences in what constitutes a "theory" in mathematics and in economics (also noted by earlier contributors to the theory, e.g. Gale (1960)). The abundance of different vector spaces does not mean that the theory of vector spaces is a failure; on the contrary, it means that it emphasizes the importance of the theory. Perhaps the general equilibrium theory should be considered as establishing a framework for the treatment of allocation problems in society, investigating what can be concluded in broad generality and where specific assumptions will be needed in order to obtain definite descriptions of what will happen.

The book consists of 16 chapters plus an introductory chapter, which sumps up some basic facts of microeconomics and introduces the notation to be used throughout. Then Chapters 1 and 2 deal with equilibria in exchange economies and economies with production, where we have included sections on linear economic models and international trade. The classical welfare theory is considered in Chapter 3, followed by a chapter on solutions inspired from game theory. This takes us to a treatment of large economies in Chapter 5, involving largeness on the side of agents as well as largeness in the sense of on infinity of commodities. The following chapters take us back to standard topics of equilibrium theory, namely uniqueness and stability in Chapters 6 and 7, where we touch upon some of the more advanced topics in this field. In Chapter 8, the standard competitive market is replaced by imperfect competition in some way or the other, and Chapter 9 deals with questions of incentives and implementation in the context of allocation of commodities. Then Chapter 10 resumes the discussion of Chapter 3, dealing now with market failures such as externalities, public goods and increasing returns to scale.

The remaining chapters deal with equilibria over time and under uncertainty. Markets for future delivery as well as temporary equilibria are introduced in Chapters 11 and 12 deals with overlapping generations. In Chapter 13, we introduce uncertainty and contingent commodity markets together with a brief discussion of incompleteness of markets and of sunspot equilibria, Chapter 14 discusses financial markets, and in Chapter 15, we consider equilibria with incomplete markets first with financial assets and then with real assets. The concluding Chapter 16 takes up a topic largely neglected in all previous chapters, discussing the role of money in general equilibrium.

The text consists of material which has been used in teaching, together with newer material extending the scope of the text and improving its up-to-dateness. I am grateful to Bodil O. Hansen for his valuable advise and suggestions for improvement during the process of writing this text.

About the Author

 **Hans Keiding** is a Professor of economics at the University of Copenhagen. He has published in general equilibrium and in game theory as well as in other fields of economics. He is the author of several textbooks, such as *Game Theory: A Comprehensive Introduction* (2015) and *Theoretical Health Economics* (2018).

Contents

Preface   v

Introduction   1

    1   The Commodity Space   1
    2   Consumers   3
    3   Producers 10
    4   Markets 14
    5   Correspondences 17
    6   Exercises 19

1.   Exchange Economies   21

    1   Economies and Allocations in the Economy 21
    2   Existence of Equilibria 27
    3   The Survival Condition and Hierarchic Equilibria . . . 36
    4   Admissible Preferences 40
    5   Fixed Point Theorems for General Equilibrium
       Theory 45
    6   Exercises 50

2.   Economies with Production   53

    1   Walras Equilibria in Economies with Production . . . 54
    2   Linear Models of Production and Growth 59
    3   International Trade and Factor Price Equalization . . . 73
    4   Exercises 80

xii   *Contents*

3. Welfare Theorems and Market Failures   83

    1   Pareto-optimality   83
    2   Pareto-optimality and Social Optimum   98
    3   Quantifying the Lack of Efficiency   105
    4   Separation Theorems for Convex Sets   108
    5   Exercises   111

4. Cooperative Game Theory and Equilibria   115

    1   The Core of an Economy   115
    2   Allocations Induced by Other Game Theoretic
        Solutions   125
    3   Fuzzy Games and Equilibria   135
    4   Exercises   138

5. Large Economies   141

    1   Economies with Many Agents   141
    2   Economies with Many Commodities   158
    3   Large-square Economies   171
    4   Exercises   172

6. Uniqueness and Stability of Equilibria   175

    1   The Uniqueness Problem   175
    2   Stability   186
    3   Excess Demand Functions   190
    4   Exercises   196

7. The Equilibrium Manifold and Probabilistic
    Equilibrium Theory   199

    1   The Equilibrium Manifold   199
    2   Probabilistic Equilibrium Theory   207
    3   Exercises   231

8.	General Equilibrium and Imperfect Competition		233
	1	The Subjective Demand Approach	233
	2	Equilibrium with Objective Demand	239
	3	Exercises	248
9.	Incentives and Mechanisms in General Equilibrium		251
	1	Implementing Economic Equilibria	251
	2	Strategic Market Games	253
	3	Hurwicz's Theorem and the Walras Correspondence	256
	4	Allocation Processes and Information	258
	5	The VCG Mechanism and Equilibrium Allocations	266
	6	Exercises	271
10.	Market Failures		275
	1	Introduction: The Classical Market Failures	275
	2	External Effects	276
	3	Public Goods	284
	4	Natural Monopolies	291
	5	Exercises	295
11.	Time and General Equilibrium		299
	1	Time and Economic Activity	299
	2	Temporary Equilibrium	305
	3	Exercises	310
12.	Overlapping Generations Economies		313
	1	The Overlapping Generations (OLG) Model	313
	2	Pareto Optimality of Equilibria Over Time	317
	3	Indeterminacy of Walras Equilibrium in the OLG Model	324
	4	Cycles in the OLG Model	328
	5	Exercises	331

xiv   *Contents*

13. **Uncertainty and General Equilibrium**   333

    1   Introduction 333
    2   Contingent Commodities 334
    3   Equilibrium of Plans, Prices and Price Expectations . . 337
    4   Sunspot Equilibria 345
    5   Exercises 351

14. **Financial Markets**   355

    1   One-Period Models of Financial Equilibrium 355
    2   Dynamic Models of Financial Markets 370
    3   Exercises 374

15. **Economies with Incomplete Markets**   377

    1   Equilibria in Economies with Financial Assets 377
    2   Pseudoequilibria in Economies with Real Assets . . . 380
    3   Exercises 395

16. **General Equilibrium and Money**   399

    1   Money and Decentralized Exchange 399
    2   Cybercurrencies and Blockchain 409
    3   Money and Memory 412
    4   Exercises 418

Bibliography   421

Index   429

Introduction

This chapter briefly reviews some of the basic concepts of general equilibrium theory. It draws on standard microeconomics, in particular consumer theory and the theory of the firm, and in addition, it introduces some of the mathematical tools to be used repeatedly in the following chapters, such as correspondences and the associated concepts of continuity.

1. The Commodity Space

The general equilibrium model deals with *markets* and with exchange of *commodities* in such markets. A first step in developing such a theory will be to establish the notion of commodities, the objects which are exchanged between agents or allocated to agents. This is at the same time trivially simple, given that we are used to handling commodities in our private lives, and somewhat complicated, since we need a notion which can also be used in applications of our theory which does not immediately suggest themselves. As a consequence, we place more effort in emphasizing what distinguishes one type of commodity from another than in explaining what exactly constitutes a commodity.

First of all, commodities may be material goods or they may be in the form of services delivered. In any case, the quality of the good or service delivered should be fully specified, so that two different qualities of a good will be two different commodities. What, then,

1

constitutes the "quality" of goods and services, or rather, how can we determine that they are different? The general answer is that as soon as the difference matters to one agent in the economy, it causes the commodities to be different.

Such differences will show up in many of the models to be considered, so here we mention only a few. One of the important cases is that of *dated commodities* — if the time of delivery matters, then this is a relevant criterion for distinguishing among commodities, so that the same good delivered at different dates will be considered as different commodities, each with a price of its own. This corresponds to what is encountered in real-life trading in forward contracts and futures. We return to this when dealing with allocation over time in Chapter 11. In the same direction, delivery contingent on a uncertain event in the future will give rise to distinct commodities, something which will be developed further in Chapter 13.

With this notion of a commodity, it comes as no surprise that the number of distinct commodities must be quite large. We assume that there are l different commodities, which means that the number is finite. As we proceed, we shall see that the assumption of finitely many commodities will be too restrictive in some situations, so we shall also consider models with infinitely many commodities. However, the insights gained from models satisfying our assumption will be useful even there.

We shall assume several properties of commodities. One of the more fundamental is that commodities can be available — at least in principle — in all possible quantities; in particular, we assume that they are arbitrarily divisible. Consequently, a collection, or, in the terminology to be used from now, *bundle* of commodities is an element of $\mathbb{R}^l$, the l-dimensional Euclidean space consisting of vectors $x = (x_1, \ldots, x_l)$ with l real coordinates. We say that the *commodity space* is $\mathbb{R}^l$. As mentioned above, we shall occasionally use commodity spaces other than $\mathbb{R}^l$, and this will be duly emphasized in each case.

Since we model the commodity space as $\mathbb{R}^l$, l-dimensional Euclidean space, it will be endowed with the Euclidean norm $\| \cdot \|$

Introduction 3

(defined by $\|x\| = (x_1^2 + \cdots + x_l^2)^{1/2}$) and an inner product $\cdot$, whereby

$$x \cdot x' = \sum_{h=1}^{l} x_h x'_h.$$

We shall use the following notational conventions: The positive orthant in $\mathbb{R}^l$, that is the set $\{x \in \mathbb{R}^l \mid x_h \geq 0, h = 1, \ldots, l\}$, is denoted $\mathbb{R}^l_+$. We shall use the notation $\geq$ on vectors in $\mathbb{R}^l$ defined by

$$x \geq x' \Leftrightarrow x - x' \in \mathbb{R}^l_+,$$

so that $x \geq x'$ is equivalent to $x_h \geq x'_h$ for all h. In some contexts, it will be convenient to also use the notation $\mathbb{R}^l_{++}$ for $\{x \in \mathbb{R}^l \mid x_h > 0, \text{ all } h\}$, and $\mathbb{R}^l_-$ for $-\mathbb{R}^l_+$, the set of vectors having all non-positive coordinates.

2. Consumers

In this section, we briefly review the microeconomic theory of the consumer, leaving out topics which are of minor relevance for general equilibrium. Also, we concentrate on the fundamental aspects, avoiding details which rely on special functional forms or differentiability properties of utility functions.

2.1 *Consumption sets*

The role of the consumer in a model of general economic equilibrium is self-evident: the consumer must carry out a certain consumption plan, and the choices made according to this plan should reflect the conditions of the market. Formally, a consumption plan is a commodity bundle $x = (x_1, \ldots, x_l)$, where the kth coordinate specifies the amount of commodity k made available to the consumer in question.

Not every commodity bundle may be eligible as a consumption plan. On the intuitive level, one would exclude vectors with negative coordinates. However, it may be convenient to allow for negative consumption of some commodities, for example when labor delivered by the consumer is considered as negative consumption of the commodity "leisure". We introduce the *consumption set* X as the set of all commodity bundles which are feasible as consumption plans for the consumer in question. Thus, X is a subset of $\mathbb{R}^l$, and its elements $x \in X$ are called *feasible* consumption plans.

What does a consumption set look like? Clearly, this will depend on the model at hand, but most of the consumption sets that occur in general equilibrium models have some basic properties of well-behavedness, which we state and comment upon as follows.

Assumption 0.1. A consumption set X is well-behaved if it has the following properties:

(a) X is non-empty,
(b) X is closed,
(c) X is convex,
(d) X is bounded from below, i.e. there is $b \in \mathbb{R}^l$, such that $x_h \geq b_h$, $h = 1, \ldots, l$, for all $x \in X$,
(e) X is upper comprehensive, i.e. $X + \mathbb{R}^l_+ \subseteq X$ (if $x^1 \in X$ and $x^2 \in \mathbb{R}^l$ satisfies $x^1_h \leq x^2_h$, for all h, then $x^2 \in X$).

As always, assumptions should be assessed in terms of their restrictiveness. Most of the items in (a)–(e) seem rather harmless, but in any case, we take a closer look.

Property (a) needs no comment. If X is empty, there are no consumption plans to consider. Part (b) on closedness means that if a sequence $(x^n)_{n=1}^\infty$ from X is convergent, then the limit must belong to X as well. Obviously, this property cannot be verified empirically, it is technical in the sense that it facilitates the presentation of the results, which otherwise could be more cumbersome and less intuitive. The convexity of X means that if $x^1, x^2 \in X$ are feasible consumptions, and λ is a real number with $0 \leq \lambda \leq 1$, then $\lambda x^1 + (1 - \lambda)x^2$, which is a weighted (vector) average of the two feasible consumption plans, should also be feasible.

How reasonable is this assumption of convexity? Clearly, it implies that all commodities are divisible into arbitrarily small units, so if we want to model the (unfortunately very realistic) feature of *indivisibility* of certain goods, then (c) cannot be fulfiled. In most, indeed almost all, situations considered in what follows, we consider such indivisibility as a phenomenon of secondary importance in relation to the main features of the model, so that Assumption 0.1 may be taken as at least approximately satisfied.

Part (d) of the assumption puts a bound on the amount of negative consumption which can be sustained by the consumer, and this does not seem to be controversial. Finally, part (e) implies that the consumer is technically able to consume more of all commodities no matter which feasible consumption plan we are actually considering. Allowing for a wide interpretation of consumption, possibly including the act of throwing away commodities, the present assumption is seen to be essentially one of *free disposal*, namely the assertion that disposal is technically feasible. It may be too expensive or generally undesirable to throw away goods, but this is another matter to be considered in due course.

2.2 Consumer preferences

The consumer must choose from the consumption set X. This choice is assumed to be guided by a *preference relation* on the consumption set. Formally, a preference relation is a set of pairs $(x, x') \in X \times X$, with the interpretation that the consumption bundle x' is considered as better than the consumption bundle x. It turns out to be convenient to represent the preference relation as a *preference correspondence* $P : X \rightrightarrows X$, which to each consumption bundle $x \in X$ assigns the set $P(x)$ of all consumption bundles preferred to x (for more about correspondences, see Section 5). For $P : X \rightrightarrows X$ a preference correspondence, we get the underlying preference relation as the graph of P, the set $\text{Graph}(P) = \{(x, x') \mid x' \in P(x)\}$.

In contrast to what was the case for the consumption set, there is no standard set of assumptions which together define a well-behaved preference correspondence. Different properties may be needed depending on the problems considered. However, some

6   *Theory of General Economic Equilibrium*

classes of preferences show up quite often, the one introduced here is an example of such a class.

Assumption 0.2. A preference correspondence P is standard if Graph P is open (in $X \times X$) and for all $x \in \mathbb{R}_+^l$,

(i) $x \notin P(x)$ (irreflexivity),
(ii) $P(x)$ is non-empty (non-satiation),
(iii) $P(x)$ is convex.

The class of standard preference correspondences is denoted $\mathcal{P}^K$. The property of irreflexivity is the formal counterpart of what is intuitively understood by one bundle being preferred to, or considered better than, another bundle, and as such it causes no trouble. The same holds for the non-satiation property, which is not always needed. The convexity assumption, on the other hand, is one which we would rather be without, there is no compelling reason why a convex combination of two bundles preferred to x should also be preferred to x. As we shall see in the chapters to follow, the assumption is needed for most of our results.

Apart from the properties satisfied by standard preferences, we may occasionally need some additional ones, restricting the generality of our treatment but allowing stronger results. Indeed, traditional general equilibrium theory was developed under the assumption that the preference relation of the consumer can be described by a utility function: Let $\succsim$ be the relation on X defined from the preference correspondence P as

$$\succsim = \{(x, x') \in X \times X \mid x' \in P(x) \text{ or } x \notin P(x')\},$$

(in the sequel, we write $x' \succsim x$ instead of $(x, x') \in \succsim$) with the interpretation that $x' \succsim x$ if x' is at least as good as x. The relation $\succsim$ is described by a utility function $u : X \to \mathbb{R}$ if

$$x' \succsim x \Leftrightarrow u(x') \geq u(x),$$

so that statements about preferences can be translated to statements about values of the function u.

The basic result about preferences represented by utility functions is due to Debreu (1959): A preference relation $\succsim$ on a subset X of $\mathbb{R}^l$ can be represented by a continuous utility function u if

(a) $\succsim$ is a preorder, i.e. $\succsim$ satisfies the conditions

(i) $\forall x \in X, x \succsim x$ (irreflexivity),
(ii) for $x, x', x'' \in X$, if $x'' \succsim x'$ and $x' \succsim x$, then $x'' \succsim x$ (transitivity),

(b) for $x, x' \in X$, either $x \succsim x'$ or $x' \succsim x$ ($\succsim$ is complete),
(c) for all $x \in X$, the upper and lower level sets ($\{x' \in X \mid x' \succsim x\}$ and $\{x' \in X \mid x \succsim x'\}$, respectively) are closed in X.

We shall not use utility representations of preferences except for examples and some particular cases.

2.3 Consumer demand

Given a consumer with characteristics (X, P) consisting of a consumption set X and a preference correspondence P, a preferred choice from any subset X' of X is an element $x^0 \in X'$ which is *maximal for P on X'* in the sense that $P(x^0) \cap X' = \emptyset$.

Theorem 0.1. *Let $X' \subset X$ be non-empty, convex and compact, and let P be a standard preference correspondence. Then there is an element $x^0 \in X'$ which is maximal for P on X'.*

The proof exploits a fixed-point argument for correspondences, and it is postponed to Chapter 1. We note that the convexity assumption on X' can be dispensed with if the preference relation derived from P has a continuous utility representation, since in this case the problem reduces to finding a maximum of a continuous function on a compact set. Fortunately, the subsets X' occurring in equilibrium analysis are typically convex.

Indeed, if the consumer has access to a market given by a price system $p \in \mathbb{R}^l_+ \setminus \{0\}$ and is endowed with an income w to be used for purchases in the market, then the relevant choice set is the *budget set*

$$\gamma(p, w) = \{x \in X \mid p \cdot x \leq w\},$$

8 *Theory of General Economic Equilibrium*

Box 1. Differentiable utility and demand: In some cases to be considered later, it will be assumed that consumer demand can be expressed as a differentiable function of prices and income. This will be the case if the preferences can be represented by utilities with suitable properties.

Assume that the utility function u describing the preferences is twice continuously differentiable (C^2) in every point $x \in$ int X and that the matrix

$$\begin{pmatrix} u'' & u' \\ (u')^t & 0 \end{pmatrix},$$

has full rank $l + 1$ (here, u'' is the matrix of second partial derivatives of u, u' is the gradient of u, and the notation $\cdot^t$ stands for the transposition).

If x maximizes utility under the budget constraint $p \cdot x = w$, then it satisfies the first order conditions for a constrained maximum,

$$u' - \lambda p = 0,$$
$$p \cdot x - w = 0,$$

where $\lambda \neq 0$ is a Lagrangian multiplier. Writing this equation system as

$$F(x, \lambda, p, w) = 0,$$

we see that under the above assumption on u, we may apply the Implicit Function Theorem to get that the solution (x, λ) can be written locally as a C^1 function f of (p, w) with

$$Df = -(D_{(x,\lambda)}F)^{-1} D_{(p,w)}F,$$

where

$$D_{(x,\lambda)}F = \begin{pmatrix} u'' & p \\ p^t & 0 \end{pmatrix} = \begin{pmatrix} u'' & \lambda^{-1}u' \\ (\lambda^{-1}u')^t & 0 \end{pmatrix},$$

has full rank. This means that in particular ξ is a C^1 function for all (p, w) such that $p \in$ int X and $w > \min\{p \cdot x \mid x \in X\}$.

The reasoning above may be used also to obtain demand functions that are C^∞ given that the utility functions are C^∞ and satisfy the rank condition.

(here $\cdot$ denotes the interior product in $\mathbb{R}^l$, so that $p \cdot x = \sum_{h=1}^{l} p_h x_h$). The set of maximal elements for P in the budget set is the *demand* at prices p and income w,

$$\xi(p, w) = \{ x \in \gamma(p, w) \mid P(x) \cap \gamma(p, w) = \emptyset \}.$$

Introduction 9

We collect some useful results about the consumer's demand in a theorem. For this, we introduce the set

$$M_w = \left\{ (p, w, x) \in \mathbb{R}_{++}^l \times \mathbb{R} \times X \,\middle|\, w > \min_{x \in X} p \cdot x \right\}$$

of combinations of price, income and consumption for which the value of the bundle exceeds its minimum over the consumption set.

Theorem 0.2. *Assume that X satisfies Assumption 0.1 and P satisfies Assumption 0.2.(i)–(ii), and let $p \in \mathbb{R}_+^l$. If the minimum-wealth condition $w \geq \min_{x \in X} p \cdot x$ is satisfied, then the following hold:*

(i) *If $p \in \mathbb{R}_{++}^l$, then the demand set $\xi(p, w)$ is non-empty.*

(ii) *If P satisfies local non-satiation ($P(x) \cap U \neq \emptyset$ for each $x \in X$ and each neighborhood U of x), then there is $x^0 \in \xi(p, w)$ with $p \cdot x^0 = w$.*

(iii) *For all $(p, w, x) \in M_w$, if $p \cdot x = w$ and $P(x) \cap \gamma(p, w) \neq \emptyset$, then $P(x)$ contains elements x' of X with $p \cdot x' < w$.*

(iv) *The set $\{(p, w, x) \mid x \in \xi(p, w)\}$ is a closed subset of M_w.*

Proof. (i) Follows immediately from Theorem 0.1. To show (ii), we note that if $P : X \rightrightarrows X$ is standard, then so is also the correspondence $\widehat{P} : X \rightrightarrows X$ defined by

$$\widehat{P}(x) = \{\lambda x + (1 - \lambda)x' \mid x' \in P(x), 0 < \lambda \leq 1\}.$$

Moreover, each neighborhood of x contains a point $x'' \in \widehat{P}(x)$ ($\widehat{P}$ has the property of local non-satiation). Applying Theorem 0.1 to $\widehat{P}$, we get that there is $x^0 \in X$ such that $\widehat{P}(x^0) \cap \gamma(p, w) = \emptyset$, and by our construction, we must have that $p \cdot x^0 = w$. Since $P(x) \subseteq \widehat{P}(x)$, we have that $x^0 \in \xi(p, w)$.

The statement (iii) is an immediate consequence of the openness of $P(x)$. For (iv), we use that the sets $\{(p, w, x) \mid p \cdot x > w\}$ and $\{(p, w, x) \mid P(x) \cap \{x' \mid p \cdot x' < w\} \neq \emptyset\}$ are open in M_w and that $\{(p, w, x) \mid x \in \xi(p, w)\}$ is the complement of their union. $\square$

The properties stated in Theorem 0.2 are standard in the microeconomic theory of the consumer. In particular, property (ii) is known as the equivalence of preference maximization and expenditure minimization. Property (iv) is usually proved by reference to Berge's maximum theorem, to be mentioned in Section 5, but can also be established without reference to correspondences. The special case of differentiable utilities and demand is considered in Box 1.

In the following chapters, we shall make extensive use of the demand $\xi(p, w)$ of a consumer at given prices p and income w. In the context of general equilibrium, the income w is usually derived from some other components of the model. The simplest case is the one where the consumer is endowed with an initial bundle of goods $\omega \in \mathbb{R}_+$ which is sold in the market, giving the amount $w = p \cdot \omega$. If $x \in \xi(p, p \cdot \omega)$, then the change from initial to final bundle of goods is the *net trade* of the consumer, and we can define the *excess demand* of the consumer as

$$\zeta(p) = \{x - \omega \in \mathbb{R}^l \mid x \in \xi(p, p \cdot \omega)\}. \tag{1}$$

If the set $\xi(p, p \cdot \omega)$ contains only a single element, we may consider ξ and ζ as functions, which under the assumptions of Theorem 0.2 (iv) will be continuous.

3. Producers

Our treatment of producers in this introductory chapter will be similar to that of consumers in Section 2, we describe the fundamental characteristics of a productive unit in the economy. We begin our analysis by describing a producer, and then we introduce the producer's optimal choices at given prices.

3.1 *Production sets*

In general, a production may be described by two commodity bundles, a bundle $a = (a_1, \ldots, a_l)$ of inputs and a bundle $b = (b_1, \ldots, b_l)$ of outputs of the l commodities. Although this approach will be used at some instances in the chapters to follow, for most of the purpose of our theory, it suffices to consider the net production $y = b - a$. From now on, a *production* $y = (y_1, \ldots, y_l)$ is a commodity bundle where the kth coordinate specifies the net production of commodity k, $k = 1, \ldots, l$. It follows that the kth coordinate is negative if the production process uses more of commodity k than is produced — that is to say, if k is a (net) input in the production — and it is positive if k is a (net) output.

We characterize the producer by the *production set* $Y \subset \mathbb{R}^l$ consisting of all productions which are technically feasible. For $l = 2$, one can visualize the shape of Y directly as shown in Box 2.

In the present general context, there is a natural partial order defined on productions: If $y^1, y^2 \in Y$ are two feasible productions with $y_k^2 \geq y_k^1$ for all k, then y^2 yields as much output as y^1 and uses no more input than y^1, so from an economic point of view, y^2 is at least as good as y^1 and possibly better. When choosing from Y, it seems unreasonable to select a production which is dominated in the above sense. This leads to the following definition.

Box 2. Production sets in low dimensions: For a two-dimensional illustration of production sets, one has to keep input or output of most commodities fixed, leaving two to be subject to variation. In Fig. 1, the left panel shows the case where commodity 1 is input and commodity 2 is output, feasible productions are all points below the graph. To the right, both commodities are output. The northeast boundary is called a *transformation curve*.

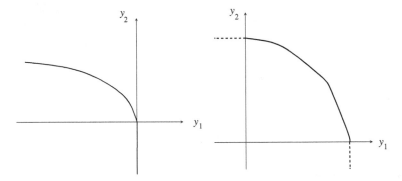

Fig. 1. Production sets illustrated in two dimensions.

We leave it to the reader to sketch the case where both commodities are inputs.

Definition 0.1. A production $y \in Y$ is efficient if there is no other $y' \in Y$ and $y' \geq y$ implies $y' = y$.

Since efficient productions suggest themselves as possible choices in a context where resources are scarce, it would be convenient that such productions exist, indeed that each feasible production should be dominated by an efficient production. This property will be achieved if the production set satisfies the following standard assumptions of well-behavedness.

Assumption 0.3. The production set $Y \subset \mathbb{R}^l_+$ satisfies the following assumptions:

  (i) Y is closed,
 (ii) $0 \in Y$ (possibility of inaction),
(iii) $Y + \mathbb{R}^l_- \subseteq Y$ (free disposal),
 (iv) for each $y \in Y$, the set $\{y' \in Y \mid y' \geq y\}$ is bounded,
  (v) Y is convex.

It is straightforward that the boundedness assumption (iv) together with closedness of Y (i) guarantees the existence of efficient productions dominating any $y \in Y$. Also, boundedness (iv) together with (ii) secures that there is an upper bound to free production.

The convexity assumption (v) is one which is rather restrictive, since it rules out increasing returns to scale. We return to this problem later, considering models where some but not all of the items (i)–(v) are satisfied.

A production set exhibits constant (*non-decreasing, non-increasing*) *returns to scale* if for all $y \in Y$ and $\lambda \geq 0$ ($\lambda \geq 1$, $\lambda \in [0,1]$), $\lambda y \in Y$. The presence of non-decreasing returns to scale may lead to violation of Assumption 0.3, which poses some problems to the workings of competitive markets, so it will be ignored for the time being and considered separately in Chapter 10.

If Y_1 and Y_2 are production sets, and production plans are chosen in both, $y_j \in Y_j$, $j = 1, 2$, then $y_1 + y_2$ expresses the production plan obtained if y_1 and y_2 are carried out simultaneously. The set $Y_1 + Y_2$,

Introduction 13

defined as

$$Y_1 + Y_2 = \left\{ y = y_1 + y_2 \,\middle|\, y_1 \in Y_1, y_2 \in Y_2 \right\} \qquad (2)$$

(known as the Minkowski sum of the two sets Y_1 and Y_2), gives us all such simultaneous productions.

3.2 _Production functions_

If the production set Y is convex, then there is an obvious candidate for such a function, namely the indicator function $\delta(\cdot|Y)$ which takes the value 0 outside Y and $+\infty$ on Y (cf. e.g. Rockafellar, 1970). However, the traditional choice of an indicator function for a production set is a function $f_Y : \mathbb{R}^l \to \mathbb{R}$ with the property that

$$f_Y(y) = \begin{cases} s \leq 0 & y \in Y, \\ 0 & y \in Y \text{ and } y \text{ efficient}, \\ r > 0 & y \notin Y. \end{cases}$$

The textbook version of a production function is however not this indicator function. It is used in cases where one commodity, say commodity 1, always occurs as an output, whether the remaining commodities are inputs. In this case, one may express the maximal obtainable output given inputs $y_{-1} = (y_k)_{k \neq 1}$ as $g(y_{-1})$, where

$$g(y_{-1}) = \max\{y_1 \mid f(y_1, y_{-1}) \leq 0\}.$$

We shall make use of the type of one-commodity production functions only occasionally, mainly in examples.

3.3 _Choice of production at given prices_

When commodities can be bought and sold in the market at prices $p \in \mathbb{R}^l_+ \backslash \{0\}$, the domination relation induced on Y by the relation $\geq$ on $\mathbb{R}^l$ can be further refined: Introducing the _profit_ derived from a production y at prices p as $p \cdot y$, one immediately gets that $y' \geq y$ implies that $p \cdot y' \geq p \cdot y$ for all $y, y' \in Y$. We also note that productions $y^0 \in Y$ are efficient if and only if they maximize the profit $p \cdot y$ over

Y for some price $p \in \mathbb{R}_+^l \setminus \{0\}$. This gives an additional explanation of the interest in efficient productions.

The assumption that producers choose profit-maximizing productions from their production set is standard in general equilibrium theory, and we shall adhere to it in most of the chapters to follow. Given that producers are not considered as individuals endowed with preferences of their own, but rather as derived agents whose decisions are based on preferences of some other agents in control of production, the assumption is somewhat unsatisfactory. It may be added that it seems quite controversial that real-world producers should be profit maximizers. Be this as it may, the assumption is one which makes the reasoning simple and transparent, and bearing in mind the role of the general equilibrium model as a benchmark rather than a realistic picture of the economy, we shall be happy with it most of the time, leaving some more fundamental reservations to later chapters.

When Y satisfies Assumption 0.3, profit maximization at different prices will describe the production set fully, at least for our purposes. Indeed, the support function $\delta^*(\cdot|C) : \mathbb{R}^l \to \mathbb{R}$ of the convex set $C \subset \mathbb{R}^l$ is defined as

$$\delta^*(p|C) = \min\{p \cdot c \mid c \in C\},$$

and $-\delta^*(-p|Y)$ will give the maximal profit for any $p \in \mathbb{R}_+^l$. The support function $\delta^*(\cdot|Y)$ is the conjugate of the indicator function $\delta(\cdot|Y)$ mentioned above, meaning that it satisfies the equation

$$\delta^*(p|Y) = \sup_{y \in \mathbb{R}^l} [p \cdot y - \delta(y|Y)],$$

and this conjugacy is exploited in the duality theory of cost and utility functions.

4. Markets

Markets play a fundamental role in general equilibrium theory. We have already considered choices from a market given by a price system, and this type of market will be a recurrent theme in the sequel. There are, however, other types of markets which are important depending on the context, and it may be useful to consider markets in a more general setting.

Before introducing general markets, we introduce the even more basic notion of an *exchange*. An exchange between two agents i and j is a vector $x_{ij} \in \mathbb{R}^l$, where the kth coordinate specifies the amount of commodity k transferred from j to i. With this interpretation, we have that to the exchange x_{ij} between i and j there always corresponds an exchange $x_{ji} = -x_{ij}$ between j and i. In practice, only two commodities are used in the exchange, so that most of the components of x_{ij} are 0.

A system of exchanges between m agents is an array $X = (x_{ij})_{i=1\ j=1}^{m\ \ m}$ of exchanges between i, j, for $i, j \in \{1, \ldots, m\}$, satisfying the condition of skew-symmetry introduced above, that is $x_{ij} = -x_{ji}$ for all $i, j \in \{1, \ldots, m\}$, so that in particular $x_{ii} = 0$ for all i. Such a system of exchanges represents a situation where each agent may have performed exchanges with all the other agents. A *repeated system of exchanges*, then, is an array $(X^s)_{s=1}^{r}$ of r systems of exchanges.

If $(X^s)_{s=1}^{r}$ is a repeated system of exchanges, and $\omega_i \in \mathbb{R}^l$ is an initial bundle possessed by agent i, then this bundle will be transformed as a result of the exchanges to

$$x_i = \sum_{s=1}^{r} \sum_{j \neq i} x_{ij}^s.$$

Using the properties of exchanges, it is easily seen that the total number of commodities is unchanged by the exchanges

$$\sum_{i=1}^{m} x_i = \sum_{i=1}^{m} \omega_i,$$

since exchanges as considered here do not use up or create resources.

Exchanges may be subject to additional constraints, such as being subject to a price system $p \in \mathbb{R}^l \setminus \{0\}$, so that $p \cdot x_{ij} \leq 0$ for all i and j (which by skew-symmetry implies that $p \cdot x_{ij} = 0$). Other constraints may occur as well, and this is where general markets enter the picture.

A *market* is a subset $\mathcal{M}^0$ of the commodity space, here taken to be $\mathbb{R}^l$. We may think of the market as being controlled by a *market agent*, who puts $\mathcal{M}^0$ at the disposal of the other agents for a single or repeated choice of exchange. If agents $i_1, \ldots, i_n$ have chosen exchanges $x_{i_j,\mu}$ from $\mathcal{M}^0$, the endowment of the market agent is modified by $\sum_{j=1}^n x_{i_j,\mu}$.

The market $\mathcal{M}^0$ is *anonymous* if all agents are allowed to select elements of $\mathcal{M}^0$, and as many times as desired, so that if $m \in \mathcal{M}^0$, then agents can obtain all net trades λm for $\lambda \in \mathbb{N}$. If the commodities are divisible, and we allow for the possibility that coalitions of agents with r members can trade on behalf of its members and split the trade arbitrarily among its members, then the set of trades open to an agent, denoted $\mathcal{M}$, is the convex cone spanned by $\mathcal{M}^0$.

Further structure on the set of trades obtainable through a market $\mathcal{M}^0$ is obtained if the market sustains an equilibrium (which at present does not need to be specified in detail), since in this case the market agent should be able to substantiate the sum of exchanges chosen, and assuming that the market agent has no initial endowment, this means that $\mathcal{M}$ must intersect the negative orthant,

$$\mathcal{M} \cap \mathbb{R}^l_- \neq \emptyset. \tag{3}$$

Another equilibrium property is that of *no arbitrage:* it should not be possible to attain a non-zero vector from $\mathbb{R}^l_+$ by a suitable trade from the market,

$$\mathcal{M} \cap \mathbb{R}^l_+ = \{0\}. \tag{4}$$

The conditions (3) and (4) are rather weak, and the markets $\mathcal{M}$ are not necessarily of the type usually considered, which are those sustained by a price system,

$$\mathcal{M}_p = \{x \in \mathbb{R}^l_+ \mid p \cdot x = 0\}. \tag{5}$$

For this, one needs some additional assumption of competition between market agents, ruling out that points in $\mathbb{R}_- \setminus \{0\}$ could occur

Introduction 17

in equilibrium, so leaving 0 as the only feasible vector of aggregate trades in equilibrium, and if non-trivial equilibria occur, then $\mathcal{M}$ must be a vector subspace.

The final step toward the form (5) of the market is an assumption on the dimension of the subspace $\mathcal{M}$, namely that dim $\mathcal{M} = l - 1$; one says that there are *complete markets*. Taken together, the assumptions made so far give us the desired form of markets. We shall return to markets incomplete in later chapters, but for most of the theory to be considered, markets will have the form (5).

5. Correspondences

Correspondences, or set-valued mappings, entered the toolbox of mathematical economists with the general equilibrium model. We have encountered them in the form of preference correspondences, and they also show up when considering consumer demand at given prices and income, since the set of preference-maximal elements of the budget set need not be a singleton. And there will be numerous other instances as we proceed.

In other fields of mathematics, for example in optimization theory, correspondences occur under other names, most often as *multimaps*, and the term "correspondence" seems now slightly outdated, but we have chosen to keep it, thus honoring the time-long connection with general equilibrium. At the present introductory state, we restrict ourselves to a general introduction and to the notion of continuity for correspondences.

Let X and Y be topological spaces. For most applications, we may think of X and Y as subsets of Euclidean spaces of suitable dimension. A *correspondence* is a map φ taking elements $x \in X$ to subsets $\varphi(x)$ of Y; we use the notation $\varphi : X \rightrightarrows Y$ for such a correspondence. The *graph* of φ is the set

$$\text{Graph } \varphi = \{(x, y) \in X \times Y \mid y \in \varphi(x)\}. \tag{6}$$

For purposes of analysis, one needs a notion of continuity of a correspondence, a property which guarantees that the set $\phi(x)$ does not change too much when x is changed. The notion of continuity for functions does not immediately carry over to correspondences, and instead one has to operate with several notions, each covering a specific part of what is usually understood as continuity.

Definition 0.2. Let $\phi : X \rightrightarrows Y$ be a correspondence.

(i) φ is upper hemicontinuous (uhc) at $x \in X$ if for each open set V in Y with $\varphi(x) \subset V$, there is an open neighborhood U of x such that $\varphi(x') \subset V$ for all $x' \in U$.

(ii) φ is lower hemicontinuous (lhc) at $x \in X$ if for each open set V in Y with $V \cap \varphi(x) \neq \emptyset$, there is an open neighborhood U of x such that $\varphi(x') \cap V \neq \emptyset$ for all $x' \in U$.

(iii) φ is continuous at x if φ is both upper and lower hemicontinuous at x.

(iv) φ is uhc (lhc, continuous) if it is uhc (lhc, continuous) at all $x \in X$.

Upper hemicontinuity of a correspondence at a point means that the value set does not expand too much when moving to neighboring points, and lower hemicontinuity means that it does not collapse. Taken together, a continuous correspondence has values which do not change too abruptly when the point is changed slightly. The two partial notions of continuity do not necessarily go together, and we shall consider several important cases of correspondences which are lower but not upper hemicontinuous.

The continuity notions in Defintion 0.2 have a relation to open- or closedness properties of the graph of a correspondence as defined in (6).

Lemma 0.1. *Let $\varphi : X \rightrightarrows Y$ be a correspondence. Then the following hold:*

(i) *If Graph φ is open, then φ is lhc.*

(ii) *If Graph φ is closed and Y is compact, then φ is uhc.*

Proof. (i) is straightforward and left to the reader.

To prove (ii), we assume to the contrary that there is $x \in X$ and an open set V in Y containing $\varphi(x)$ such that all neighborhoods of x

contain a point x' with some $y' \in \varphi(x')$ not belonging to V. Thus, we may construct a sequence $(x_n, y_n)_{n=0}^{\infty}$ with $y_n \in \varphi(x_n)$, $y_n \notin V$, and $x_n \to x$. By compactness of Y, we may subtract a subsequence $(x_{n_v}, y_{n_v})_{v=0}^{\infty}$ such that $(y_{n_v})_{v=0}^{\infty}$ converges to some $y \in Y \backslash V$. By closedness of Graph φ, we have that $y \in \varphi(x)$, a contradiction. $\qquad\square$

6. Exercises

(1) Suppose that a consumer has access to only two commodities, namely (1) rice today and (2) rice tomorrow. There is a given consumption $\bar{x}_1$ of rice today without which the consumer does not survive until tomorrow. Determine the consumption set of this consumer and consider whether it satisfies Assumption 0.1. If not, what can be done to reestablish well-behavedness of the consumption set? (The example is from Debreu (1959)).

(2) Let $X = \mathbb{R}_+^l$ and suppose that the preference correspondence $P \in \mathcal{P}^K$ has the following properties:

   (i) for all $x \in X$, $x \in \mathrm{cl}\, P(x)$,
   (ii) for all $x, y \in X$, either $x \in \mathrm{cl}\, P(y)$ or $y \in \mathrm{cl}\, P(x)$ (P is complete),
   (iii) for $x, y, z \in X$, $y \in \mathrm{cl}\, P(x)$ and $z \in \mathrm{cl}\, P(y)$ implies that $x \notin \mathrm{cl}\, P(z)$ (P is transitive).

   Suppose that P has the following monotonicity property: If $x', x \in X$ and $x' \geq x$ and $x' \neq x$, then $x' \in P(x)$. Show that P can be represented by a continuous utility function.

(3) Let (X, P) be a consumer with $X = \mathbb{R}_+^l$, and assume that $\xi(p, w)$ is non-empty for all $p \in \mathbb{R}_{++}^l$ and $w \in \mathbb{R}_+$. Show that ξ is homogeneous of degree 0 in p and w, i.e. that $\xi(\lambda p, \lambda w) = \xi(p, w)$ for all $(p, w) \in \mathbb{R}_{++}^l \times \mathbb{R}_+$.

   Suppose that P satisfies $P(\lambda x) = \lambda P(x)(= \{\lambda x' \mid x' \in P(x)\})$ for all $x \in X$ and $\lambda > 0$. Show that the demand correspondence $\xi(p, w)$ is positively homogeneous in w in the sense that

   $$\xi(p, \lambda w) = \lambda \xi(p, w) = \{\lambda x \mid x \in \xi(p, w)\},$$

   for all $w \in \mathbb{R}_+$ and $\lambda \in \mathbb{R}_+$.

Show that the excess demand is homogeneous of degree zero: $\zeta(\lambda p) = \zeta(p)$ for $\lambda > 0$ whenever $\zeta(p)$ is defined.

(4) (Lexicographic preferences) Define $P : \mathbb{R}_+^l \rightrightarrows \mathbb{R}_+^l$ by

$$P(x) = \{x' \in \mathbb{R}_+^l \mid x_1' > x_1 \text{ or } x_1' = x_1, x_2' > x_2 \text{ or}$$
$$\cdots x_h' = x_h, h \le l - 1, x_l' > x_l\}.$$

Does P belong to $\mathcal{P}^K$? Can P be represented by a continuous utility function?

(5) Let X be a consumption set satisfying Assumption 0.1 and suppose that the preference correspondence P can be represented by a continuous utility function. Let $p \in \mathbb{R}_{++}^l$ be a price system, and assume that the wealth w of the consumer satisfies $w \ge \min\{p \cdot x \mid x \in X\}$. Show that there is a maximal element for P in the budget set $\{x \in X \mid p \cdot x \ge w\}$.

(6) Suppose that $Y \subset \{y \in \mathbb{R}^l \mid y_h \le 0, h = 2, \ldots, l\}$ (only the first commodity can be an output) and that it can be described by a production function of the form

$$y_1 \le \prod_{h=2}^{l} (-y_h)^{\alpha_h}$$

with $\alpha_h \in \mathbb{R}_+$, $h = 2, \ldots, l$. Show that Y exhibits constant (non-decreasing, non-increasing) returns to scale if $\sum_{h=2}^{l} \alpha_h = 1$ ($\sum_{h=2}^{l} \alpha_h > 1$, $\sum_{h=2}^{l} \alpha_h < 1$).

Find for $p \in \mathbb{R}_{++}^l$ the profit-maximizing production(s) in Y.

(7) Let Y be a production set which has constant returns to scale. Let $p \in \mathbb{R}^l$ be a price system and assume that there is some $y^0 \in Y$ which maximizes profits on Y, so that $p \cdot y^0 \ge p \cdot y$ for all $y \in Y$. Show that $p \cdot y^0 = 0$.

(8) Suppose that the production set $Y \subset \mathbb{R}^l$ satisfies Assumption 0.3 and that, in addition,

$$Y + Y \subseteq Y.$$

Show that Y exhibits constant returns to scale.

Chapter 1

Exchange Economies

In the Introduction chapter, we have considered the concepts needed for our theory of general economic equilibrium. It is now time to apply the concepts to the model. For the time being, we leave the producers out of the story, postponing their introduction to Chapter 2. In this chapter, we consider those economies where the purpose of economic interaction is to distribute (or redistribute) the initial endowment among the consumers. The reason that we begin with this seemingly trivial part of economic activity is that most of the theoretical problems will already be present at this level, and the subsequent introduction of production into the economy will cause only minor disturbances.

In Section 1, we introduce the main equilibrium concept, that of a Walras equilibrium, and consider the question of the existence of the Walras equilibria. This will take us into a discussion of the existence proofs and of the assumptions on preferences and endowments that will make such proofs possible. At the end of this chapter, a short section provides an introduction on the necessary mathematical tools.

1. Economies and Allocations in the Economy

In the model of this chapter, we shall have a finite number of commodities, l, m number of consumers who receive commodity bundles in $\mathbb{R}^l$, and n number of producers carrying through a net output vector in $\mathbb{R}^l$. There are some given initial endowments of commodities.

1.1 Walras equilibria: Definitions

Our first task in the discussion of exchange economies is to fix certain terminologies. Since a model of an economy without production, from now on an *exchange economy*, arises when we specify its consumers, we may refer to the economy as

$$\mathcal{E} = ((X_i, P_i)_{i=1}^m, \omega).$$

Here, the finite set of consumers, as described in the Introduction chapter, is indexed by $i = 1, \ldots, m$, and the total initial endowment is $\omega \in \mathbb{R}_+^l$. Occasionally, it will be convenient to use the notation $M = \{1, \ldots, m\}$ for the index set, and the economy is then written as $\mathcal{E} = ((X_i, P_i)_{i \in M}, \omega)$. There is no explicit reference to the number of commodities in the model, which is a natural number l, large or small. Since much of what is said would be trivial if $l = 1$, we may think of l as being greater or equal to 2.

In some cases, the total initial endowment ω is all that matters, but for many of our considerations, in particular for those dealt with in this chapter, we need to know how this endowment is allocated initially to the consumers in the form of bundles ω_i, for $i = 1, \ldots, m$, with $\sum_{i=1}^m \omega_i = \omega$. When this information is available, we speak about a *private ownership economy* $\varepsilon = (X_i, p_i, \omega_i)_{i=1}^m$.

The aim of the general equilibrium theory is to study allocations arising from economic activity in this world of l commodities. The final consequences of such an activity will be an *allocation*, which is an assignment of a commodity bundle to each consumer. Formally, an allocation is a map x from M to $\mathbb{R}_+^l$, usually written as $x = (x_1, \ldots, x_m)$. The allocation is *individually feasible* if

$$x_i \in X_i, \quad i = 1, \ldots, m, \tag{1}$$

so that each consumer can survive with the bundle assigned to him/her, and it satisfies *aggregate feasibility* if

$$\sum_{i=1}^m x_i \le \omega, \tag{2}$$

so that the total amount assigned to the consumers of each commodity does not surpass the endowment. The allocation is *feasible* if it satisfies both (1) and (2). The set of feasible allocations in $\mathcal{E}$ is denoted $F(\mathcal{E})$.

In general, the set of feasible allocations is a subset of ml-dimensional Euclidean space and, as such, quite difficult to visualize. However, for economies with only two commodities and with two consumers having well-behaved consumption sets and preferences described by utilities, feasible allocations can be represented in the two-dimensional plane using the *Edgeworth box* shown in Box 1.

Markets in the economy $\mathcal{E}$ are defined by prices $p \in \mathbb{R}^l_+ \setminus \{0\}$. Given the price p, individual budget sets are determined as

$$\gamma_i(p) = \{x_i \in X_i \mid p \cdot x_i \leq \omega_i\}.$$

Box 1. The Edgeworth box: This very useful construction allows us to illustrate general concepts of equilibrium theory in the simple setting of a two-consumer–two-commodity economy.

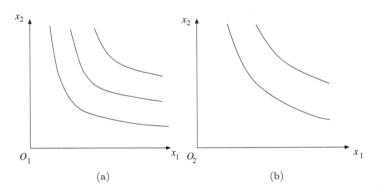

Fig. 1. (a) Toward the Edgeworth box; (b) Consumers and their preferences.

Plotting the initial endowment ω into consumer 1's diagram to the point O_2 and attaching the diagram for consumer 2 top-down from this point, we get the (Edgeworth) box in Fig. 2.

(Continued)

Box 1. *(Continued)*

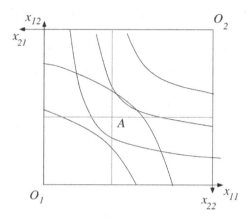

Fig. 2. The Edgeworth box: The diagram for consumer 2 has been attached to that for consumer 1 at the point representing the total endowments.

Let A be an arbitrary point in the box. Reading the coordinates of the point in each of the consumers' coordinate systems, we get bundles (x_{11}, x_{12}) and (x_{21}, x_{22}) which clearly sum to ω. Conversely, every feasible allocation obviously defines a point in the Edgeworth box.

The price p determines an equilibrium if there are bundles x_i for consumers $i = 1, \ldots, m$, which (1) are maximal for P_i in the budget set and (2) constitute a feasible allocation. An equilibrium of this type is called a *Walras equilibrium* (Box 2).

Definition 1.1. Let $\mathcal{E} = (X_i, P_i, \omega_i)_{i=1}^{m}$ be a private ownership economy. An array (x, p), where $x = (x_1, \ldots, x_m)$ is an allocation and $p \in \mathbb{R}^l \setminus \{0\}$ a price, is a Walras equilibrium if

(i) $x \in F(\mathcal{E})$,
(ii) for $i = 1, \ldots, m$, $P(x_i) \cap \{x_i' \in X_i \mid p \cdot x_i' \leq p \cdot \omega\} = \emptyset$.

The equilibrium conditions state that the individual decisions about consumption are *compatible* in the sense that the aggregate demand for each commodity is no greater than the available supply.

> **Box 2. Walras equilibrium in the Edgeworth box:** In Fig. 3, we illustrate the Walras equilibrium in the economy depicted in the Edgeworth box.
>
>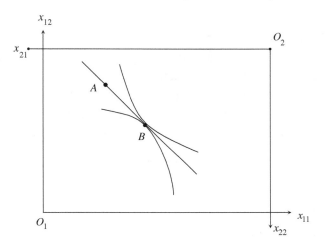
>
> Fig. 3. A Walras equilibrium illustrated in the Edgeworth box.
>
> The initial endowment corresponds to the point A. In the point B, the preferred bundles for consumer 1 are all above the line through A and B, and similarly, the preferred bundles for consumer 2 are all below this line. Prices p_1 and p_2 of the two goods such that the slope of the line equals $-p_1/p_2$ together with the bundles in B will define a Walras equilibrium.

1.2 Excess demand and Walras' law

In the definitions of a Walras equilibrium given above, we have emphasized the simultaneous satisfaction of individual optimality and aggregate feasibility, an approach which is natural in the given context where consumers have preferences that are neither necessarily complete nor transitive. Traditionally, the equilibrium conditions have been stated using demand functions or demand correspondences. If consumers in $\mathcal{E} = (X_i, P_i, \omega_i)_{i=1}^m$ have excess demand functions $\zeta_i(p)$, as defined in Eq. (1) in the Introduction, then aggregate excess demand at the price p is

$$\zeta(p) = \sum_{i=1}^{m} \zeta_i(p),$$

defined for all prices p such that each consumer has a well-defined demand given these prices and the income $p \cdot \omega_i$. Under standard assumptions of monotonicity of preferences, this is the case whenever p belongs to $\mathbb{R}^l_{++}$, which is the interior of $\mathbb{R}^l_+$.

The excess demand function for the economy $\mathcal{E}$ has some properties which are consequences following from Theorem 0.2.

Lemma 1.1. *The excess demand function* $\zeta : \mathbb{R}^l_{++} \to \mathbb{R}^l$ *for the exchange economy* $\mathcal{E}$ *satisfies*

  (i) ζ *is continuous,*
 (ii) ζ *is positively homogeneous of degree zero in the sense that* $\zeta(\lambda p) = \zeta(p)$ *for all* $p \in \mathbb{R}^l_{++}$ *and* $\lambda > 0$,
(iii) $p \cdot \zeta(p) = 0$ *for all* $p \in \mathbb{R}^l_{++}$.

The property (iii) is known as *Walras' law*. It is a consequence of the optimizing behavior inherent in the definition of demand, whereby (under some assumption of non-satiation) the consumer will use all the available income in the optimum. As presented, Walras' law can be seen as a constraint that must be satisfied for all prices, also prices for supply and demand are not in balance. This restricts the independence of the markets: If there is equilibrium in $l - 1$ of the l markets, then there is also equilibrium in the last, lth, market. This observation is useful when considering the search for a Walras equilibrium as the problem of solving the l equations

$$\zeta_h(p) = 0, \quad h = 1, \dots, l \tag{3}$$

in prices $p_1, \dots, p_l$. Since the excess demand functions are homogeneous of degree zero, the price of one commodity may be chosen arbitrarily. This particular commodity, say commodity 1, is then used as a *numeraire*, a basis for the measurement of relative prices, with $p_1 = 1$.

Needless to say, this argument does not guarantee the existence of a solution to the system (3), the more so, since we demand that prices should be positive (or at least non-negative). We shall not use Walras' law explicitly when we turn to the question of whether or not the economy $\mathcal{E}$ has an equilibrium. It does, however, play some role in the argumentation, since it must be satisfied in the equilibrium, so that violations of Walras' law may be taken as an indication of the unbalancedness of at least some market.

In may cases, we shall also consider a slightly weaker version of the Walras equilibrium, known as a *quasi-equilibrium*. The only difference between the two equilibrium concepts is the condition for individual optimization, which is replaced by a condition of expenditure minimization.

Definition 1.2. A quasi-equilibrium in the private-ownership economy $\mathcal{E} = (X_i, P_i, \omega_i)_{i=1}^m$ is an array (x, p), where x is an allocation and p is a price, such that

(i) $x \in F(\mathcal{E})$,
(ii) for $i = 1, \ldots, m, p \cdot x_i = \inf \left\{ p \cdot x_i' \,\middle|\, x_i' \in P(x_i) \right\}$.

The connection between quasi-equilibria and Walras equilibria is the same as what we discussed in connection with Theorem 0.2.(ii) in the Introduction of this book.

Lemma 1.2. *Let (x, p) be a quasi-equilibrium in $\mathcal{E}$ and assume that each consumer i satisfies the (no-)minimum-wealth condition*

$$p \cdot x_i > \min \left\{ p \cdot x_i' \,\middle|\, x_i' \in X_i \right\}$$

Then (x, p) is a Walras equilibrium. $\square$

2. Existence of Equilibria

In this section and in several occasions in the sections and chapters that follow, we shall consider the existence of equilibria, in the present case, Walras equilibria, and later, other types of equilibria. Since this will take us into rather lengthy considerations which may be formal and seemingly unrelated to the underlying economic content of the models, some introductory comments may be called for.

2.1 Why existence proofs?

What is called an existence proof is basically a consistency check. A model is set up, in the present case, a model of individual optimizing behavior in a market, and the fundamental purpose of such a model should be to obtain a tool for studying the consequences of changes in the initial conditions, that is, to explain the pricing of commodities obtained in the market by the underlying conditions of individual tastes and scarcity of resources. But before any such analyses can be carried out, one must be sure that the model makes sense and that it does not contain postulates that conflict with each other. Finding the properties of equilibria in a model where there can be no equilibria is not a very promising activity.

Historically, this consistency check of the model has taken different forms. An intuitive approach, at least in the simple version of the model where demand functions are available, takes as point of departure that prices which do *not* yield an equilibrium must change in some systematical way, and with suitable assumptions on the reaction of overall demand to changes in prices, it may be shown that there is a price for which excess demand is zero. Unfortunately, this approach needs assumptions on derived concepts (such as aggregate excess demand) rather than on the basic data of the economy, and even so, these assumptions are restrictive.

Instead of this, we consider what is now the standard approach to existence proofs, as initiated by several authors in the 1950s, see, e.g. Arrow and Debreu (1954). From a formal point of view, they differ from the above intuitional approaches by the use of fixed-point theorems in one version or another, see Section 5. In the

Exchange Economies

following section, In what follows, we consider several approaches to the problem of proving existence of a Walras equilibrium.

2.2 Types of existence proofs

The existence of a Walras equilibrium can be shown in many ways, all of which involve a fixed-point theorem, in our case that of Kakutani, at some step. A statement and proof of this theorem is given in Section 5 of this chapter.

We limit ourselves to showing the existence of *quasi-equilibria* (Box 3). However, once a quasi-equilibrium is shown to exist, one gets to the Walras equilibrium using standard methods (cf. Lemma 1.2).

2.2.1 The simultaneous optimization method

We are now ready to prove the existence of a quasi-equilibrium. The proof given below follows Gale and Mas-Colell (1975).

Theorem 1.1. *Let $\mathcal{E} = (X_i, P_i, \omega_i)_{i=1}^m$ be an economy, and assume that*

(1) *the set $F(\mathcal{E})$ is non-empty and compact,*
(2) *for each consumer $i \in M$, X_i satisfies Assumption 0.1, P_i satisfies Assumption 0.2, and $\omega_i \in X_i$.*

Then there exists a quasi-equilibrium in $\mathcal{E}$.

Proof. Let $K \subset \mathbb{R}^l$ be a compact set such that $(x_1, \ldots, x_m) \in F(\mathcal{E})$ implies $x_i \in K$, each $i \in M$. Define $\widetilde{X}_i = X_i \cap K$, $i \in M$, and let $\widetilde{X} = \prod_{i \in M} \widetilde{X}_i$. Let $e = (1, 1, \ldots, 1)$ be the diagonal vector in $\mathbb{R}^l$ and let $\Delta = \{p \in \mathbb{R}_+^l \mid p \cdot e = 1\}$. Then $\widetilde{X}$ and Δ are non-empty, convex and compact.

For each i, define $\widetilde{P}_i : \widetilde{X}_i \rightrightarrows \widetilde{X}_i$ by

$$\widetilde{P}_i(x_i) = \left\{ \lambda x_i + (1 - \lambda) x_i' \,\middle|\, \lambda \in]0, 1[, x_i' \in P_i(x_i) \right\},$$

then $\widetilde{P}_i$ is lhc, convex-valued and irreflexive in the sense that $x_i \notin \widetilde{P}_i(x_i)$, and if $\widetilde{P}_i(x_i) \neq \emptyset$, then it intersects every neighborhood of x_i (local non-satiation).

Next, define for each i the budget correspondence $\gamma_i : \widetilde{X} \times \Delta \rightrightarrows \widetilde{X}_i$ by

$$\gamma_i(x, p) = \left\{ x_i' \in \widetilde{X}_i \,\middle|\, p \cdot x_i' < \frac{1}{2} p \cdot x_i + \frac{1}{2} p \cdot \omega_i \right\},$$

and the modified preference correspondence $\varphi : \widetilde{X} \times \Delta \rightrightarrows \widetilde{X}_i$ by

$$\varphi_i(x, p) = \begin{cases} \gamma_i(x, p) \cap \widetilde{P}_i(x_i) & p \cdot x_i \leq p \cdot \omega_i, \\ \gamma_i(x, p) & \text{otherwise.} \end{cases}$$

By construction, $x_i \notin \varphi_i(x, p)$. Moreover, φ_i is lhc with convex values (since γ_i has an open graph and $\widetilde{P}_i(x_i)$ is lhc and both have convex values).

Finally, define the correspondence $\varphi_0 : \widetilde{X} \times \Delta$ by

$$\varphi_0(x, p) = \left\{ p \in \Delta \,\middle|\, p \cdot \sum_{i \in M} x_i > p \cdot \sum_{i \in M} \omega_i \right\}.$$

Then φ_0 has convex values and an open graph. Moreover, $p \notin \varphi_0(x, p)$ for all $p \in \Delta$.

Applying Theorem 1.7 (in Section 5), we get the existence of an array $(x_1^*, \ldots, x_m^*, p^*)$ such that $\varphi_i(x^*, p^*) = \emptyset$ for $i \in M \cup \{0\}$. We note that $p^* \cdot x_i^* \leq p^* \cdot \omega_i$ for each $i \in M$, since otherwise, $\varphi_i(x^*, p^*) = \gamma_i(x^*, p^*) \neq \emptyset$. Thus, $p \cdot \sum_{i \in M} x_i^* \leq p^* \cdot \sum_{i \in M} \omega_i$, and $\sum_{i \in M}(x_i^* - \omega_i) \in \mathbb{R}_-^l$, since otherwise, by separation of convex sets, there would be some $p' \in \varphi_0(x^*, p^*)$, contradicting that $\varphi_0(x^*, p^*) = \emptyset$. We conclude that $x^* \in F(\mathcal{E})$.

Since $p^* \cdot x_i^* \leq p^* \cdot x_i^*$, each $i \in M$, we have that $\varphi_i(x^*, p^*) = \gamma_i(x^*, p^*) \cap \widetilde{P}_i(x_i^*)$. Then either $\gamma_i(x^*, p^*) = \emptyset$, in which case $p^* \cdot x_i^* = p^* \cdot \omega_i = \inf\{p^* \cdot x_i \mid x_i \in X_i\}$, so that trivially $x_i' \in P_i(x_i^*)$ implies $p^* \cdot x_i' \geq p^* \cdot x_i^*$, or we have $\gamma_i(x^*, p^*) \neq \emptyset$. But then, the construction of $\widetilde{P}_i$ implies that

Exchange Economies

Box 3. Existence of Walras equilibria in Cobb–Douglas economies: Let $\mathcal{E} = (X_i, P_i, \omega_i)_{i=1}^{m}$ be an economy where $X_i = \mathbb{R}_+^l$ and the preferences are described by utility functions of the form

$$u(x) = x_1^{c_1} \cdots x_l^{c_l}$$

for $x \in \mathbb{R}_+^l$, with parameters $c_1, \dots, c_l \in \mathbb{R}_{++}$, known as Cobb–Douglas utility functions. Clearly, any parameter vector $c' = \lambda c$ with $\lambda > 0$ gives rise to the same preferences as c, so we may normalize the parameter vectors in any convenient way.

If prices $p = (p_1, \dots, p_l)$ are given, the consumer i with parameter vector $c_i = (c_{i1}, \dots, c_{i,l})$ and endowment ω_i will have a demand for commodity h, which can be found as

$$\xi_{ih}(p, p \cdot \omega_i) = \frac{c_{ih}}{\sum_{k=1}^{l} c_{ik}} \frac{p \cdot \omega_i}{p_h},$$

(cf. Exercise 3 in Introduction chapter) and a Walras equilibrium must satisfy the equations

$$\sum_{i=1}^{m} \frac{c_{ih}}{\sum_{k=1}^{l} c_{ik}} \frac{p \cdot \omega_i}{p_h} = \sum_{i=1}^{m} \omega_{ih}, \quad h = 1, \dots, l-1 \tag{4}$$

(where the lth equation follows from Walras' law).

A particular case will be useful at a later stage: Suppose that all consumers have the same endowment, namely $\omega_i = (1, \dots, 1)$ for $i = 1, \dots, m$. If we normalize the prices so that $\sum_{h=1}^{l} p_h = 1$, then (4) can be rewritten as

$$\sum_{i=1}^{m} \frac{c_{ih}}{\sum_{k=1}^{l} c_{ik}} \frac{1}{p_h} = m, \quad h = 1, \dots, l-1,$$

so that

$$\frac{1}{m} \sum_{i=1}^{m} \frac{c_{ih}}{\sum_{k=1}^{l} c_{ik}} = p_h, \quad h = 1, \dots, m, \tag{5}$$

which gives an explicit formula for the equilibrium prices. Conversely, if p are equilibrium prices for an economy with Cobb–Douglas preferences and endowments $\omega_i = (1, \dots, 1)$, then (5) can be used to find the unique (except for normalization) parameter vector c for which a given price vector will be a Walras equilibrium price.

$p^* \cdot x_i^* = p^* \cdot \omega_i$ and

$$p^* \cdot x_i' \geq p^* \cdot \omega_i \tag{6}$$

for all $x_i' \in \widetilde{X}_i \cap \widetilde{P}_i(x_i^*)$. By the definition of $\widetilde{X}_i$ and $\widetilde{P}_i$, we have that (6) must hold for all $x_i' \in P_i(x_i^*)$. $\qquad\square$

2.2.2   The demand correspondence approach

The classical existence proofs exploit the notion of a demand correspondence: For each $i \in M$, define the (quasi-)demand correspondence $\xi_i : \Delta \rightrightarrows X_i$ by

$$\xi_i(p) = \left\{ x_i \in X_i \,\middle|\, p \cdot x_i = p \cdot \omega_i \text{ and } x_i' \in P_i(x_i) \text{ implies } p \cdot x_i' \geq p \cdot \omega_i \right\}.$$

Assuming $\zeta(p) = \sum_{i \in M} \xi_i(p) - \sum_{i \in M} \{\omega_i\}$ gives us a correspondence $\zeta : \Delta \rightrightarrows \mathbb{R}^l$, known as the excess demand correspondence, and a quasi-equilibrium is obtained if $\zeta(p)$ intersects $\mathbb{R}^l_-$. That this will actually happen can be established through the Gale–Debreu–McKenzie–Nikaido (GDMN) lemma:

Lemma 1.3 (GDMN). *Let $K \subset \mathbb{R}^l$ be non-empty, convex and compact, and let $\zeta : \Delta \rightrightarrows K$ be a closed correspondence with non-empty, convex values, such that $p \cdot x = 0$ for all $x \in \zeta(p)$. Then there is $p^0 \in \Delta$ such that $\zeta(p^0) \cap (\mathbb{R}^l_-) \neq \emptyset$.*

Proof. Define $\varphi_0 : K \times \Delta \rightrightarrows \Delta$ by

$$\varphi_0(x, p) = \left\{ p' \in \Delta \,\middle|\, p' \cdot x > 0 \right\}$$

and apply Theorem 1.7 to the correspondences ζ and φ_0. $\qquad\square$

As noted by Uzawa (1962), the GDMN lemma implies Kakutani's fixed point theorem: We define an *abstract excess demand correspondence* as a correspondence $\hat{\zeta} : \mathbb{R}^l_+ \backslash \{0\} \rightrightarrows \mathbb{R}^l$ with the following properties:

(i) $\hat{\zeta}$ is upper hemicontinuous with convex values,
(ii) $\hat{\zeta}(p) = \hat{\zeta}(\lambda p)$ for all p (homogeneity of degree zero),
(iii) $p \cdot z = 0$ for all $z \in \hat{\zeta}(p)$ (Walras' law).

Exchange Economies 33

An equilibrium for the abstract excess demand correspondence is a point p^0, such that $\hat{\zeta}(p^0) = 0$.

Lemma 1.4 (Uzawa). *Let* $\Phi : \Delta \rightrightarrows \Delta$ *be an upper hemicontinuous correspondence with non-empty, closed and convex values. If every abstract excess demand correspondence has an equilibrium, then* Φ *has a fixed point.*

Proof. Define the correspondence $\hat{\zeta} : \mathbb{R}_+^l \setminus \{0\} \rightrightarrows \mathbb{R}_+^l$ by

$$\hat{\zeta}(p) = \left\{ \frac{x}{p \cdot x} \,\middle|\, x \in \Phi\left(\frac{p}{p \cdot e}\right) \right\} - \frac{1}{p \cdot p}\, p,$$

where $e = (1, \ldots, 1)$. It is easily checked that $p \cdot z = 0$ for each $z \in \hat{\zeta}(p)$, $p \in \mathbb{R}_+^l \setminus \{0\}$, so that $\hat{\zeta}$ has property (iii) of an abstract excess demand correspondence. The other two properties follow almost directly from our construction, and it follows from our assumption that there is $p^0 \in \Delta$ such that $p^0 \in \Phi(p^0)$, so Φ indeed has a fixed point. $\qquad\square$

It should be noted that although the aggregate excess demand derived from an exchange economy is an abstract excess demand correspondence, the converse may not be true: An abstract excess demand correspondence may not necessarily come from utility maximizing consumers in an exchange economy. We return to this problem in Chapter 8.

The demand correspondence approach will work if the correspondences ξ_i are closed with non-empty, convex values. However, as noted in the Introduction of this book, convexity of $\xi_i(p)$ does not follow from our general assumptions on preferences, so further assumptions, such as, e.g. transitivity, must be made. To adapt the demand correspondence approach to the general situation, an approach introduced by Shafer (1974) may be used: At each $x_i \in X_i$, choose a (local) utility function $u(\cdot, x_i)$ such that $P_i(x_i) = \{x_i' \mid u_i(x_i', x_i) > u_i(x_i, x_i)\}$. Now one may define demand and excess demand (depending on utility and through utilities on bundles) and apply the GDMN lemma. We shall exploit a similar idea below and postpone a detailed treatment.

2.2.3 The Pareto-frontier approach

Still another line of argument proceeds through the set of Pareto-optimal allocations. We sketch briefly the concepts involved, since Pareto-optimality will be developed in more detail in Chapter 3: A feasible allocation $(x_1, \ldots, x_m)$ is Pareto-optimal if there is no other feasible allocation $(x'_1, \ldots, x'_m)$ such that $x'_i \in \operatorname{cl} P_i(x_i)$ for each $i \in M$ and $x'_j \in P_j(x_j)$ for some $j \in M$.

As we shall see in Chapter 3, if $(x_1, \ldots, x_m)$ is Pareto-optimal, then there is $p \in \Delta$ such that

$$p \cdot z \geq 0 \quad \text{for } z \in \sum_{i \in M} \operatorname{cl} P_i(x_i)$$

$$p \cdot z \leq 0 \quad \text{for } z \in \left\{ \sum_{i \in M} \omega_i \right\} + \mathbb{R}^l_-.$$

This fact may be used in an existence proof: If $(x_1, \ldots, x_m)$ is Pareto-optimal and p is a price vector with these properties, then $(x_1, \ldots, x_m, p)$ is a quasi-equilibrium if $p \cdot x_i = p \cdot \omega_i$ for each $i \in M$.

To carry through the argument, one needs some convexity properties of the set of Pareto-efficient allocations, and it seems to work best when preferences are defined by utility functions. For an existence proof using this approach, see Arrow and Hahn (1971) and Negishi (1960).

2.2.4 The social utility function approach

A variety of the argument used above combines the ideas of Shafer (1974) with a classical result for a linear-preferences economy due to Gale (1960).

Lemma 1.5. *Let $\mathcal{E}$ be such that $P_i = \left\{ (x_i, x'_i) \in \mathbb{R}^l_+ \times \mathbb{R}^l_+ \,\middle|\, p^i \cdot x'_i > p^i \cdot x_i \right\}$, all $i \in M$, for some $p^i \in \Delta$, and $\sum_{i \in M} \omega_i \in \mathbb{R}^l_{++}$. Let $\theta_1, \ldots, \theta_m$ be positive numbers with $\sum_{i \in M} \theta_i = 1$.*

If $x = (x_1, \ldots, x_m)$ is a feasible allocation which maximizes the function

$$(p^1 \cdot x_1)^{\theta_1} \cdots (p^m \cdot x_m)^{\theta_m}$$

over all feasible allocations, and p is defined by $p_h = \max_{i \in M}(p_h^i \theta_i)/(p^i \cdot x_i)$,
then (x, p) *satisfies*

(i) $p \cdot x_i = \theta_i$,
(ii) $x_i' \in P_i(x_i)$ *implies* $p \cdot x_i' > \theta_i$,

for each $i \in M$.

For a proof of Lemma 1.5, the reader is referred to Exercise 6. To exploit Lemma 1.5 in the general situation of Theorem 1.1, we approximate the preferences at each x_i by suitable linear preferences. For $i \in M$, define P_i° by

$$P_i^\circ(x_i) = \left\{ p \in \Delta \,\middle|\, p \cdot x_i' \geq p \cdot x_i, \text{ all } x_i' \in P_i(x_i) \right\}.$$

Since P_i is lower hemicontinuous, P_i° is closed. Also, P_i° has convex values. However, $P_i^\circ(x_i)$ may be empty, so we must assume that $P_i(x_i) \cap \mathbb{R}_-^l = \emptyset$ for all $x_i \in \mathbb{R}_+^l$, i.e. a (weak) monotonicity assumption on preferences. Furthermore, we assume that $\omega_i \in \text{int } \mathbb{R}_+^l$ for each $i \in M$.

Let $\Delta' = \left\{ p \in \mathbb{R}_+^l \,\middle|\, p \cdot \sum_{i \in M} \omega_i = 1 \right\}$ and define the set $B \subset \mathbb{R}_+^m$ by

$$B = \left\{ (\beta_1, \dots, \beta_m) \,\middle|\, \exists p \in \Delta' : \beta_i = p \cdot \omega_i, i = 1, \dots, m \right\}.$$

Furthermore, let the correspondences φ_k for $k = 1, 2, 3, 4$ be defined by

$$\varphi_1(x_1, \dots, x_m) = \left\{ p = (p^1, \dots, p^m) \in \Delta^m \,\middle|\, p^i \in P_i^\circ(x_i), i \in M \right\},$$

$$\varphi_2(p^1, \dots, p^m, \beta_1, \dots, \beta_m) = \left\{ x = (x_1, \dots, x_m) \,\middle|\, x \in \mathcal{F}, \right.$$

$$\left. x \text{ maximizes } (p^1 \cdot x_1')^{\beta_1} \cdots (p^m \cdot x_m')^{\beta_m} \text{ on } \mathcal{F} \right\},$$

$$\varphi_3(p) = \{ \beta = (\beta_1, \dots, \beta_m) \mid \beta_i = p \cdot \omega_i, i \in M \},$$

$$\varphi_4(x, p, \beta) = \begin{cases} \left\{ p \in \Delta' \,\middle|\, \exists \lambda > 0, p_h = \lambda \max_{i \in M}(p^i \cdot x_i)^{-1} p_h^i \beta_i, h \in M \right\} \\ \qquad\qquad\qquad\qquad\qquad\qquad \text{if } p^i \cdot x_i > 0, i \in M, \\ \Delta' \qquad\qquad\qquad\qquad\qquad\qquad\quad \text{otherwise.} \end{cases}$$

Each of the correspondences φ_k is closed with non-empty convex values. Applying Theorem 1.7, we get a fixed point of the product correspondence from $\mathcal{F} \times B \times \Delta^m \times \Delta'$ to itself. Using Lemma 1.5, it is easily verified that the fixed point defines a quasi-equilibrium in $\mathcal{E}$.

3. The Survival Condition and Hierarchic Equilibria

In the preceding sections, we have considered quasi-equilibria and neglected the final step, taking us from a quasi-equilibrium to a genuine Walras equilibrium, to a later consideration, to which we turn to now.

The basic reasoning for this final step is simple enough, stated here as a lemma.

Lemma 1.6. *Let* $\mathcal{E} = (X_i, P_i, \omega_i)_{i=1}^m$ *be an economy and* (x, p) *a quasi-equilibrium in* $\mathcal{E}$. *If* $P_i(x_i)$ *is open and*

$$p \cdot x_i > \inf \{p \cdot x_i' \mid x_i' \in X_i\}, \tag{7}$$

for $i = 1, \ldots, m$, *then* (x, p) *is a Walras equilibrium.*

Proof. Suppose on the contrary there is some i such that $p \cdot x_i' \le p \cdot x_i$ for some $x_i' \in P_i(x_i)$. By (7), there is some $x_i^0 \in X_i$, such that $p \cdot x_i^0 < p \cdot x_i$, and by convexity of X_i together with openness of $P_i(x_i)$, we get that the line segment $[x_i^0, x_i']$ intersects $P_i(x_i)$, indicating that there would be $x_i'' \in P_i(x_i)$ with $p \cdot x_i'' < p \cdot x_i$, which is a contradiction. $\square$

The condition (7) in Lemma 1.6 is known as the *minimum-wealth condition*. Unfortunately, it is formulated on derived concepts (quasi-equilibria) rather than on the basic concepts of $\mathcal{E}$, so the logical next step is to search for conditions on the primitives of $\mathcal{E}$ which will guarantee that (7) is satisfied at all quasi-equilibria.

3.1 *Irreducibility and resource-relatedness*

The simplest way of avoiding minimum-wealth problems is to assume that each consumer can survive with his/her initial endowment and still leave something unused, that is $\omega_i \in \text{int} \, X_i$ for each i.

In many cases, in particular when other aspects are more important for the study of equilibria, this interiority condition will be assumed without further comments. In a broader perspective, however, the assumption that every consumer can survive without trading is unsatisfactory, for example in models of interpersonal trade, and several different approaches have been undertaken in order to obtain minimum wealth in a more general setting. What matters in the social context of commodity trading is not that each agent can satisfy her own needs without trade but rather that each agent has something that is desired by the other agents. One way of expressing this is through the condition of irreducibility formulated by McKenzie (1959).

The economy $\mathcal{E}$ is *irreducible* if for every feasible allocation $x \in \mathcal{F}(\mathcal{E})$ with $\sum_{i=1}^{m} x_i = \sum_{i=1}^{m} \omega_i$ and every partition of $M = \{1, \ldots, m\}$ into non-empty sets M_1 and M_2, there is an allocation x' such that

$$\sum_{i \in M_1} (x'_i - \omega_i) + \sum_{i \in M_2} (x'_i - x_i) = 0, \tag{8}$$
$$x'_i \in P_i(x_i) \text{ for each } i \in M_2.$$

In other words, the initial endowment of any group is desired by its complement: M_2 would be better off after some trade with the complement if M_2 is endowed with $(x_i)_{i \in M_2}$ and M_1 with its original endowment.

The irreducibility assumption has the consequence that if one consumer satisfies the minimum wealth condition in a quasi-equilibrium, then so do all consumers.

Lemma 1.7. *Let (x, y, p) be a quasi-equilibrium in an irreducible economy $\mathcal{E}$. If $P_i(x_i)$ is open and $p \cdot x_i > \inf \{p \cdot x'_i \mid x'_i \in X_i\}$ for some $i \in M$, then $p \cdot x_i > \inf \{p \cdot x'_i \mid x_i \in X_i\}$ for all $i \in M$.*

Proof. Suppose that $\{i \in M \mid p \cdot x_i = \inf p \cdot x'_i \mid x'_i \in X_i\} \neq \emptyset$ and denote this set by M_1. By irreducibility, there is an allocation x' satisfying (8), and from $p \cdot \sum_{i \in M_1} (x'_i - \omega_i) \geq 0$, we get that $p \cdot x'_i \leq p \cdot x_i$ for each $i \in M_2$, contradicting the result in Lemma 1.6. $\square$

38   Theory of General Economic Equilibrium

When the economy is irreducible, all consumers have something in their endowment that is interesting for some other consumers. An alternative formulation of this interconnectedness of the consumers uses resource relatedness as introduced in Arrow and Hahn (1971). If x is an allocation belonging to a quasi-equilibrium in $\mathcal{E} \doteq (X_i, P_i, \omega_i)_{i=1}^m$ (or more generally, a Pareto-optimal allocation in $\mathcal{E}$, cf. Chapter 3), then $i^0 \in M$ is *resource related* to $i^1 \in M$ if there is an allocation x' and a bundle $x''_{i^1} \in X_{i^1}$ for i^1 such that

$$\sum_{i \in M} (x'_i - \omega_i) + (x''_i - \omega_i) = 0,$$

$$x'_{i^0} \in P_{i^0}(x_{i^0}), x'_i \in \text{cl}\, P_i(x_i), \text{ all } i \in M. \tag{9}$$

Thus, consumer i^0 can be made better off, and no one is worse off, if the endowment of agent i^1 was to be duplicated. If all consumers are resource related, then again one may deduce that if one consumer satisfies the minimum-wealth condition, so do all the other consumers, cf. Exercise 5.

For the final step, assuring that quasi-equilibria are actually Walras equilibria, we need an aggregate survival condition. We formulate the result which follows immediately from Lemma 1.7 as a theorem.

Theorem 1.2. *Let $\mathcal{E} = (X_i, P_i, \omega_i)_{i=1}^m$ be an irreducible economy satisfying the aggregate interiority condition $\sum_{i \in M} \omega_i \in \sum_{i \in M} X_i$. If $\mathcal{E}$ has a quasi-equilibrium, then there exists a Walras equilibrium in $\mathcal{E}$.* $\square$

3.2   Hierarchic equilibria

When the minimum-wealth condition does not hold, a quasi-equilibrium may fail to be Walrasian. For the consumers who attain the minimal value of feasible consumption bundles at the equilibrium prices, the quasi-equilibrium bundles may not be maximal for their preferences in the budget set. In such situations, the consumers split into two disjoint sets M_0 and M_1, where M_1 consists of all the consumers with minimum-wealth bundles, and M_0 are those getting

Exchange Economies 39

a bundle with value above minimum-wealth. The consumers in M_1 may be thought of as being so poor that they cannot buy anything at the equilibrium prices. However, they still trade with each other, as is shown in the following case of a reducible economy.

Theorem 1.3. *Let $\mathcal{E} = (X_i, P_i, \omega_i)_{i=1}^m$ be an economy with $\omega_i \in X_i$, all i, and suppose that there is $p \in \Delta$ and a partition $\{M_0, M_1\}$ of $M = \{1, \ldots, m\}$ such that for each $i \in M_0$,*

$$p \cdot \omega_i \geq \min_{x_i \in X_i} p \cdot x_i,$$

whereas for $i \in M_1$,

$$p \cdot \omega_i = \min_{x_i \in X_i} p \cdot x_i \text{ and } \{x_i \in \gamma_i(p) \mid \gamma_i(p) \cap P_i(x_i) \neq \emptyset\} \text{ is open.}$$

Then there is an allocation $x \in F(\mathcal{E})$ and a non-trivial linear form p^1 on $\operatorname{Ker} p = \{x \in \mathbb{R}^l \mid p \cdot x = 0\}$, such that either

(i) *(x, p) is a Walras equilibrium or*
(ii) *(x, p) is a quasi-equilibrium, and for each $i \in M_1$, $p^1 \cdot (x_i - \omega_i) = 0$, and $x' \in P_i(x_i)$, $p \cdot (x'_i - \omega_i) = 0$ implies $p^1 \cdot (x'_i - \omega_i) \geq 0$.*

Proof. For each $i \in M_1$, let $\widetilde{X}_i)\{x_i \in X_{|}p \cdot x_i = p \cdot \omega_i\}$, $\widehat{X}_i = \widetilde{X}_i + \{\lambda p \mid \lambda \geq 0\}$, and define $\widehat{P}_i : \widehat{X}_i \rightrightarrows \widehat{X}_i$ by

$$\widehat{P}_i(\hat{x}_i) = \begin{cases} P_i(\widetilde{x}_i) \cap \widetilde{X}_i + \{\lambda p \mid \lambda \geq 0\} & P_i(\widetilde{x}_i) \cap \widetilde{X}_i \neq \emptyset, \\ \{\hat{x}_i\} + \{\lambda p \mid \lambda > 0\} & \text{otherwise,} \end{cases}$$

where $\hat{x}_i \in \widehat{X}_i$, $\hat{x}_i = \widetilde{x}_i + \lambda p$ for some $\lambda \geq 0$. Further, for each $n \in \mathbb{N}$, let $\omega_i^n = \omega_i + \frac{1}{n}p$.

For each n, the economy $((X_i, P_i, \omega_i)_{i \in M_0}, (\widehat{X}_i, \widehat{P}_i, \omega_i^n)_{i \in M_1})$ satisfies the assumptions of Theorem 1.1 and thus has a quasi-equilibrium $(\hat{x}^n, q^n)$. If $q^n = p$ for some n, then (x, p), where $x_i^n = \hat{x}_i^n$ for $i \in M_0$ and $x_i = \hat{x}_i^n - \frac{1}{n}p$ for $i \in M_1$, is a Walras equilibrium. If $q^n \neq p$ for all n, then each q^n defines a non-trivial linear form $\widehat{q}^n$ on $\operatorname{Ker} p$, and w.l.o.g. we may assume that $\|\widehat{q}\| = 1$.

Let $\widetilde{x}_i^n \in \widetilde{X}_i$ be such that $x^n = \widetilde{x}^n + \lambda p$. Then the sequence $((\hat{x}_i^n)_{i \in M_0}, (\widetilde{x}_i^n)_{i \in M_1}, \widetilde{q}^n)_{n \in \mathbb{N}}$ has an accumulation point (x, p^1). It can be easily verified that x and p^1 satisfy the conclusions of the theorem.

$\square$

In the situation considered, the commodities were split into hierarchies, so that consumers having endowment in the higher hierarchy can buy those in the lower hierarchy, whereas endowment in the lower hierarchies does not permit trade in the higher one. This generalizes to hierarchies with more than two layers. For the general version of the hierarchical equilibrium, it is more convenient to focus on net trades rather than bundles. Thus, we consider an exchange economy $(Z_i, P_i)_{i \in M}$, where $Z_i \subset \mathbb{R}^l$ is a set of individually feasible net trades and $P_i : Z_i \rightrightarrows Z_i$ describes the preferences over net trades of consumer $i \in M$.

Definition 1.3. Let $\mathcal{L} = (L^1, \ldots, L^k)$ with $L^k \subset L^{k-1} \subset \cdots \subset L^1 = \mathbb{R}^l$ be a chain of subspaces of $\mathbb{R}^l$ where for each j there is i such that 0 belongs to the relative interior of $Z_i \cap L^j$. A *hierarchic equilibrium* relative to $\mathcal{L}$ is an array $(z_1, \ldots, z_m, (p^j)_{j=1}^k)$, where $z_i \in Z_i, i = 1, \ldots, m$, and $p^j \in \Delta, j = 1, \ldots, k$, satisfying the following conditions:

(i) $\sum_{i=1}^m z_i = 0$,
(ii) For $i = 1, \ldots, m$, if $\min_{z_i' \in Z_i} p^j \cdot z_i' < 0$, then $z_i \in L^j, p^j \cdot z_i = 0$ and
$P_i(z_i) \cap \{z_i' \in Z_i \cap L^j \mid p^j \cdot z_i \le 0\} = \emptyset$.

Intuitively, in a hierarchic equilibrium, consumers and commodities are split into classes, so that members of classes with higher index are too poor to buy all the commodities available to those in classes with a lower index. Hierarchic equilibria have been studied by several authors, cf. Danilov and Sotskov (1990), and Florig (2002).

4. Admissible Preferences

In the previous sections, we have shown that quasi-equilibria exist for a rather large class of exchange economies specified by consumption sets, preferences and endowments of their consumers.

Exchange Economies 41

So far, we have considered *sufficient conditions* on the characteristics of the consumers, allowing us to carry through an existence proof. It can easily be verified that the sufficient conditions are not necessary in the sense that no equilibria would exist in economies not satisfying these conditions. In this section, the first step is taken toward establishing necessary conditions for the existence of equilibria in exchange economies. We focus on preferences and therefore keep the other characteristics entirely standard, so that all consumption sets are $\mathbb{R}_+^l$ and survival conditions on endowments are satisfied throughout.

A correspondence $\varphi : A \rightrightarrows B$, where A and B are subsets of Euclidean spaces, is said to have a *closed selection* if there is a closed correspondence $\psi : A \rightrightarrows B$ with $\emptyset \neq \psi(x) \subseteq \varphi(x)$ for all $x \in A$.

Definition 1.4. A set $\mathcal{P}$ of preferences is admissible if

(a) each economy $\mathcal{E} = (P_i, \omega_i)_{i=1}^m$ over $\mathcal{P}$ has a quasi-equilibrium,
(b) for fixed $P_1, \ldots, P_m \in \mathcal{P}$, the correspondence $\Pi[P_1, \ldots, P_m]$: $\left(\mathbb{R}_+^l \setminus \{0\}\right)^m \rightrightarrows \Delta$ assigning to each $(\omega_1, \ldots, \omega_m)$ the set of quasi-equilibrium prices for $(P_i, \omega_i)_{i=1}^m$ has a closed selection (there is a closed correspondence $\psi : \left(\mathbb{R}_+^l \setminus \{0\}\right)^m \rightrightarrows \Delta$ with $\emptyset \neq \psi(x) \subseteq \Pi[P_1, \ldots, P_m]$ for all $x \in A$.

Thus, the class $\mathcal{P}$ is admissible if quasi-equilibria always exist when preferences are taken from $\mathcal{P}$, and the equilibrium price correspondence is continuous in the weak sense of admitting a closed selection.

In what follows we consider some examples of admissible sets of preferences. The class $\mathcal{P}^K$ satisfying Assumption 0.2 was introduced in Introduction chapter, and we used it in Sec., so that it satisfies part (a) of the condition for admissibility. It is easy to check that (b) holds as well.

For the next class considered, we introduce a lemma.

Lemma 1.8. *Let $\mathcal{P}$ be an admissible set of preferences, and let $\mathcal{P}^M$ be the set of $\mathcal{P}$-majorized preferences, that is for each $P \in \mathcal{P}^M$ there is $P' \in \mathcal{P}$,* *such that $P(x) \subseteq P'(x)$ for all $x \in \mathbb{R}^l_+ \setminus \{0\}$. Then $\mathcal{P}^M$ is an admissible set of preferences.*

Proof. Let $P_1, \ldots, P_m \in \mathcal{P}^M$, and for each i, let $P'_i \in \mathcal{P}$ be such that $P_i(x) \subseteq P'_i(x)$, all $x \in \mathbb{R}^l_+ \setminus \{0\}$. Then $(P'_i, \omega_i)_{i=1}^m$ has quasi-equilibria and

$$\Pi[P'_1, \ldots, P'_m](\omega_1, \ldots \omega_m) \subseteq \Pi[P_1, \ldots, P_m](\omega_1, \ldots, \omega_m)$$

for each $(\omega_1, \ldots, \omega_m) \in (\mathbb{R}^l_+ \setminus \{0\})^m$. Since $\Pi[P'_1, \ldots, P'_m]$ has a closed selection, we have $\Pi[P_1, \ldots, P_m]$. $\qquad\square$

The sets $\mathcal{P}^K$ and $\mathcal{P}^{KM} = (\mathcal{P}^K)^M$ will be used repeatedly in what follows. We mention some other conditions on preferences which in the present setup yield the same sets of preferences:

(1) In McKenzie (1981), preferences P satisfy the conditions that $P(x)$ and $P^{-1}(x) = \{y \in \mathbb{R}^l_+ \mid x \in P(y)\}$ are open, and $x \notin \operatorname{conv} P(x)$ for each $x \in \mathbb{R}^l_+$. The preference $(\operatorname{conv} P)$ defined by

$$(\operatorname{conv} P)((x) = \operatorname{conv} P(x), \ x \in \mathbf{R}^l_+$$

belongs to $\mathcal{P}^K$: Choose $y \in (\operatorname{conv}, P)(x)$; then $y \in \operatorname{ri} \operatorname{conv} (\{y_1, \ldots, y_n\})$ for some $y_1, \ldots, y_n \in P(x)$, and since all the points $y_1, \ldots, y_n$ belong to $P(x')$ for x' in some neighborhood of x, so does a whole neighborhood of y, so that $P \in \mathcal{P}^K$.

(2) Borglin and Keiding (1976) considered preferences P which are *locally majorized* by members of $\mathcal{P}^K$ in the following sense: For each $x \in \mathbb{R}^l_+$ there is a neighborhood U_x of x and a preference $P_x \in \mathcal{P}^K$ such that $P(x') \subseteq P_x(x')$ for $x' \in U_x$. That such preferences belong to $\mathcal{P}^{KM}$ is formulated as a lemma.

Lemma 1.9. *Let $\mathcal{P}$ be locally majorized by members of $\mathcal{P}^K$. Then $\mathcal{P} \subset \mathcal{P}^{KM}$.*

Proof. Since $\mathbb{R}^l_+$ is paracompact, the covering $(U_x)_{x \in \mathbb{R}^l_+}$ has a locally finite closed refinement (see, e.g. Willard, 1970), i.e. there is an index set A and a family $\{F_\alpha \mid \alpha \in A\}$ such that

(a) each $x \in \mathbb{R}^l_+$ has a neighborhood meeting only finitely many of the sets F_α, $\alpha \in A$,

(b) each F_α is contained in some U_x,

(c) $\cup_{\alpha \in A} F_\alpha = \mathbb{R}^l_+$.

Assign to each $\alpha \in A$ an $x(\alpha) \in \mathbb{R}_{++}^l$, such that $F_\alpha \subset U_{x(\alpha)}$, and let $P_\alpha = P_{x(\alpha)}$. Define now the correspondence $\overline{P} : \mathbb{R}_+^l \rightrightarrows \mathbb{R}_+^l$ by

$$\overline{P}(x) = \bigcap_{\alpha : x \in F_\alpha} P_\alpha(x).$$

Then $P(x) \subseteq \overline{P}(x)$ since $P(x) \subseteq P_\alpha(x)$ for each α with $x \in F_\alpha$ and $x \notin \overline{P}(x)$. It remains to show that $\overline{P}$ is open in $\mathbb{R}_+^l \times \mathbb{R}_+^l$.

For $x \in \mathbb{R}_+^l$, let V_x be a neighborhood of x only finitely meeting many of the closed sets F_α. Then there is a neighborhood $W_x \subseteq V_x$ of x such that $\{\alpha \mid x' \in F_\alpha\} \subseteq \{\alpha \mid x \in F_\alpha\}$ for all $x' \in W_x$. It follows that

$$\bigcap_{\alpha : x \in F_\alpha} P_\alpha(x') \subseteq \overline{P}(x')$$

for all $x' \in W_x$. Since for fixed x, the correspondence $\cap_{\alpha : x \in F_\alpha} P_\alpha$ has an sssopen graph, we conclude that $\mathcal{P}$ has an open graph. Thus, $\overline{P} \in \mathcal{P}^K$ and $P \in \mathcal{P}^{KM}$. $\square$

We now return to abstract sets $\mathcal{P}$ of admissible preferences. Our goal is to establish results of the following type: If (1) $\mathcal{P}$ contains certain well-behaved preferences and (2) each $P \in \mathcal{P}$ satisfies a suitable monotonicity condition, then $\mathcal{P}$ is a subset of $\mathcal{P}^{KM}$, which then may be considered as a "best possible" set of preferences to be used in general equilibrium models (with a finite commodity space). There is a trade-off between conditions (1) and (2), as will be seen in what follows.

Define the dual correspondence $P^\circ : \mathbb{R}_+^l \rightrightarrows \Delta$ of a preference P by

$$P^\circ(x) = \{p \in \Delta \mid p \cdot x' > p \cdot x, \text{ all } x' \in P(x)\}.$$

Lemma 1.10. *Let P be a preference such that for each $x \in \mathbb{R}_+^l \backslash \{0\}$, $P^\circ(x) \neq \emptyset$ and for some neighborhood U_x of x, the restriction of P° to U_x has a closed selection. Then, $P \in \mathcal{P}^{KM}$.*

Proof. For each $x \in \mathbb{R}_+^l \backslash \{0\}$, let $\varphi_x : U_x \rightrightarrows \Delta$ be a closed selection of P° restricted to U_x, and define $P_x : U_x \rightrightarrows \mathbb{R}_+^l$ by

$$P_x(x') = \left\{ x'' \in \mathbb{R}_+^l \mid p \cdot x'' > p \cdot x', \text{ all } p \in \varphi_x(x') \right\}.$$

Then $P_x(x')$ is convex, $x' \notin P_x(x')$, and $P(x') \subseteq P_x(x')$ for all $x' \in U_x$.

Furthermore, P_x has an open graph in $U_x \times \mathbb{R}^l_+$. It suffices to show that $P_x(x')$ and $P_x^{-1}(x')$ are both open sets. Thus, let $y \in P_x(x')$; then $\min\{p \cdot y \mid p \in \varphi_x(x')\}$ is attained and is equal to some $k > p \cdot x'$. If $\varepsilon < k - p \cdot x'$, then for each y' with $\|y - y'\| \le \varepsilon$ and each $p \in \Delta$, we have

$$|p \cdot (y - y')| \le \|p\| \, \|y - y'\| \le \varepsilon.$$

Thus, $y' \in P_x(x')$, so $P_x(x')$ is open.

Next, let $y \in P_x(x')$ and suppose that there is a sequence $(x_n)_{n \in \mathbb{N}}$ in U_x with $x_n \to x'$ such that $y \notin P_x(x_n)$. Then for each n, there is $p_n \in \varphi_x(x_n)$ with $p_n \cdot y \le p_n x_n$. W.l.o.g., we may assume that the sequence $(p_n)_{n \in \mathbb{N}}$ converges to some p'. We then have that $p' \cdot y \le p' \cdot x'$, and also, since φ_x is closed, $p' \in \varphi_x(x')$, a contradiction showing that $P_x^{-1}(y)$ must be open. Now, reasoning as in the proof of Lemma 1.9 (and noting that $\mathbb{R}^l_+ \backslash \{0\}$ is paracompact), we get the existence of a correspondence $\overline{P} : \mathbb{R}^l_+ \backslash \{0\} \rightrightarrows \mathbb{R}^l_+$ such that $\overline{P}(x)$ is convex, $x \notin \overline{P}(x)$ and $P(x) \subseteq \overline{P}(x)$, all $x \in \mathbb{R}^l_+ \backslash \{0\}$, and $\overline{P}$ has an open graph.

It remains only to extend $\overline{P}$ to an element of $\mathcal{P}^K$ by a suitable definition of $\overline{P}(0)$. Let U_e be an open convex neighborhood of $e = (1, \ldots, 1)$, such that $p \cdot x > 0$ for all $p \in \Delta$, $x \in U_e$. Then $\{x\} + U_e \subset \overline{P}(x)$ for all $x \in \mathbb{R}^l_+ \backslash \{0\}$, and setting $\overline{P}(0) = U_e$ makes $\overline{P}$ an element of $\mathcal{P}^K$.

$\square$

Let $\mathcal{P}^{cl}$ be the set of ('classical') preferences P that can be represented by continuous, strictly quasi-concave and monotonic utility functions. Then $\mathcal{P}^{cl}$ is admissible, and it may be considered to be a rather natural subset of any admissible set $\mathcal{P}$. We restrict attention to preferences that satisfy the following boundary condition: If $x \in \text{bd}\,\mathbb{R}^l_+$ and $p \in \Delta$, $p \cdot x = 0$, then $p \in P^\circ(x)$. Note that if P is an arbitrary preference, then the preference P^* defined by

$$P^*(x) = \begin{cases} P(x) & x \in \text{int}\,\mathbb{R}^l_+ \\ \{y \in P(x) \mid p \cdot y > 0, \text{ all } p \in \Delta \text{ with } p \cdot x = 0\} & \text{otherwise.} \end{cases}$$

satisfies this boundary condition, and P^* is equivalent to P in the sense that the set of quasi-equilibria is unaffected if any P_i is replaced by P_i^*.

Theorem 1.4. *Let $\mathcal{P}$ be an admissible set of preferences, $m \geq 2$, which contain the elements of $\mathcal{P}^{cl}$ satisfying the boundary condition, and for which each $P \in \mathcal{P}$ is weakly monotonic in the following sense: For each $x \in \mathbb{R}_+^l$, if there is $x' \leq x$ and $p \in P^\circ(x')$ with $p \cdot x = p \cdot x'$, then $p \in P^\circ(x)$. Then $\mathcal{P} \subseteq \mathcal{P}^{KM}$.*

Proof. Let $P \in \mathcal{P}$ and $\bar{x} \in \mathbb{R}_+^l \setminus \{0\}$. Define $\widetilde{P} \in \mathcal{P}^{cl}$ as the preference represented by the utility function

$$u(x) = \min\{x_1, \ldots, x_l\},$$

for $x \in \mathbb{R}_+^l$. Let p be a quasi-equilibrium price for the economy $((P, \bar{x}), (\widetilde{P}, e) \ldots, (\widetilde{P}, e))$. If $p \in \text{int}\,\Delta$, then the excess demand of each agent $(\widetilde{P}, e)$ is 0, so $p \in P^\circ(x)$. If $p \in \text{bd}\,\Delta$, then the excess demand of $(\widetilde{P}, e)$ is ≥ 0, and there must be $x' \leq \bar{x}$ with $p \cdot x' = p \cdot \bar{x}$ and $p \in P^\circ(x')$. By the weak monotonicity condition, $p \in P^\circ(\bar{x})$.

Thus, for all $\bar{x} \in \mathbb{R}_+^l \setminus \{0\}$, we have that

$$\Pi[P, \widetilde{P}, \ldots, \widetilde{P}](\bar{x}, e, \ldots, e) \subseteq P^\circ(\bar{x}) \neq \emptyset,$$

and the theorem follows from Lemma 1.10. $\square$

5. Fixed Point Theorems for General Equilibrium Theory

In economic equilibrium analysis, existence theorems are proved by various fixed point theorems for continuous functions or correspondences satisfying suitable continuity assumptions. The simplest, at least conceptually, of these is Brouwer's fixed point theorem stating that a continuous function, mapping a convex, compact subset of $\mathbb{R}^l$ into itself, leaves at least one point fixed. Usually, one starts by proving this theorem, but it is as easy to prove a fixed point theorem for correspondences directly, since this is what we use. The Brouwer fixed point theorem is then obtained as a corollary.

An n-simplex s_n is the convex hull of a finite set of $n+1$ points in a topological vector space V, i.e. $s_n = \text{conv}(\{0, 1, \ldots, n+1\})$. The points $v_0, v_1, \ldots, v_n$ are called the vertices of s_n. A face of s_n is the convex hull of some of the vertices of s_n. Clearly, a face is either empty or a

q-simplex for some $q \leq n$, and the vertices are those faces of s_n that are 0-simplexes. The restriction of the topology of V to s_n induces a topology on s_n, which is metrizable since s_n is homeomorphic to a subset of some finite-dimensional Euclidean space.

A subdivision of s_n is a finite set sd of q-simplexes, $q \leq n$, such that

(i) s_n is the union of the n-simplexes of sd,
(ii) every two simplexes in sd intersect in a common face, which might be empty,
(iii) every face of a simplex in sd is a simplex in sd.

By the mesh of a subdivision, we understand the number $\max\{\mathrm{diam}(s') \mid s' \in \mathrm{sd}\}$. We leave it to the reader to check that there exist subdivisions of arbitrarily small mesh.

A labeled subdivision (sd, Λ) is a subdivision sd together with a map Λ from the set of vertices (= 0-simplexes) in sd to the set $\{0, 1, \ldots, n\}$, such that

$$v \in \mathrm{conv}\left(\left\{v_{i_0}, \ldots, v_{i_k}\right\}\right) \Rightarrow \Lambda(v) \in \{i_0, \ldots, i_k\}.$$

An n-simplex $s' \in \mathrm{sd}$ is completely labeled if the label set of its vertices is $\{0, 1, \ldots, n\}$.

The existence of completely labeled simplexes is asserted by the following result known as Sperner's lemma (Sperner, 1928).

Lemma 1.11. *In a labeled subdivision* (sd, Λ) *of an n-simplex s_n, there are an odd number of completely labeled simplexes.*

Proof. By induction on the number of points $n + 1$ in the simplex s_n, for $n + 1 = 1$, the simplex s_n is a point and the result is trivial. Suppose that it holds for every simplex with n points, $n \geq 2$. For every n-simplex s'_n in sd, let $\gamma(s'_n)$ denote the number of faces of s'_n with label set $\{0, \ldots, n\}$. Then

$$\gamma(s'_n) = \begin{cases} 1 & \text{if } s'_n \text{ is completely labeled,} \\ 0 \text{ or } 2 & \text{otherwise,} \end{cases}$$

Exchange Economies 47

so if Γ denotes the number of completely labeled simplexes, we have that

$$\Gamma \equiv \sum \gamma(s_n') \pmod 2,$$

where summation runs over all n-simplexes in sd. In $\sum \gamma(s_n')$, all $(n-1)$-simplexes except those on the boundary of s_n are counted twice, and all the $(n-1)$-simplexes on the boundary with level set $\{1,\ldots,n\}$ must be on the face $\hat{v}_{n+1} = \text{conv}(\{v_0,\ldots,v_{n-1}\})$. By the induction hypothesis, the number of such $(n-1)$-simplexes is odd, so Γ must be an odd number. $\qquad\square$

From Sperner's lemma, we derive a first result about fixed points of correspondences.

Lemma 1.12. *Let* $\varphi : s_n \rightrightarrows s_n$ *be a correspondence such that*

(i) $\varphi(x)$ *is convex for all* $x \in s_n$,
(ii) *Graph* $\varphi = \{(x,y) \in (s_n)^2 \mid y \in \varphi(x)\}$ *is open in* $(s_n)^2$,
(iii) *for all* $x \in s_n$, $\varphi(x) \cap \{v_0,\ldots,v_n\} \neq \emptyset$.

Then φ *has a fixed point, i.e. there is* $x^0 \in s_n$ *such that* $x^0 \in \varphi(x^0)$.

Proof. Suppose to the contrary that $x \notin \varphi(x)$ for all $x \in s_n$. Choose a number $m \in \mathbb{N}$ and let sd_m be a subdivision of s_n with $\text{mesh}(\text{sd}_m) \leq \dfrac{1}{m}$. Construct a labeled subdivision (sd_m, Λ_m) as follows: For every vertex v of sd_m, let $\Lambda_m(v)$ be a number $r \in \{0,1,\ldots,n\}$, such that $v_r \notin \varphi(v)$. By convexity of $\varphi(v)$ and the assumption that $v \notin \varphi(v)$, there is such a number r. Moreover, we can choose r such that $v \in \text{conv}\{v_{i_0},\ldots,v_{i_k}\}$ implies that $r \in \{i_0,\ldots,i_k\}$, so (sd_m, Λ_m) is actually a labeled subdivision. By Lemma 1.11, there is at least one completely labeled simplex s_n^m.

Now, letting $m \to \infty$ and using compactness of s_n, we get a sequence $(s_n^m)_{n=1}^{\infty}$ of n-simplexes converging to a point $x^0 \in s_n$. Using the way we labeled the simplexes and the fact that the complement of Graph φ is a closed set, it is seen that for $i = 0,\ldots n$, $v_i \notin \varphi(x^0)$. But this contradicts property (iii) of φ, so φ must have a fixed point. $\quad\square$

48 *Theory of General Economic Equilibrium*

We can now prove the following fixed point theorem of Fan (1969).

Theorem 1.5. *Let $K \subset V$ be non-empty, convex and compact, and let $\varphi : K \rightrightarrows K$ be a correspondence such that*

(i) *$\varphi(x)$ is non-empty and convex for all $x \in K$,*
(ii) *Graph φ is open in $K \times K$.*

Then φ has a fixed point.

Proof. For all $x \in K$, there is a point y_x in $\varphi(x)$ and a neighborhood U_x of x such that $y_x \in \varphi(z)$ for all $z \in U_x$. The family $\{U_x \mid x \in X\}$ covers K, and by compactness, a finite subfamily $U_{x_0}, \ldots, U_{x_n}$ covers K. Let $s_n = \text{conv}(\{y_{x_0}, \ldots, y_{x_n}\})$ and consider the correspondence $\varphi' : s_n \rightrightarrows s_n$ defined by $\varphi'(x) = \varphi(x) \cap s_n$ for $x \in s_n$. Then φ' satisfies the assumptions in Lemma 1.12, and consequently, it has a fixed point, which is also a fixed point for φ. $\qquad\square$

Using the above mentioned result, we can state and prove Kakutani's fixed point theorem (Kakutani, 1941).

Theorem 1.6. *Let $K \subset V$ be non-empty, convex and compact, and let $\varphi : K \rightrightarrows K$ be a correspondence such that*

(i) *$\varphi(x)$ is non-empty, closed and convex for all $x \in K$,*
(ii) *φ is upper hemicontinuous.*

Then φ has a fixed point.

Before proving Theorem 1.6, we introduce a lemma.

Lemma 1.13. *Let K and φ be as in Theorem 1.6, and denote by Ψ the set of correspondences $\psi : K \rightrightarrows K$ such that*

(i) *$\varphi(x)$ is convex and $\varphi(x) \subset \psi(x)$ for all $x \in K$,*
(ii) *Graph ψ is open in $K \times K$.*

Then for all $(x, y) \in K \times K$ with $y \notin \varphi(x)$, there is $\psi \in \Psi$ such that $(x, y) \in \text{cl Graph } \psi$.

Proof. If $y \notin \varphi(x)$, then there is a convex neighborhood U of 0 in V such that $y \notin \text{cl}\,[\varphi(x) + U]$. By upper hemicontinuity of φ, there is a neighborhood W of x such that $\varphi(z) \subset \varphi(x) + U$ for $z \in \text{cl}\,W$. Define $\psi : K \rightrightarrows K$ by

$$
\psi(x) = \begin{cases} \varphi(x) + U & z \in \text{cl}\,W, \\ K & \text{otherwise.} \end{cases}
$$

It is readily seen that $(x, y) \notin \text{cl}\,\text{Graph}\,\psi$ and that $\psi \in \Psi$. $\qquad\square$

Now we can prove the theorem.

Proof of Theorem 1.6. Let $\mathcal{F}$ be the family of closed sets

$$
\mathcal{F} = \{\text{cl}\,\text{Graph}\,\psi \mid \psi \in \Psi\} \cup \text{Diag}(K),
$$

where Ψ is the set defined in Lemma 1.13 and $\text{Diag}(K) = \{(x, x) \mid x \in K\}$. By Theorem 1.5, every finite subfamily of $\mathcal{F}$ has a non-empty intersection. By compactness of $K \times K$, so has $\mathcal{F}$. Let (x^0, x^0) belong to $\cap \mathcal{F}$. By Lemma 1.13, $x^0 \notin \varphi(x^0)$ would imply $(x^0, x^0) \notin \text{cl}\,\text{Graph}\,\psi$ for some $\psi \in \Psi$, a contradiction, so $x^0 \in \varphi(x^0)$. $\qquad\square$

We use a version of Kakutani's fixed-point theorem, which exploits an idea of Gale and Mas-Colell (1975). For this purpose, we introduce some terminology: Let $K_1 \subset \mathbb{R}^{l_1}$, $K_2 \subset \mathbb{R}^{l_2}$ be non-empty sets; a correspondence $\varphi : K_1 \times K_2 \rightrightarrows K_2$ has property GM if

(i) the set $\{(x_1, x_2) \in K_1 \times K_2 \mid \varphi(x_1, x_2) \neq \emptyset\}$ is open,
(ii) there is a closed correspondence $\psi : K_1 \times K_2 \rightrightarrows K_2$ with convex values such that $x_2 \in \psi(x_1, x_2)$ implies $x_2 \in \varphi(x_1, x_2)$.

We may then state and prove the following theorem:

Theorem 1.7. *Let $K_i \subset \mathbb{R}^{l_i}$ be non-empty, convex and compact, and let $\varphi_i : K_1 \times \cdots \times K_r \rightrightarrows K_i$ be a correspondence with convex values and an open graph, $i = 1, \ldots, r$. Then there is $(x_1, \ldots, x_r) \in \prod_{i=1}^r K_i$, such that for each i, either $x_i \in \varphi_i(x_1, \ldots, x_r)$ or $\varphi_i(x_1, \ldots, x_r) = \emptyset$.*

Proof. For each i, define $\widehat{\psi}_i : \prod_{i=1}^r K_i \rightrightarrows \prod_{i=1}^r K_i$ by

$$\widehat{\psi}_i(x_1,\ldots,x_r)$$

$$= \begin{cases} \left\{(x_1',\ldots,x_r') \in \prod_{i=1}^r K_i \,\big|\, x_i' \in \psi(x_1,\ldots,x_r)\right\} & \varphi(x_1,\ldots,x_r) \neq \emptyset, \\ K_i & \text{otherwise,} \end{cases}$$

where ψ_i is the correspondence given by property GM. Then each $\widehat{\psi}_i$ is closed with non-empty, convex values. Define $\widehat{\psi}$: $\prod_{i=1}^r K_i \rightrightarrows \prod_{i=1}^r K_i$ by $\widehat{\psi}(x_1,\ldots,x_r) = \prod_{i=1}^r \widehat{\psi}_r(x_1,\ldots,x_r)$ and apply Theorem 1.6. $\qquad\square$

Remark 1.1. The conclusion of Theorem 1.7 holds if some of the correspondences φ_i are also instead upper hemicontinuous with non-empty closed and convex values, cf. Exercise 6.

6. Exercises

(1) Investigate the Walras equilibria in the Edgeworth box for an economy where the consumers have linear preferences, so that $P_i(x_i) = \{x_i' \mid q_i \cdot (x_i' - x_i) > 0\}$ for some $q_i \in \mathbb{R}_+^2$.

(2) Consider an exchange economy with two consumers and two goods. Consumption sets are $\mathbb{R}_+^2$, and preferences are described by utility functions

$$u_1(x_{11}, x_{12}) = \ln x_{11} + x_{12}, \quad u_2(x_{21}, x_{22}) = x_{21} + 2\sqrt{x_{22}},$$

for the two consumers. The initial endowments are $\omega_1 = (4, 6)$, $\omega_2 = (6, 4)$.

Find a Walras equilibrium in this economy.

(3) Consider an economy with two consumers and two goods. The consumers have consumption set $\mathbb{R}_+^2$ and preferences described by utility functions

$$u_1(x_{11}, x_{12}) = \min\{x_{11}, x_{12}\}, \; u_2(x_{21}, x_{22}) = x_{21}^2 + x_{22}^2,$$

Exchange Economies 51

and the initial endowments are $\omega_1 = (1,2)$, $\omega_2 = (2,1)$. Show that there is no Walras equilibrium in this economy.

(4) Prove the Brown–Matzkin theorem: There are price-endowment pairs $(p,(\omega_i)_{i=1}^m)$ and $(p',(\omega_i')_{i=1}^m)$, such that for no collection of preference relation $(P_i)_{i=1}^m$ on $\mathbb{R}_+^l$ representable by utility functions and satisfying monotonicity it would be the case that p is a Walras equilibrium price for $(\mathbb{R}_+^l, P_i, \omega_i)_{i=1}^m$ and p' is a Walras equilibrium price for $(\mathbb{R}_+^l, P_i, \omega_i')_{i=1}^m$.

(5) Let $\mathcal{E} = (X_i, P_i, \omega_i)_{i=1}^m$ be an economy where all consumers are resource related (cf. Section 3.1). Show that if one consumer satisfies the minimum, wealth condition, so do all the other consumers.

(6) Prove Lemma 1.5: Let $\mathcal{E}$ be such that $P_i = \{(x_i, x_i') \in \mathbb{R}_+^l \times \mathbb{R}_+^l \mid p^i \cdot x_i' > p^i \cdot x_i\}$, all $i \in M$, for some $p^i \in \Delta$, and $\sum_{i \in M} \omega_i \in \text{int } \mathbb{R}_{++}^l$. Let $\theta_1, \dots, \theta_m$ be positive numbers with $\sum_{i \in M} \theta_i = 1$.

If $x = (x_1, \dots, x_m)$ is a feasible allocation that maximizes the function

$$(p^1 \cdot x_1)^{\theta_1} \cdots (p^m \cdot x_m)^{\theta_m}$$

over all feasible allocations, and p is defined by $p_h = \max_{i \in M}(p_h^i \theta_i)/(p^i \cdot x_i)$, then (x,p) satisfies (i) $p \cdot x_i = \theta_i$, and (ii) $x_i' \in P_i(x_i)$ implies $p \cdot x_i' > \theta_i$ for each $i \in M$.

(7) Prove the statement in Remark 1.1.

(8) Let $\mathcal{E} = (X_i, P_i, \omega_i)_{i=1}^m$ be an exchange economy where $X_i = \mathbb{R}_+^l$, P_i satisfies Assumption 0.2, and $\omega_i \in \mathbb{R}_{++}^l$ for each i.

Show that there is a preference correspondence P on $\{v \in \mathbb{R}^{m+1}l_+ \mid v_h \leq K\}$ for some $K > 0$ satisfying Assumption 0.2 such that

$$P(x_1, \dots, x_m, p) = \emptyset \Leftrightarrow (x_1, \dots, x_m, p) \text{ is a Walras equilibrium in } \mathcal{E}. \tag{10}$$

Which properties of the preference correspondence P in addition to those of Assumption 0.2 are necessary for (10) to hold?

Chapter 2

Economies with Production

In this chapter, we extend the model introduced in the previous chapter so as to take account of production: Commodities are not just available, but are produced using other commodities. This gives rise to a specific economic activity, production, which in our model is carried out by specific agents, *producers*, who select the bundles of commodities used as inputs and outputs. Producers are treated as derived agents, i.e. they are not themselves individuals endowed with preferences on bundles or allocations, rather they represent the preferences of some or all consumers in a particular way to be specified in the model.

The presence of production poses some new problems, and one might fear that all that was done in the previous chapter for economies without production must be redone to take care of production. Fortunately, this is not the case: As long as we are concerned with market behavior under perfect competition, the new agents pose few if any new problems, so that the results obtained for exchange economies can be extended with almost no pain to cover the new situation.

Thus, the present chapter might be a very brief one, at least as long as our scope is only the classical perfect-competition model of production and exchange. Since we have a wider perspective on the theory of general equilibrium, we consider in more detail what could be regarded as special cases of the general model, namely, the von Neumann model of growth and the Heckscher–Ohlin theory of international trade.

1. Walras Equilibria in Economies with Production

1.1 *Production economies and their Walras equilibria*

In order to define an economy with production, we need to add *producers* (cf. Introduction chapter) to the model of an exchange economy. The producers are defined by their *production sets* $Y_j \subset \mathbb{R}^l_+$ for $j = 1, \ldots, n$, given that we have n producers in the model. An economy with production is therefore an array

$$\mathcal{E} = \left((X_i, P_i)_{i=1}^m, (Y_j)_{j=1}^n, \omega \right),$$

specifying the characteristics of both consumers and producers. Allocations in $\mathcal{E}$ are arrays $(x, y) = (x_1, \ldots, x_m, y_1, \ldots, y_n)$, where for each i, x_i is a consumption bundle for the consumer i, and similarly, y_j is a (net) production for each producer j, $j = 1, \ldots, n$. The allocation (x, y) is *feasible* if it is *individually feasible*,

$$x_i \in X_i, \quad i = 1, \ldots, m, \quad y_j \in Y_j, \quad j = 1, \ldots, n, \tag{1}$$

so that consumption and production can be carried out, and *aggregate feasible*,

$$\sum_{i=1}^m x_i \leq \sum_{j=1}^n y_j + \omega, \tag{2}$$

that is, if for each commodity, what is consumed is no more than what is available from initial endowment not used by producers and from production. The set of feasible allocations in $\mathcal{E}$ is written as $F(\mathcal{E})$. In the case where the initial endowment is owned by the consumers, we have that $\omega = \sum_{i=1}^m \omega_i$.

In order to introduce the Walras equilibrium for an economy with production, we need a rule for determining the way in which producers' profits are transferred to consumers. The tradition prescribes a linear function, so that each consumer i receives a share θ_{ij} of the profits earned by producer j: If prices are given by $p \in \mathbb{R}^l$ and producers have chosen production plans $y_j \in Y_j$, $j = 1, \ldots, n$, then consumer i gets an income

$$w_i(p) = p \cdot \omega_i + \sum_{j=1}^{n} \theta_{ij} p \cdot y_j. \tag{3}$$

The profit shares θ_{ij} satisfy the condition

$$\sum_{i=1}^{m} \theta_{ij} = 1$$

for $j = 1, \ldots, n$, so that the the profits $p \cdot y_j$ are fully distributed among consumers. For the interpretation as shares, it seems natural to assume also that $\theta_{ij} \geq 0$ for all i and j, even if this is not strictly necessary. We define an *economy with private ownership* as

$$\mathcal{E} = \left((X_i, P_i, \omega_i)_{i=1}^{m}, (Y_j)_{j=1}^{n}, (\theta_{ij})_{i=1 \, j=1}^{m \, n} \right),$$

where all the ingredients have been explained above.

Definition 2.1. Let $\mathcal{E} = ((X_i, P_i, \omega_i)_{i=1}^{m}, (Y_j)_{j=1}^{n}, (\theta_{ij})_{i=1 \, j=1}^{m \, n})$ be an economy with private ownership. An array (x, y, p), where (x, y) is an allocation and $p \in \mathbb{R}^l$ a price, is a Walras equilibrium if

  (i) for $i = 1, \ldots, m$, x_i is maximal for P_i in the budget set $\{x_i' \in X_i \mid p \cdot x_i' \leq p \cdot \omega_i + \sum_{j=1}^{n} \theta_{ij} p \cdot y_j\}$,
  (ii) for $j = 1, \ldots, n$, $p \cdot y_j \geq p \cdot y_j'$, all $y_j' \in Y_j$,
  (iii) $(x, y) \in F(\mathcal{E})$.

As for economies without production, we may also consider the slightly weaker concept of a quasi-equilibrium, which differs only in the conditions for individual optimality of consumers, where maximum for the preference relation in the budget set is replaced by expenditure minimization in the preferred set.

Definition 2.2. Let $\mathcal{E} = ((X_i, P_i, \omega_i)_{i=1}^{m}, (Y_j)_{j=1}^{n}, (\theta_{ij})_{i=1 \, j=1}^{m \, n})$ be an economy with private ownership. An array (x, y, p), where (x, y) is an

56   *Theory of General Economic Equilibrium*

allocation and $p \in \mathbb{R}^l$ a price, is a Walras equilibrium if

(i) for $i = 1, \ldots, m, p \cdot x_i = \inf\{ p \cdot x_i' \mid x_i' \in P_i(x_i)\}$,
(ii) for $j = 1, \ldots, m, p \cdot y_j \geq p \cdot y_j'$ for all $y_j' \in Y_j$,
(iii) $(x, y) \in F(\mathcal{E})$.

1.2   *The existence problem in economies with production*

The question of existence of (quasi-)equilibria in production economies can be settled rather easily using our work in Chapter 1, since from a formal point of view, producers can be transformed to consumers of a particular type. In order to perform this reduction of the problem, we must however accept that consumer incomes are defined as arbitrary continuous functions, rather than just linear functions, of the price system. This should come as no surprise given the definition of consumer income in (3), and the existence proofs for exchange economies' work are easily adapted to this situation (cf. Exercise 3 in Chapter 1).

Lemma 2.1. *Let $\mathcal{E} = ((X_i, P_i, \omega_i)_{i=1}^{m}, (Y_j)_{j=1}^{n}, (\theta_{ij})_{i=1\,j=1}^{m\ \ n})$ be an economy with production, where for each producer j,*

(i) *Y_j is convex,*
(ii) *Y_j satisfies the free-disposal assumption $\{y_j\} - \mathbb{R}_+^l \subseteq Y_j$ for each $y_j \in Y_j$.*

Then there is an exchange economy $\widetilde{\mathcal{E}}$ with $m + n$ consumers, of which the first m are identical with the consumers in $\mathcal{E}$, and a system of income transfers, such that (x, y, p) is a quasi-equilibrium in $\mathcal{E}$ if and only if $(x_1, \ldots, x_m, \tilde{x}_1, \ldots, \tilde{x}_n, p)$, with $\tilde{x}_j = -y_j$ for $j = 1, \ldots, n$, is a quasi-equilibrium in $\widetilde{\mathcal{E}}$.

Proof. Let K be a number so large that $(x_1, \ldots, x_m, y_1, \ldots, y_n) \in F(\mathcal{E})$ implies that $|y_{jh}| < K$ for $j = 1, \ldots, n$, $h = 1, \ldots, l$, and define for $j = 1, \ldots, n$ the consumer $(\widetilde{X}, \widetilde{P})$, where the consumption set $\widetilde{X}_j$ is defined by

$$\widetilde{X}_j = \{ y_j \mid y_j \in Y_j, |y_{jh}| < K, h = 1, \ldots, l\}, \tag{4}$$

and the preferences $\widetilde{P}_j : \widetilde{X}_j \rightrightarrows \widetilde{X}_j$ by

$$\widetilde{P}_j(\tilde{x}_j) = \{\tilde{x}_j\} + \mathbb{R}^l_+.$$

The initial endowment of the consumer $(\widetilde{X}_j, \widetilde{P}_j)$ is chosen as 0. Let the incomes of consumers $i = 1, \ldots, m$ be given by (3) and for the new consumers $j = 1, \ldots, n$ by

$$w_j(p) = \inf \{ p \cdot \tilde{x}_j \mid \tilde{x}_j \in \widetilde{X}_j < 0 \}.$$

Let (x, y, p) be a quasi-equilibrium in $\mathcal{E}$. Then $(x, y) \in F(\mathcal{E})$, so that $x_i \in X_i$ for $i = 1, \ldots, m$ and $y_j \in Y_j$, and by the definition of $\widetilde{X}_j$, we get that $\tilde{x}_j \in \widetilde{X}_j$ for $j = 1, \ldots, n$. Moreover,

$$\sum_{i=1}^{m} x_i + \sum_{j=1}^{n} \tilde{x}_j = \sum_{i=1}^{m} x_i - \sum_{j=1}^{n} y_j \le \sum_{i=1}^{m} \omega_i,$$

so that $(x_1, \ldots, x_m, \tilde{x}_1, \ldots, \tilde{x}_n) \in F(\widetilde{\mathcal{E}})$. For $i = 1, \ldots, m$, the individual optimization condition (ii) in Definition 2.1 is identical to the individual optimization condition in $\mathcal{E}$ under the budget constraint given by $w(p)$, and since y_j is maximal for $p \cdot y'_j$ on Y_j, it is maximal on any subset of Y_j containing y_j, so that $p \cdot \tilde{x}'_j$ attains its minimum on $\widetilde{X}_j$ at $\tilde{x}_j$, giving the value $p \cdot \tilde{x}_j = w_j(p)$, which is the individual optimization condition in the quasi-equilibrium of $\widetilde{\mathcal{E}}$.

Conversely, let $(x, \tilde{x}, p)$ be a quasi-equilibrium in the exchange economy $\widetilde{\mathcal{E}}$. Then feasibility of $(x, \tilde{x})$ in $\widetilde{\mathcal{E}}$ translates to feasibility of (x, y) in $\mathcal{E}$ in the same way as above, and similarly, the individual optimality for the m consumers $(X_i, P_i), i = 1, \ldots, m$, in $\widetilde{\mathcal{E}}$ carries over to individual optimality of the consumers in $\mathcal{E}$. To show that also condition (iii) in Definition 2.1 is satisfied, we note that $y_j = -\tilde{x}_j$ maximizes $p \cdot y_j$ on the set

$$\{y_j \mid y_j \in Y_j, |y_{jh}| < K, h = 1, \ldots, l\},$$

and since Y is convex, it must maximize $p \cdot y'_j$ on all of Y, which is condition (iii). $\qquad\square$

Using the result in Lemma 2.1, we immediately get the following result on the existence of quasi-equilibria in economies with production.

Theorem 2.1. *Let $\mathcal{E} = ((X_i, P_i, \omega_i)_{i=1}^m, (Y_j)_{j=1}^n, (\theta_{ij})_{i=1\,j=1}^{m\ \ n})$ be a production economy with private ownership, where*

(A1) *the set $F(\mathcal{E})$ is non-empty and compact,*

(A2) *for each i, X_i satisfies Assumption 0.1 and P_i satisfies Assumption 0.2,*

(A3) *for each j, Y_j is closed, convex, and satisfies $Y_j - \mathbb{R}_+^l \subseteq Y_j$.*

Then there is a quasi-equilibrium in $\mathcal{E}$.

From the existence of quasi-equilibria, we may deduce existence of Walras equilibria when we add assumptions guaranteeing that expenditure minimization entails preference maximization. Using Lemma 1.6 (which holds also when there are producers in the economy), we get the following corollary.

Corollary. *Let $\mathcal{E} = ((X_i, P_i, \omega_i)_{i=1}^m, (Y_j)_{j=1}^n, (\theta_{ij})_{i=1\,j=1}^{m\ \ n})$ be a production economy with private ownership satisfying the assumptions (A1)–(A3) in Theorem 2.1, and assume that $w_i(p)$ as defined in (3) satisfies*

$$w_i(p) > \inf \{p \cdot x_i' \mid x_i' \in X_i\}$$

for all $p \in \Delta$, $i = 1, \ldots, m$. Then there is a Walras equilibrium in $\mathcal{E}$.

1.3 Aggregate production and feasibility

In our formulation of the theorem, the condition (A1) asserts that there are feasible allocations in $\mathcal{E}$, which clearly is a precondition for the existence of a quasi-equilibrium, and that the set of feasible allocations is bounded, so that arbitrarily large inputs or outputs cannot be sustained by the available resources. Since (A1) is not formulated in terms of original data of $\mathcal{E}$, we need to elaborate somewhat upon this assumption, searching for conditions on production sets, consumption sets and endowment which will guarantee that the set of feasible allocations is (a) non-empty and (b) bounded. Part (a) is a

Economies with Production 59

survival assumption of the same type of that treated in the previous chapter, so we concentrate on condition (b).

In the case of an exchange economy, boundedness of the set of feasible allocations is a straightforward consequence of the lower boundedness of consumption sets X_i for $i = 1, \ldots, m$. In the case of an economy with production, the obvious counterpart would be upper boundedness of production sets, but this conflicts with the interpretation of production sets as a collection of technologically feasible productions, ruling out production under constant returns to scale where any feasible production may be scaled up or down arbitrarily.

Let $Y = \sum_{j=1}^{n} Y_j$ be the aggregate production set. What matters in the present context is not that Y itself should have an upper bound, rather we want that its intersection with $\{\omega\} - \sum_{i=1}^{m} X_i$ should be bounded, and with the usual assumptions on consumption sets, this means that at $Y \cup \mathbb{R}_+^l$ must be bounded from above, so that there is a bound to the possibility of aggregate-free production.

Turning then to the individual production sets, we note that even if $Y_j \cap \mathbb{R}_+^l = \{0\}$ for each j, so that no free production is possible in any firm, aggregate production may exhibit free production: Let $l = 2$, and suppose that there are two producers with production sets

$$Y_1 = \{(y_{11}, y_{12}) \mid y_{11} \leq 0, y_{12} \leq -2y_{11}\},$$
$$Y_2 = \{(y_{21}, y_{22}) \mid y_{22} \leq 0, y_{21} \leq -2y_{22}\},$$

then every $y \in \mathbb{R}_+^2$ can be written as $y = y_1 + y_2$ with $y_1 \in Y_1, y_2 \in Y_2$. Thus, if we allow for production sets with no upper bound, we must make sure that these production sets cannot combine in a way as to allow for unbounded free production, so that boundedness assumptions in one way or another cannot be avoided.

2. Linear Models of Production and Growth

In this section, we consider economies where production has a linear structure, so that if two production plans are both technically

60 *Theory of General Economic Equilibrium*

feasible, then so is their sum. This linearity has far-reaching consequences, some of which are considered in the following section.

2.1 *The simple linear production model*

We consider an economy with l goods, each of which can be produced using the other ones. Assume that to produce one unit of the hth commodity, one needs $a_{hk} \geq 0$ units of the kth commodity. Then the production of good h can be described by an l-dimensional vector $(a_{h1}, \ldots, a_{hl})$, where $a_{hh} = 0$, often referred to as a *process* or *activity*. Assuming constant returns to scale, we get that if x_h units of good h are needed, then the use of good k is $x_h a_{hk}$. We assume that each of the goods considered is needed as input somewhere, so that for each k, there is h, such that $a_{hk} > 0$.

Writing the l processes corresponding to each of the goods as columns in an $(l \times l)$ matrix A, it is seen that producing the output array $x = (x_1, \ldots, x_l)$ of goods, the input demand is Ax. The surplus available is then $x - Ax = (I - A)x$, where I is the $(l \times l)$ unit matrix.

Definition 2.3. The matrix A is *productive* if there is $\overline{x} \in \mathbb{R}_+^l$ such that $(I - A)\overline{x} \in \mathbb{R}_{++}^l$, i.e. $\overline{x}$ is positive in all coordinates.

If A is productive, then the economy can create a positive surplus of each of the l commodities. A more specific demand would be that the economy should be able to establish a given surplus vector $c \in \mathbb{R}_+^l$, so that the equation

$$(I - A)x = c \tag{5}$$

has a solution $x \in \mathbb{R}^l$.

Finally, we may investigate whether the economy can sustain balanced growth in the sense that for some $\mu < 1$ and $x \in \mathbb{R}_+^l$, the net output at the production x is μx, or equivalently, whether there is $\lambda > 0$ and $x \in \mathbb{R}_+^l$, such that

$$Ax = \lambda x. \tag{6}$$

Economies with Production 61

There are some obvious interrelations between the concepts introduced.

Lemma 2.2. *If A is productive and $(I - A)x \in \mathbb{R}_+^l$ for some $x \in \mathbb{R}^l$, then x is non-negative, i.e. $x \in \mathbb{R}_+^l$.*

Proof. Clearly, $\bar{x} \in \mathbb{R}_{++}^l$, where $\bar{x}$ is the vector given in Definition 2.3, since for each h, we have that $\bar{x}_h = \sum_{k=1}^l a_{hk}\bar{x}_k$, and $a_{hk} > 0$ for some k by our assumptions on A. Assume now that $(E - A)x \in \mathbb{R}_+^l$ but $x \notin \mathbb{R}_+^l$, and let

$$\theta = \max_h \left\{ -\frac{x_h}{\bar{x}_h} \right\},$$

w.l.o.g. we may assume that $\theta = \frac{x_1}{\bar{x}_1}$. Then $\theta > 0$, and if $x' = x + \theta\bar{x}$, then $x' \in \mathbb{R}_+^l$ with $x_1' = 0$. However,

$$x + \theta\bar{x} > Ax + A\theta\bar{x} = Ax' \geq 0,$$

from which $x_1' > 0$, a contradiction. $\qquad\square$

Lemma 2.3. *If A is productive, then $I - A$ is regular.*

Proof. If $(I - A)x = 0$, then also $(I - A)(-x) = 0$, and from Lemma 2.2, we obtain that $x \geq 0$ and $-x \geq 0$, so that $x = 0$. $\qquad\square$

We combine the above results to the following.

Theorem 2.2. *If A is productive, then Eq. (5) has one and only one solution $x \geq 0$.*

Moreover, the following holds:

Lemma 2.4. *The following are equivalent*:

(a) *Equation (5) has a unique solution $x \geq 0$,*
(b) *$I - A$ is regular and $(I - A)^{-1}$ is non-negative (that is, every element in $(I - A)^{-1}$ is ≥ 0).*

Proof. The implication $(b) \Rightarrow (a)$ is trivial. $(a) \Rightarrow (b)$: The matrix $I - A$ is regular by Lemma 2.3, and we have that $(I - A)^{-1} \geq 0$ for each $c \geq 0$. Choosing $c = e_i = (0, \ldots, 0, 1, 0, \ldots, 0)$ (the ith unit vector), for $i = 1, \ldots, l$, we obtain (b). $\qquad\square$

2.1.1 The Perron–Frobenius theorem

The problem about the possibility of sustained growth turns out to be a question of existence of eigenvalues and eigenvectors under non-negativity constraints. Here, one may use the following theorem of Perron and Frobenius.

Theorem 2.3 (Perron–Frobenius). *Let A be an $(l \times l)$-matrix with non-negative elements. Then the following hold*:

(i) *A has at least one non-negative eigenvalue. To the largest of these eigenvalues, λ, corresponds a non-negative eigenvector.*
(ii) *The matrix $\mu I - A$ has an inverse with non-negative elements if and only if $\mu > \lambda$.*
(iii) *If $Ay \geq \mu y$ for some $\mu \in \mathbb{R}$ and $y \geq 0$, $y \neq 0$, then $\lambda \geq \mu$.*
(iv) *$\lambda \geq |\omega|$ for every eigenvalue ω of A.*

Proof. Let

$$M(A) = \{\rho \mid \rho I - A \text{ has a non-negative inverse}\}.$$

Choose $x > 0$ and $\rho > 0$ large enough, so that $\rho x > Ax$, that is, $(\rho I - A)x \geq 0$ or alternatively, $(I - \rho^{-1}A)x \geq 0$. Then the matrix $\rho^{-1}A$ is productive, and $I - \rho^{-1}A$ has a non-negative inverse, meaning that $\rho \in M(A)$. Clearly, $\eta > \rho$ implies that $\eta \in M(A)$, and $\rho \in M(A)$ implies that $\rho > 0$. Also, $M(A)$ is an open set: ρ belongs to $M(A)$ whenever $\rho x > Ax$ for some $x \geq 0$, and then $\rho'x > Ax$ for ρ' sufficiently close to ρ.

Let $\lambda = \inf M(A)$. Then $\lambda \geq 0$ and $\lambda \notin M(A)$. We show that there is $x \in \mathbb{R}_+^l$, such that $Ax = \lambda x$ (so that λ is indeed an eigenvector of A): Choose some $c > 0$ and let $y(\rho) = (\rho I - A)^{-1}c$ for $\rho \in M(A)$. Let $\rho, \eta \in M(A)$ with $\rho \geq \eta$. Then

$$(\rho I - A)(y(\rho) - y(\eta)) = c - \rho y(\eta) + Ay(\eta)$$
$$= c - \rho y(\eta) + \eta y(\eta) - \eta y(\eta) + Ay(\eta)$$
$$= c - \rho y(\eta) + \eta y(\eta) - c = (\eta - \rho)y(\eta),$$

so that

$$y(\rho) - y(\eta) = (\eta - \rho)(\rho I - A)^{-1}y(\eta) \geq 0. \tag{7}$$

Choose a sequence $(\rho_n)_{n\in\mathbb{N}}$ in $M(A)$ decreasing towards λ. From (7), it is seen that $y(\rho_n)$ is non-decreasing in all coordinates. It follows that also $\sigma_n = e \cdot y(\rho_n)$ is non-decreasing (here, e is the diagonal unit vector $e = (1, 1, \ldots, 1)$ in $\mathbb{R}^l$). There are two possibilities:

(1) The sequence $(\sigma_n)_{n=1}^{\infty}$ has an upper bound. Then $y(\sigma_n)$ converges to some y, and since $(\sigma_n I - A)y(\rho_n) = c$, we get that $(\lambda I - A)y = c > 0$, meaning that $\lambda \in M(A)$, a contradiction.

(2) The sequence $(\sigma_n)_{n\in\mathbb{N}}$ goes to $+\infty$. Let $x_n = \frac{y(\rho_n)}{\sigma_n}$. Then all x_n belongs to a compact set, so we may assume that $x_n \to x$. From $(\rho_n I - A)x_n = \frac{1}{\sigma_n}c$, we get that $(\lambda I - A)x = 0, x \geq 0$. This concludes the proof of (i). (ii) is straightforward.

For (iii), suppose that $\mu > \lambda$. Then there is a non-negative inverse of $\mu I - A$. From $0 \geq (\mu I - A)y$, we then obtain that $y \leq 0$, a contradiction. Finally, (iv) is shown as follows: Let $z \in \mathbb{C}^l$ be an eigenvector associated with an eigenvalue ω, so that $Az = \omega z$. Then

$$\sum_{h=1}^{l} a_{kh}z_h = \omega z_k, \quad k = 1, \ldots, l.$$

Taking absolute values $|\cdot|$, we get that

$$\sum_{h=1}^{l} a_{kh}|z_h| \geq |\omega||z_k|, \quad k = 1, \ldots, l,$$

or $A|z| \geq |\omega||z|$, where $|z| = (|z_1|, \ldots, |z_l|)$. Using (iii) gives the conclusion (iv). $\qquad\square$

The eigenvalue λ is called the *dominating root* of A and is denoted $\lambda(A)$. We look for a condition on A which will assure that $\lambda(A) \geq 0$.

Definition 2.4. The matrix A is decomposable if there is a proper non-empty subset J of $\{1, \ldots, l\}$ such that $a_{ij} = 0$ for $i \notin J, j \in J$.

Lemma 2.5. *The matrix A is decomposable if there is a real number $\mu \geq 0$ and a non-negative vector x with some, but not all coordinates are equal to 0, such that $Ax \leq \mu x$. Otherwise, A is indecomposable.*

Proof. Let $J = \{h \mid x_h\} > 0$, then $J \neq \emptyset$ and $J \neq \{1, \ldots, l\}$. Since $Ax \leq \mu x$, we get that $\sum_{h=1}^{l} a_{kh} x_h \leq 0$ for $k \neq J$. But each member of the sum is non-negative, so $a_{kh} = 0$ for $k \neq J, h \in J$. $\qquad \square$

Matrix A is indecomposable if A is different from the (1×1) matrix $\{0\}$ and A is not decomposable. We have the following.

Theorem 2.4. *Let A be an indecomposable matrix. Then the following holds*:

(i) $\lambda(A) \geq 0$ *and for every eigenvector $x \geq 0$ belonging to $\lambda(A)$, we have that $x > 0$.*
(ii) *The eigenvector belonging to $\lambda(A)$ is uniquely determined up to a scalar multiple.*

Proof. (i) Since $Ax = \lambda x$ and $x \geq 0$, then by Lemma 2.5, we cannot have that $x_h = 0$ for some h. If $\lambda(A) = 0$, we have that $Ax = 0$, that is, $\sum_{h=1}^{l} a_{kh} x_h = 0$ for all k, but then $a_{kh} = 0$, all k, h, and $A = 0$, which for $l \geq 2$ means that A is decomposable, and for $l = 1$ that A is the (1×1) zero matrix, and in either case we have a contradiction.

(ii) If x and y are two distinct eigenvectors belonging to $\lambda(A)$, then there is $\theta > 0$, such that $\theta x \geq y$ and $\theta x_h = y_h$ for some h. But then $\theta x - y$ is a non-negative eigenvector belonging to $\lambda(A)$, and $\theta x - y$ is non-negative with some zero coordinates, a contradiction. $\qquad \square$

2.1.2 *The Marx–Leontief model*

In the simple production model introduced in the previous section, it was seen that if A is productive, then it is possible to produce an arbitrary commodity bundle $c \geq 0$. In many cases, it is however assumed that one or more commodities (1) cannot be produced, and (2) are available in limited supply. If we assume that there is a single such commodity (the lth commodity in the following called labor),

Economies with Production 65

then we have $l - 1$ processes, about which we further assume that $a_{lh} > 0$ for $h = 1, \ldots, l - 1$ (labor is essential for producing each of the other commodities). A model of this kind is called a (simple) Leontief model. We denote by A_1 the $(l - 1) \times (l - 1)$ submatrix of A corresponding to the first $l - 1$ commodities. Assume in the following that A_1 is productive.

For $p \in \mathbb{R}_+^l$ a vector of prices with $p_l = 1$, the *profit* connected with producing one unit of commodity $h \in \{1, \ldots, l - 1\}$ is the quantity

$$\pi_h = p_h - \sum_{k=1}^{l-1} p_k a_{kh} - a_l.$$

Theorem 2.5. *There exists a price vector $p \in \mathbb{R}_+^l$ with $p_l = 1$, such that $\pi_h = 0$ for $h = 1, \ldots, l - 1$.*

Proof. We must show that there is a non-negative solution to the system of equations

$$p_h - \sum_{k=1}^{l-1} p_k a_{kh} - a_l = 0, \quad h = 1, \ldots, l - 1,$$

or, in matrix notation, to

$$p(I - A_1) = a_l, \tag{8}$$

where $a_l = (a_{l1}, \ldots, a_{l,l-1})$. Since A_1 is productive, the matrix $(I - A)$ has a non-negative inverse, and the result follows from (8) after multiplying by $(I - A_1)^{-1}$. $\qquad\square$

Remark 2.1. The price p_h of commodity h can be written as $p_h = \sum_{k=1}^{l-1} \alpha_{kh} a_k$, where α_{kh} is the (k, h)th element in $(I - A_1)^{-1}$. The prices are thus determined by the input of labor in each of the processes.

The above remark points to the Marxian labor theory of value. According to Marx, the value of a commodity is determined by the amount of labor which is used either directly as labor input or indirectly, through the production of other inputs, in its production. But the labor theory of value applies to labor as well: To find the

66   *Theory of General Economic Equilibrium*

value of labor, we must determine the quantity of goods necessary to (re-)produce the one unit of labor, in the sense that this quantity is necessary in order for the worker to survive and work.

This gives us once again l processes, where the last one produces one unit of labor through the typical composition of consumption goods needed for the worker. We are looking for a vector $v \in \mathbb{R}^l_+$, $v \neq 0$, of values with the properties

(1^*) for $h \neq l$, $v_h = \sum_{k=1}^l v_k a_{kh}$,

(2^*) $v_l \geq \sum_{k=1}^l v_k a_{kl}$.

For simplicity, we assume that all the coordinates a_{kl} in the lth process (except $a_{ll} = 0$) are positive. Then A is indecomposable and $\lambda(A) > 0$. There are three cases to consider:

Case 1° $\lambda(A) > 1$. Then A is not productive; this case is not interesting and will not be considered any further.

Case 2° $\lambda(A) = 1$. Again, A is not productive, but there is a vector $x > 0$, such that $Ax = x$. This means that the economy can subsist even without creating surplus of any commodity. This case was called *simple reproduction* by Marx. Clearly, $\lambda(A) = \lambda(A^T)$, where A^T is the transpose of A, so there is a vector $v > 0$, such that $vA = v$. Thus, (2^*) above is satisfied with equality.

Case 3°. This is what Marx calls *extended reproduction*, the economy grows at the rate $\dfrac{1}{\lambda(A)} > 1$. The matrix $I - A$ has a nonnegative inverse, and the problem of finding a value vector has solutions

$$v = d(I - A)^{-1},$$

where $d = (0, \ldots, 0, d_l)$ with $d_l > 0$. It may be noted that $v_l > \sum_{k=1}^l v_k a_{kl}$, so that more value is extracted from the workers than what was used in the production of their labor, so that *surplus value* is created.

Assume conversely that $v \in \mathbb{R}^l_+$ is a solution to (1^*) and (2^*) with strict inequality in (2^*). Let x be a positive eigenvector belonging to

$\lambda(A)$ (existing since A is indecomposable). Then

$$(1 - \lambda(A))vx = v(I - A)x > 0,$$

from which we get that $\lambda(A) < 1$.

Define the *rate of exploitation* as

$$e = \frac{v_l - \sum_{k=1}^{l-1} v_k a_{kl}}{\sum_{k=1}^{l} v_k a_{kl}}.$$

We have shown the following theorem, which can be traced back to Marx (1867).

Theorem 2.6 (Marx). *The following are equivalent:*

(i) $e > 0$,

(ii) $\dfrac{1}{\lambda(A)} > 1$ (*the economy can grow at a positive rate*).

2.2 The von Neumann model

In the previous sections, we have assumed throughout that

(1) each commodity is produced in exactly one process,
(2) every process produces only one commodity.

To avoid these restrictive assumptions, we now extend the model somewhat. It is assumed that there are n processes, with $n \geq l$. In each of the processes, some of the commodities are produced using some (possibly the same) commodities. The jth process is defined by a vector $a_j = (a_{1j}, \ldots, a_{lj})$ of input and a vector $b_j = (b_{1j}, \ldots, b_{lj})$ of output, whereby all a_{ij} and b_{ij} are non-negative. The process (a_j, b_j) is operated at the intensity $x_j \geq 0$, and then $x_j b_j \in \mathbb{R}_+^l$ is produced using inputs $x_j a_j \in \mathbb{R}_+^l$.

Formally, a *von Neumann model* is defined by two $(l \times n)$ matrices A and B (where column j corresponds to the jth process), satisfying

(i) $a_{ij} \geq 0, b_{ij} \geq 0, i = 1, \ldots, l, j = 1, \ldots, n$,
(ii) $\sum_{i=1}^{n} a_{ij} > 0, j = 1, \ldots, n$,
(iii) $\sum_{j=1}^{n} b_{ij} > 0, i = 1, \ldots, l$.

68 *Theory of General Economic Equilibrium*

The condition (ii) states that every process uses at least one commodity as input, and similarly (iii) states that every commodity can be produced in some process.

The model may be further generalized, allowing for an infinite menu of processes which might be operated. It may be noted that the input–output combinations that can be realized in the von Neumann model corresponds to the convex cone in $\mathbb{R}_+^l \times \mathbb{R}_+^l$ spanned by the vectors (a_j, b_j), $j = 1, \ldots, n$.

Definition 2.5. A von Neumann–Gale model is a closed convex cone $Z \subset \mathbb{R}_+^l \times \mathbb{R}_+^l$ with $(0, y) \notin Z$ for $y \neq 0$, such that

$$(pr_2 Z) \cap \mathbb{R}_{++}^l \neq \emptyset, \tag{9}$$

where pr_2 is projection on the second factor in $\mathbb{R}_+^l \times \mathbb{R}_+^l$.

Following the terminology introduced above, we refer to elements $(u, y) \in Z$ as processes. For a process $(u, y) \in Z$, we define the growth rate $\alpha(u, y) = \sup \{\alpha \mid \alpha u \leq y\}$.

Lemma 2.6. *The function* $\alpha : Z \to \mathbb{R} \cup \{\infty\}$ *is upper semicontinuous.*

Proof. Let $(u^n, y^n)_{n \in \mathbb{N}}$ be a sequence converging to (u, y). If $u = 0$, then $y = 0$ and $\alpha(u, y) = \infty$, so we may assume that $u \neq 0$. The sequence $(\alpha(u^n, y^n))_{n \in \mathbb{N}}$ has an accumulation point $\overline{\alpha} \in \mathbb{R} \cup \{\infty\}$, so for a suitable subsequence, we have that $\lim \alpha(u^{n_v}, y^{n_v}) = \overline{\alpha}$. Since $\alpha(u^{n_v}, y^{n_v}) u^{n_v} \leq y^{n_v}$ for all v, it must be the case that $\overline{\alpha} < \infty$ and in the limit $\overline{\alpha} u \leq y$, so that $\alpha(u, y) \geq \overline{\alpha}$. Since $\overline{\alpha}$ was an arbitrary accumulation point, we conclude that $\alpha(u, y) \geq \limsup \alpha(u^n, y^n)$. $\square$

We now exploit that an upper semicontinuous function defined on a compact set attains its maximum, and we define

$$\alpha(Z) = \{\alpha(u, y) \mid (u, y) \in Z, \|(u, y)\| = 1\}$$
$$= \{\alpha(u, y) \mid (u, y) \in Z, (u, y) \neq 0\},$$

Economies with Production   69

where the last equality is obtained from the positive homogeneity of α. The quantity $\alpha(Z)$ is called the von Neumann growth rate of Z.

Definition 2.6. An equilibrium in the von Neumann–Gale model Z is a process $(\overline{u}, \overline{y}) \in Z$, a price vector $\overline{p} \in \mathbb{R}^l_+$, and a number α, such that

(i) $\alpha \overline{u} \leq \overline{y}$,
(ii) $\overline{p} \cdot \overline{y} = \alpha \overline{p} \cdot \overline{u}$ and $\overline{p} \cdot y \leq \alpha \overline{p} \cdot u$ for all $(u, y) \in Z$,
(iii) $\overline{p} \cdot \overline{y} > 0$.

The reasoning behind this definition is as follows. The economy is assumed in each period to choose a production on the ray determined by $(\overline{u}, \overline{y})$, so that the scale of production increases by the factor α. For this to be physically possible, the amount produced in the previous period should be sufficient for the purposes of input, and this is expressed by (i). The choice of production is governed by the price $\overline{p}$. Producers are assumed to buy the commodities needed for input, for which they must pay the cost measured by the prices together with interest rates and then to sell the output. Condition (ii) says that producers maximize profits. Finally, (iii) assures non-triviality of prices and production.

Theorem 2.7. *Let Z be a von Neumann–Gale model satisfying the condition:*

$$\exists (\overline{u}, \overline{y}) \in Z : \alpha(\overline{u}, \overline{y}) = \alpha(Z), \overline{y} > 0. \tag{10}$$

Then there is an equilibrium $(\overline{u}, \overline{y}, \overline{p}, \alpha)$ in Z with $\alpha = \alpha(Z)$.

Proof. Let $C = \{y - \alpha(Z)u \mid (u, y) \in Z\} \subset \mathbb{R}^l$. Then C is a convex cone. From the definition of $\alpha(Z)$, it is seen that $C \cap \mathbb{R}^l_{++} = \emptyset$. Furthermore, $\overline{c} = \overline{y} - \alpha(Z)\overline{u} \in \mathbb{R}^l_+$, so by separation of convex sets, there exists $\overline{p} \in \mathbb{R}^l_+, \overline{p} \neq 0$ with $\overline{p} \cdot c \leq 0$ for $c \in C$. Clearly, $(\overline{u}, \overline{y}, \overline{p}, \alpha(Z))$ satisfies (i) and (ii). Condition (iii) follows from (10) and $\overline{p} \geq 0$. $\square$

2.2.1 More about Marx

In this section, we look for a generalization to von Neumann–Gale models of the result from Section 2.1.2, and as before, the lth commodity is supposed to be labor. There seems to be no obvious way of introducing a value vector v, so we shall restrict ourselves to the exploitation rate e.

Define

$$e(Z) = \max \left\{ \frac{y_l}{u_l} - 1 \,\middle|\, (u, y) \in Z, y - u \geq 0 \right\},$$

where the existence of a maximum is assured in the same way as above. Then $e(Z)$ is the cheapest way of reproducing labor. We want to find a connection between $e(Z)$ and $\alpha(Z)$. For this, we shall need the following assumption.

Assumption 2.1. For every equilibrium price p in Z, there is $(u, y) \in Z$, such that $\tilde{p} \cdot y > \tilde{p} \cdot u$, where $\tilde{p} = (p_1, \dots, p_{l-1}, 0)$.

The assumption states that there is a process which yields a positive profit if the wage is set to zero. With this assumption, we get the following result.

Theorem 2.8. Let Z be a von Neumann–Gale model satisfying Assumption 2.1. Then

$$e(Z) > 0 \Leftrightarrow \alpha(Z) > 1.$$

Proof. $\Leftarrow$: If $\alpha(Z) > 1$, then there is $(u, y) \in Z$ with $y - u \geq 0$. $\Rightarrow$: Clearly, $\alpha(Z) \geq 1$. Assume that $\alpha(Z) = 1$. The set

$$C = \{z \in \mathbb{R}^l \mid \exists (u, y) \in Z, y - u = x\}$$

will then contain an element $z \in \mathbb{R}_+^l$ with $x_l > 0$, but no elements from $\mathbb{R}_{++}^l$. Let $p \in \mathbb{R}_+^l$, $p \neq 0$, be a linear form separating C and $\mathbb{R}_+^l$, so that $p \cdot c \leq 0$ for $c \in C$ and $p \cdot z \geq 0$ for $z \in \mathbb{R}_+^l$. Then $p_l = 0$, p is an equilibrium price, and for all $(u, y) \in Z$, $p \cdot y - p \cdot u \leq 0$ contradicting Assumption 2.1. We conclude that $\alpha(Z) > 1$. $\square$

2.2.2 Spectral theory for correspondences

Let Z be a von Neumann–Gale model. Then we may define a correspondence $\varphi : \mathrm{pr}_1 Z \rightrightarrows \mathbb{R}^l_+$ by

$$\varphi(u) = \{y' \in \mathbb{R}^l_+ \mid (u, y) \in Z \text{ for some } y \geq y'\},$$

so that $\varphi(u)$ is the set of commodity bundles which can be obtained if the input bundle u is available. Then φ has the following properties:

(i) $\varphi(u_1) + \varphi(u_2) \subseteq \varphi(u_1 + u_2)$ all $u_1, u_2 \in \mathbb{R}^l_+$ (superadditivity),
(ii) $\varphi(\lambda u) = \lambda \varphi(u)$ for all $u \in \mathbb{R}^l_+$ and $\lambda > 0$ (positive homogeneity),
(iii) $\varphi(0) = \{0\}$,
(iv) φ has closed graph (namely Z),
(v) $\varphi(\mathrm{pr}_1 Z) \cap \mathbb{R}^l_{++} \neq \emptyset$.

A correspondence satisfying (i)–(v) above is said to be superlinear. Finally,

(vi) the set φ is comprehensive in the sense that $(\varphi(u) - \mathbb{R}^l_+) \cap \mathbb{R}^l_+ \subseteq \varphi(u)$ for all $u \in \mathbb{R}^l_+$.

Lemma 2.7. *Let φ be superlinear and $K \subset \mathbb{R}^l_+$ a compact set. Then $\varphi(K) = \cup_{u \in K} \varphi(u)$ is compact, in particular $\varphi(u)$ is compact for all $u \in \mathbb{R}^l_+$.*

Proof. The set $\{(u, y) \in Z \mid u \in K\}$ is closed. If the set was unbounded, there would be $(u^n, y^n) \in Z$ for each $n \in \mathbb{N}$ with $y^n \to \infty$, and we may assume that $\|y^n\| > \|y^{n-1}\| - 1$ for each n. By compactness, there is a subsequence such that u^n converges to some $u \in K$. Then $\varphi(u)$ cannot be bounded, and in particular, $\varphi(u)$ must contain elements y' with arbitrarily large norms $\|y'\|$. Using property (ii), we get that this holds also for $\varphi(u')$ for $u' = \lambda u, 0 < \lambda \leq 1$, and we obtain that $\varphi(0)$ contains elements of $\mathbb{R}^l_+$ different from 0, contradicting (iii). We conclude that $\{(u, y) \in Z \mid u \in K\}$ is bounded, which means that $\varphi(K)$ is bounded and therefore compact. $\qquad\square$

We are moving towards the following generalization of the Perron–Frobenius theorem:

Theorem 2.9. *Let $\varphi : \mathbb{R}^l_+ \rightrightarrows \mathbb{R}^l_+$ be superlinear. Then there is a real number $\lambda \geq 0$ and a non-empty, compact, convex, and comprehensive set $K \neq \{0\}$, such that $\varphi(K) = \lambda K$.*

The proof of Theorem 2.9 will follow in several steps. First of all, we note that by Lemma 2.7, φ induces a map Φ of the set

$$\mathcal{K} = \{K \subset \mathbb{R}_+^l \mid K \neq \emptyset, K \text{ compact, convex, comprehensive}\}$$

on itself, whereby $\Phi(K) = \cup_{u \in K}\varphi(u)$. $\mathcal{K}$ is a metric space with the Hausdorff metric

$$d_H(K_1, K_2) = \max\{\rho(K_1, K_2), \rho(K_1, K_2)\},$$

where $\rho(K_i, K_j) = \sup_{u \in K_i} d(u, K_j)$ and d is the usual Euclidean metric.

Lemma 2.8. *The map $\Phi : \mathcal{K} \to \mathcal{K}$ is continuous and positively homogeneous (that is $\Phi(\lambda K) = \lambda \Phi(K)$ for $\lambda > 0$).*

Proof. Positive homogeneity of Φ is straightforward, so we turn to proving continuity.

The correspondence φ is upper hemicontinuous. Let U be an open set with $\varphi(u) \subset U$, and suppose that for any neighborhood W of u, there is $v \in W$ with $\varphi \backslash U \neq \emptyset$. We may assume that the closure of W, $\overline{W}$, is compact. Choose a sequence $(u^n, y^n)_{n \in \mathbb{N}}$ with $u^n \in \overline{W}$, $y^n \in \varphi(u^n) \backslash U$, such that $u^n \to u$. For a suitable subsequence y^n converges to $y \in \varphi(u)$ (since the graph of $\varphi(u)$ is closed (property (iv)), contradicting that $y^n \notin U$ for all n.

Let $\varepsilon > 0$, and choose $G \subset \mathbb{R}_+^l$, such that $\rho(G, \Phi(K)) < \varepsilon$ and $\Phi(K) \subset G$. Then for each $u \in K$, there is δ_u, such that $\varphi(u') \subset G$ when $\|u - u'\| < \delta_u$ due to upper hemicontinuity of φ. The family of sets $B(u, \delta_u) = \{u' \mid \|u - u'\| < \delta_u\}$ for $u \in K$ is a covering of K, choosing a finite subcovering and letting δ be the smallest of the numbers δ_u associated with the sets of the subcovering, we obtain that $\rho(\Phi(K'), \Phi(K)) < \varepsilon$ for $\rho(K', K) < \delta$.

Next, we show that the correspondence φ is lower hemicontinuous. Let $y \in \varphi(u)$ and choose a sequence $(u^n)_{n \in \mathbb{N}}$ with $u^n \to u$. We may choose k points $u^{(1)}, \ldots, u^{(k)} \in \mathbb{R}_+^l$, such that u belongs to the interior of $\text{conv}(\{u^{(1)}, \ldots, u^{(k)}\})$. For n large each u^n can be written as $u^n = \sum_{i=0}^k \lambda_i^n u^{(i)}$ with $u^{(0)} = u$, and we can select the representation of u, such that $(\lambda_0^n, \lambda_1^n, \ldots, \lambda_k^n)$ converges to $(1, 0, \ldots, 0)$. Choose $y^{(i)} \in \varphi(u^{(i)})$, $y^{(0)} = y$, and let $y^n = \sum_{i=0}^k \lambda_i^n y^{(i)}$. Then $y^n \in \varphi(u^n)$ and $y^n \to y$.

Choose $\varepsilon > 0$. For each $u \in K$, we have shown that there is δ_u, such that the distance from an arbitrary point in $\varphi(u)$ to $\varphi(v)$ is $< \varepsilon$ when $\|v - u\| < \delta_u$. Using compactness, we get the existence of a $\delta > 0$, such that $\rho(K, K') < \delta$ implies that $\rho(\Phi(K), \Phi(K')) < \varepsilon$. $\qquad\square$

3. International Trade and Factor Price Equalization

In this section, we consider a special case of an economy with production which has gained classical status in economic theory, namely the Heckscher–Ohlin–Samuelson (HOS) model of international trade. Here, we have two countries producing two commodities, which can be traded internationally, using two inputs which can be traded only within the countries. The main achievements of the theory are as follows:

(a) a theorem about the *structure of international trade* (which country imports what) and as a stepping stone towards this result;
(b) the *factor price equalization theorem* stating that prices of the input commodities are the same independent of the country in which they are situated.

The background of the HOS model is the problem of identifying *comparative advantages* in international trade, determining from the given data of two economies the nature of trade flows from one economy to the other, cf. Box 1. In the HOS model, countries have access to the same technologies, so that the comparative advantages must come from other sources. If it is assumed that consumers behave in the same way in each country, what remains is the initial endowment of production factors. The structure theorem relates this initial endowment to the equilibrium trade flows.

3.1 The Heckscher–Ohlin–Samuelson model

There are two countries A and B which trade with each other in two (final) commodities, all of which are produced separately using two input commodities. We shall use the notation m and ℓ for the two

74   *Theory of General Economic Equilibrium*

Box 1. Ricardian comparative advantages: The idea of comparative advantages goes back to Ricardo (1817) and is usually presented as follows: Two countries, England and Portugal, produce the same two commodities, cloth and wine, but the labor requirements differ between the countries:

	Cloth	Wine
England	100	120
Portugal	90	80

The table shows the labor requirements per unit produced in each country. Here, the requirements are higher for England than for Portugal *in both sectors*.

But both countries will be better off trading with each other: England might transfer, say 100 units of labor from wine to cloth production, losing 100/120 units wine and gaining 1 unit cloth cheaper than producing it. This is also advantageous for Portugal, who may use the 90 units of the factor to produce 90/80 units wine instead of producing cloth.

It is seen that England, though generally less productive than Portugal, has a *comparative advantage* in the production of cloth. In equilibrium, each country exports the commodity for which they have a comparative advantage, and imports the other one.

---

production factors, traditionally referred to as "capital" and "labor". These factors are non-tradable internationally.

The technologies for producing each of the final commodities exhibit constant return to scale. Since there is only one output in each technology, they may conveniently be represented by production functions of the form

$$y_h^k \leq g_h(m_h^k, \ell_h^k), \ y_h^k, m_h^k, \ell_h^k \geq 0, \quad h = 1, 2, \ k = A, B.$$

Here, y_h^k denotes the output of commodity h in country k, whereas m_h^k and ℓ_h^k are the (numerical values of the) inputs considered as positive. The production function, g_h, is the same in each country, both countries have access to the same technology. We assume that there is at most one firm in each country producing commodity j. This is no restriction as long as firms maximize profits at given prices, cf. Exercise 2). The production functions g_h for $h = 1, 2$ satisfy some assumptions of well-behavedness, namely,

(i) g_h is homogeneous of degree 1 (constant returns to scale),
(ii) g_h is strictly quasi-concave, i.e. the level sets

$$L_h(r) = \{(m_h, \ell_h) \mid g_h(m_h, \ell_h) \le r\}$$

are strictly convex for each $r \in \mathbb{R}_+$.
(iii) g_h is monotonic, i.e. $g_h(m', \ell') \ge g_h(m, \ell)$ if $(m', \ell') \ge (m, \ell)$.

Since there is constant return to scale, if $p = (p_m, p_\ell)$ supports the level set $L_h(r)$ at (m_n, ℓ_h) (so that $p \cdot (m, \ell) \ge p \cdot (m_h, \ell_h)$ for all $(m, \ell) \in L_h(r)$), then p supports any level set at an input combination with the same factor proportion $\mu_h = \frac{\ell}{m}$.

3.1.1 *Factor price equalization*

For p, a support of $L_h(r)$ at (m_h, ℓ_h), the quantity $\pi = \frac{p_m}{p_\ell}$ is said to be a marginal rate of substitution at (m_h, ℓ_h). Clearly, marginal rates of substitution depend only on factor proportions, and from the convexity of the level sets $L_h(r)$, we get that for each h, the correspondence φ_h assigning to each factor proportion μ_h the marginal rates of substitution at any (m_h, ℓ_h) with $\mu_h = \dfrac{\ell_h}{m_h}$ is non-increasing:

$$\mu_h > \mu_h', \ \pi \in \varphi(m_h), \ \pi' \in \varphi(m_h') \Rightarrow \pi \le \pi'. \tag{11}$$

Conversely, a marginal rate of substitution gives rise to a unique factor proportion μ_h, $h = 1, 2$, in each industry. If $\mu_1 \ge \mu_2$, industry 1 is said to be more ℓ-intensive than industry 2, and vice versa for $\mu_1 \le \mu_2$. For what follows, it will be convenient to make the following assumption.

Assumption 2.2. The industries can be ranked according to factor proportions independent of the marginal rate of substitution.

The property stated in Assumption 2.2 is specific for our present case of two commodities and two factors and is not easily extended to the case of many finished goods and inputs. It is explored further in Exercise 4, but for the moment, we just accept this assumption and trace out its impacts on the model.

Assume now that the two countries trade with each other in a market for output commodities determined by the commodity prices (p_1, p_2) and that *both countries produce non-zero quantities of both goods*. If the production of commodity h in country k is y_h^k, then output value in country k is $p_1 y_1^k$ and $p_2 y_2^k$ in the two industries, $k = 1, 2$.

Let $r_1 = 1/p_1$, $r_2 = 1/p_2$ be the output corresponding to one unit of money in each industry of country k. Since equilibrium profits are zero under constant returns to scale (see Exercise 3), the value of inputs used for producing r_h is also 1, $h = 1, 2$. Since both commodities are produced, there is an *isocost* line

$$\{(m, \ell) \mid p_m^k m + p_\ell^k \ell = 1\},$$

which exactly touches each of the level sets $L(r_1)$ and $L(r_2)$ (see Fig. 1). Note that $L(r_1)$ and $L(r_2)$ do not depend on the country considered, and consequently, the factor prices (p_m, p_ℓ) can be chosen to be the same for each country. If the level sets have unique support, then this is the only possible choice.

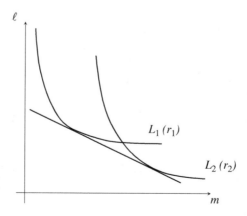

Fig. 1. Factor price equalization: Given the outputs r_1 and r_2 in the two industries, corresponding to a revenue of one unit of money, there can be at most one isocost line which is a tangent to both isoquants. Since profits are zero, factor prices are uniquely determined.

Summing up, we have obtained the following classical result (Samuelson, 1948).

Theorem 2.10. *Assume that Assumption* 2.2 *is satisfied and that both countries produce both commodities. Then the prices of both production factors are the same in the two countries.*

A crucial assumption in Theorem 2.10 is that of *no specialization:* Each country is actually producing both commodities. What happens when this is not the case can be assessed when we go somewhat further into the details of the model.

We noted above that there is a non-increasing relationship between factor proportions and marginal rates of substitution. Under the assumption of unique supports of level sets, the inverse functional relationship holds as well, so that a given marginal rate of substitution, in our case, the one defined by the common tangency condition for the level sets $L(r_1)$ and $L(r_2)$ gives rise to factor proportions μ_1 and μ_2 in each industry. The input combinations in each industry must be a point on the ray determined by the factor proportion of the industry, and the sum of the two input combinations equals the factor endowment in the country.

Let $(\overline{m}^k, \overline{\ell}^k)$, $k = A, B$, be the endowment of factors in country k and

$$(\overline{m}, \overline{\ell}) = (\overline{m}^A, \overline{\ell}^A) + (\overline{m}^B, \overline{\ell}^B)$$

the total factor endowment. Since the equilibrium allocation of inputs satisfy

$$(m_1^k, \ell_1^k) + (m_2^k, \ell_2^k) = (\overline{m}^k, \overline{\ell}^k),$$

we have that the endowment vector in each country is a sum of two input vectors, which have the direction determined by the factor intensities μ_1 and μ_2.

The situation is shown in Fig. 2 where we use an Edgeworth box in the input space. The size of the box is determined by the total endowment, and each point in the box indicates a possible distribution of total endowments between the two countries. The

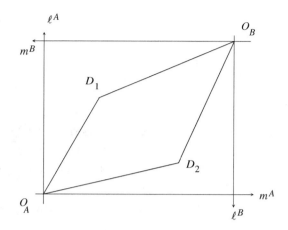

Fig. 2. The factor price equalization domain: The size of the Edgeworth box is determined by the total factor endowments, and points inside the trapezoidal figure spanned by points O_A, D_1, O_A, D_2 correspond to distributions of the factor endowments among countries at which factor price equalization will be obtained.

trapezoidal area $O_A D_1 O_B D_2$ inside the box, known as the *factor price equalization domain*, is spanned by vectors of factor inputs in each industry, inserted at both end points of the box. Points inside this area correspond to endowments where each country produces some output in each industry. If the endowment distribution is on the boundary of the area, at least one country is completely specialized in the sense that it produces only one of the output commodities.

3.1.2 The structure of international trade: Comparative advantages

Given that factor proportions can be ranked independent of factor endowments, we may reintroduce the notion of comparative advantages, now pertaining to *factor endowments*. We say that country k has a comparative advantage in the production of commodity h if the country is relatively abundantly endowed with the factor which is used intensively in industry h. Thus, country A will have a comparative advantage in the production of commodity 1 if

$$\frac{\bar{\ell}^A}{\bar{m}^A} > \frac{\bar{\ell}^B}{\bar{m}^B}, \qquad (12)$$

Economies with Production

given that as previously industry 1 is assumed to be more ℓ-intensive than industry 2.

We have that total labor endowment $\overline{\ell}^k$ for $k = A, B$ is allocated to industries as

$$\overline{\ell}^k = \ell_1^k + \ell_2^k.$$

Dividing by total endowment $\overline{m}^k$ of capital goods and rearranging, we get

$$\frac{\overline{\ell}^k}{\overline{m}^k} = \frac{\ell_1^k}{m_1^k} \frac{m_1^k}{\overline{m}^k} + \frac{\ell_2^k}{m_2^k} \frac{m_2^k}{\overline{m}^k}, \tag{13}$$

showing that the endowment ratio is a weighted average of the two factor proportions, the weights being given by the proportion of the capital goods employed in each industry.

With this terminology, the Heckscher–Ohlin structure theorem appears as a straightforward generalization of Ricardo's results.

Theorem 2.11. *Assume that in the two-country world, the consumption sectors of the countries are identical in the sense that at all commodity prices, the demands of the countries are proportional, and that Assumption 2.2 is satisfied. Then each country will produce and export the commodity in the production of which it has a comparative advantage.*

Proof. Suppose without loss of generality that industry 1 is ℓ-intensive and that country A is relatively well endowed with labor, in the sense that (12) is satisfied. There are two cases to consider:

Case 1: The endowments are such that $(\overline{m}^A, \overline{\ell}^A), (\overline{m}^B, \ell^B)$ define a point in the interior of the factor price equalization domain. Then

$$\frac{\overline{\ell}^A}{\overline{m}^A} > \frac{\overline{\ell}}{\overline{m}},$$

and for (13) to hold in both countries, it must be the case that

$$\frac{\overline{\ell}_1^A}{\overline{m}_1^A} > \frac{\overline{\ell}^B}{\overline{m}^B}.$$

Given that factor prices and therefore factor intensities are the same in the two countries, this means the proportion of output in industries 1 and 2 is larger in country A than in country B, and since the commodities are demanded in the same proportion in the two countries, the conclusion follows.

Case 2: The endowments define a point at the boundary of the factor equalization domain. This means that one industry, say industry h, is the only one operated in some country k, so that

$$\frac{\ell_h^k}{m_h^k} = \frac{\vec{\ell}^k}{\vec{m}^k}.$$

It follows that the relatively ℓ-abundant country A must produce in industry 1 if specialized, and similarly that the relatively m abundant country B must produce commodity 2 if specialized, and we have the conclusions of the theorem. $\qquad\square$

4. Exercises

(1) Find a Walras equilibrium in the economy $\mathcal{E} = ((X_i, P_i, \omega_i)_{i=1}^m, (Y_j)_{j=1}^n, (\theta_{ij})_{i=1\,j=1}^{m\ \ n})$, where $m = 3$ and for each consumer $i \in \{1, 2, 3\}$,

   (i) $X_i = \mathbb{R}_+^2$,

   (ii) P_i is represented by the utility function $u_i(x_{i1}, x_{i2}) = x_{i1}^{1-\frac{i}{4}} x_{i2}^{\frac{i}{4}}$,

   (iii) $\omega_i = \left(\frac{i}{10}, 1 - \frac{i}{10}\right)$,

   where $n = 2$ with

   $$Y_1 = \{(y_{j1}, y_{j2}) \in \mathbb{R}_+ \times \mathbb{R}_- \mid y_{j1} \le (-y_{j2})^{\frac{1}{2}}\},$$

   $$Y_2 = \{(y_{21}, y_{21}) \in \mathbb{R}_+ \times \mathbb{R}_- \mid y_{21} \le (-y_{22})^{\frac{1}{4}}\},$$

   and $\theta_{ij} = \frac{1}{3}$ for each i and j.

(2) Let Y_1, Y_2 be production sets in $\mathbb{R}^l$, and let

   $$Y = Y_1 + Y_2 = \{y \in \mathbb{R}^l \mid y = y_1 + y_2, y_1 \in Y_1, y_2 \in Y_2\}$$

   be the (Minkowski) sum of Y_1 and Y_2.

Let $p \in \mathbb{R}_+^l$ be a price system, and let $y_j^0 \in Y_j$, $j = 1, 2$. Show that y_j^0 maximizes $p \cdot y$ on Y_j, $j = 1, 2$, if and only if $y_1^0 + y_2^0$ maximizes $p \cdot y$ on $Y_1 + Y_2$.

(3) Consider a linear production model with matrix:

$$A = \begin{pmatrix} 0.15 & 0.25 \\ 0.70 & 0.05 \end{pmatrix}.$$

Check whether A is productive, and if so, find the maximal growth rate.

(4) A von Neumann model is given by the two matrices:

$$A = \begin{pmatrix} 2 & 3 & 1 \\ 4 & 0 & 4 \\ 3 & 2 & 2 \end{pmatrix} \quad B = \begin{pmatrix} 6 & 0 & 1 \\ 4 & 4 & 3 \\ 4 & 5 & 5 \end{pmatrix}.$$

Find an equilibrium.

(5) (Ergodic Markov chains) The Perron–Frobenius theorem can also be used in contexts which have no relation to production theory. A finite *Markov chain* is defined by a finite set $\{s_1, \ldots, s_n\}$ of states and an $(n \times n)$ matrix $P = (p_{ij})_{i=1}^n {}_{j=1}^n$, of *transition probabilities*, where p_{ij} denotes the probability of moving from state s_i to state s_j.

A Markov chain is irreducible if it is possible (in the sense of non-zero transition probability) to go from one state to any other state possibly in several steps. Show that a Markov chain is irreducible if its matrix of transition probabilities is indecomposable.

A Markov chain with matrix P of transition probabilities is ergodic with limiting distribution $x^* \in \Delta = \{x \in \mathbb{R}_+^n \mid \sum_{h=1}^n x_h = 1\}$ if $x^* P = x^*$. Show that if the Markov chain is irreducible, then it is ergodic with $x_h^* > 0$ for $h = 1, \ldots, n$.

(6) Prove the following classical results in the Heckscher–Ohlin framework:

Rybzinski's theorem: If a country's endowment with one production factor is increased, and the relative prices of goods remain unchanged, then the production of the good which uses this factor intensively is increased more than proportionally, and the production of the other good is reduced.

The Stolper–Samuelson theorem: Suppose that the production of one good is increased while that of the other one is reduced (possibly as a result of a tariff on imports of the first good). Then the income of the owners from the factor used intensively in the production of the first good increases, while the income from the other factor diminishes.

Chapter 3

Welfare Theorems and Market Failures

1. Pareto-optimality

By now, we have considered equilibria in exchange economies and economies with production, leaving the justification of our interest in market equilibria for later consideration, to which we turn now. For reasons which will become obvious as we proceed, we shall work in a context of economies where the ownership of producers and initial resources has not been specified. Formally, an economy is an array $\mathcal{E} = ((X_i, P_i)_{i=1}^m, (Y_j)_{j=1}^n, \omega)$ consisting of m consumers (X_i, P_i), $i = 1, \ldots, m$, n producers Y_j, $j = 1, \ldots, n$, and an initial endowment $\omega \in \mathbb{R}^l$. Allocations in $\mathcal{E}$ are (as before) arrays $(x, y) = (x_1, \ldots, x_m, y_1, \ldots, y_n) \in \mathbb{R}^{lm} \times \mathbb{R}^{ln}$, and allocations are feasible if they satisfy (1) and (2) in Chapter 2. The set of feasible allocations in $\mathcal{E}$ is denoted $F(\mathcal{E})$.

We have previously singled out some special allocations in $F(\mathcal{E})$ resulting from the interaction of agents in a market. At present, we take another approach, considering them more or less "good" or desirable for society. As is well-known, deciding upon "goodness" or social desirability of allocations is no easy matter, but it is still possible to reduce substantially the set of allocations which *a priori* can be considered as reasonable outcomes in a given economy.

To see this, suppose that

$$(x_1^0, \ldots, x_m^0, y_1^0, \ldots, y_n^0), \quad (x_1^1, \ldots, x_m^1, y_1^1, \ldots, y_n^1)$$

are feasible allocations. If every consumer in the economy is better off in the first allocation than in the second, then there would be no reason to implement the second allocation. Actually, the same might be said even if all consumers were only as well off (rather than better off) in the first as in the second allocation, as long as at least one consumer is better off. Consequently, we are left with those allocations which cannot be ruled out by pairwise comparisons of the above type.

Formally, we introduce the notion of a Pareto-optimal allocation,

Definition 3.1. An allocation $(x_1^0, \ldots, x_m^0, y_1^0, \ldots, y_n^0)$ in the economy

$$\mathcal{E} = \left((X_i, P_i)_{i=1}^m, (Y_j)_{j=1}^n, \omega \right),$$

is *Pareto-optimal* if

(i) $(x_1^0, \ldots, x_m^0, y_1^0, \ldots, y_n^0)$ is feasible,
(ii) there is no feasible allocation $(x_1^1, \ldots, x_m^1, y_1^1, \ldots, y_n^1)$ such that

$$x_i^1 \in \mathrm{cl}\, P_i(x_i^0), \quad \text{all } i, x_{i^0}^1 \in P_i(x_{i^0}^0), \quad \text{some } i^0 \in \{1, \ldots, m\}. \tag{1}$$

The statement in (1) may seem unfamiliar, involving the closure of the preferred set $P_i(x_i)$. In the case where preferences P_i can be described by utility functions u_i, the expression reduces to the usual one, stating that $u_i(x_i^1) \geq u_i(x_i^0)$ for all i with strict inequality for some i. It is easily seen that in any case, strictly monotone preferences together with the openness of $P_i(x_i^0)$ will eliminate the need for considering the closure of $P_i(x_i^0)$ instead of $P_i(x_i^0)$: If $x_{i^0}^1 \in P_i(x_{i^0}^0)$, then reducing $x_{i^0}^0$ with a small amount of some good and distributing this amount among the other consumers will yield an allocation $(x_1^2, \ldots, x_m^2, y_1^1, \ldots, y_n^1)$ with $x_i^2 \in P_i(x_i^0)$ for all i.

Returning again to the special case of preferences described by utility functions, if $(x_1^0, \ldots, x_m^0, y_1^0, \ldots, y_n^0)$ is Pareto-optimal, and $(x_1^1, \ldots, x_m^1, y_1^1, \ldots, y_n^1)$ is a feasible allocation, such that $u_i(x_i^1) > u_i(x_i^0)$ for some consumer i, then it follows from the definition that there must be another consumer r such that $u_r(x_r^1) < u_r(x_r^0)$. In other words,

Welfare Theorems and Market Failures 85

in a Pareto-optimal allocation we cannot make one consumer better off without making another consumer worse off.

1.1 Optimality and efficiency

In order to get some intuition for the set of Pareto-optimal allocations, it is useful to study special cases, and here the Edgeworth box plays an important role. As it is seen in Box 1, there are many Pareto-optimal allocations, and some of them may be rather unappealing, such as the allocations assigning everything to one consumer, leaving nothing to the others.

The Pareto-optimality of such extremely unequal divisions of the total endowment between the consumers is not a special feature of the above example, but a property which holds rather generally. Indeed, suppose that consumption sets are $\mathbb{R}^l_+$ and preferences P_i have the following monotonicity property:

$$[x_i - (\mathbb{R}^l_+ \setminus \{0\})] \cap \operatorname{cl} P_i(x_i) = \emptyset \tag{2}$$

for all $x_i \in \mathbb{R}^l_+$ (saying that no more of all goods and less of some good will never be preferred). If the aggregate endowment ω_h of each commodity h is non-zero, then every allocation $(0, \ldots, \omega, \ldots, 0)$ with ω in the ith component, 0 otherwise (so that individual i gets everything, the remaining individuals nothing), is Pareto-optimal: If $(x_1, \ldots, x_m)$ is another feasible allocation, then $x_i \leq \omega$ and $x_i \neq \omega$, so that $x_i \notin \operatorname{cl} P_i(\omega)$. Since at least one individual is worse off in $(x_1, \ldots, x_m)$ than in $(0, \ldots, \omega, \ldots, 0)$, the latter allocation is Pareto-optimal. It may easily be checked that the monotonicity property in 2 is satisfied if preferences are strictly monotone ($x'_i \geq x_i$ and $x'_i \neq x_i$ implies $x'_i \in P_i(x_i)$) and the preferred sets $P_i(x_i)$ are convex, for all $x_i \in \mathbb{R}^l_+$.

1.1.1 Efficiency and equity

For all its advantages, the concept of Pareto-optimality is seen to have its drawbacks as well, at least if we want allocations in society to have some properties of "fairness" or equal treatment of its

Box 1. Pareto-optimal in the Edgeworth box: In the special case where feasible allocations can be illustrated in an Edgeworth box, the set of Pareto-optimal allocations can be given a neat representation.

Consider a point A in the box. The set of points between the indifference curves of each of the two agents through A (or, alternatively, the intersection of the two upper level sets at A) gives all feasible allocations making each agent as well off as it is in A. Also, the points in this set other than A itself and the point A' make at least one agent better off. Since this set of points is non-empty, A is not Pareto-optimal (Fig. 1).

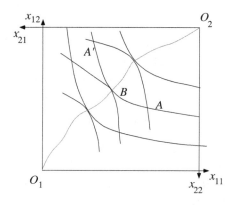

Fig. 1. Pareto-optimal in the Edgeworth box.

On the other hand, the point B, and indeed every point where the two indifference curves are tangent to each other, are Pareto-optimal (the tangency condition is in general neither necessary nor sufficient for Pareto-optimality, but this need not distract us at present). Connecting all such points, we get the *contract curve*, which by construction is the set of all points representing a Pareto-optimal allocation.

We can exhibit right away some particular Pareto-optimal allocations. The end points of the contract curve coincide with the corners O_1 and O_2 of the box: An allocation giving everything to one consumer and nothing to the other one is Pareto-optimal. Indeed, every other feasible allocation must give less of some commodity to the consumer who originally got everything, thereby making him worse off.

Box 2. The Koopmans diagram: When production matters, the Edgeworth box cannot be used, and instead, the following construction is used.

Following the tradition in economic literature, we identify the two agents (one consumer and one producer) as the single person Robinson Crusoe in the double role of both producer and consumer.

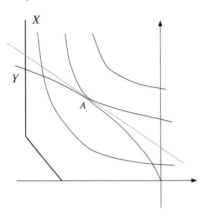

Fig. 2. The Koopmans diagram.

The production set of Robinson's firm (Y) is shown in Fig. 2, where the consumption set of Robinson the consumer is also indicated. This consumption set X contains some bundles with the first coordinate negative, corresponding to a supply of this commodity which will be some type of labor. We assume that there is no initial endowment on the island — everything must be produced.

In this economy, a feasible allocation is illustrated by a point such as B which belongs to both Y and X. The point A corresponds to an allocation which is not only feasible but also Pareto-optimal, since there is no feasible allocation as to which is better for the one and only consumer in the economy.

Let X^0 be the set of bundles that are at least as good for the consumer as that of the Pareto-optimal allocation. Then the sets Y and X^0 meet only in A, and we may draw a line through A such that X^0 is above and Y is below this line (for this, we use the convexity of X^0 and Y). As done previously, we let the price system be defined in such a way that $p = (p_1, p_2)$ is a normal to the separating line.

(Continued)

Theory of General Economic Equilibrium

Box 2. *(Continued)*
Now, considering Robinson the producer, we see that profits at the price system p are maximized in A, and the separating line acts as an isoprofit line corresponding to the maximal profit.

---

members. Phrased alternatively, the property of Pareto-optimality is a rather weak one, since the set of Pareto-optimal allocations contains quite unreasonable elements. However, this weakness is also in some sense a force: working with Pareto-optimality, we get a set of useful results to be applied to particular allocations which are Pareto-optimal *and* have some other properties depending on the context.

1.1.2 *Egalitarian and egalitarian equivalent allocations*

Having argued that Pareto-optimal allocations may violate some ideas of equality and fairness which, though rarely stated explicitly, are important in everyday thinking about distribution in society, we digress briefly on possible formalizations of such ideas. For simplicity, we confine our discussion to an exchange economy $\mathcal{E} = ((X_i, P_i)_{i=1}^m, \omega)$, where $X_i = \mathbb{R}_+^l, i = 1, \ldots, m$.

An example of an allocation achieving absolute equality in society is the *egalitarian* allocation $(\frac{1}{m}\omega, \ldots, \frac{1}{m}\omega)$. This allocation undoubtedly appeals to egalitarian sentiments, but unfortunately, it is in general not Pareto-optimal, and as we shall argue in the sequel, this implies that some agents have incentives to change the allocation, by exchanges or otherwise. The egalitarian allocation, therefore, is in a sense a highly unstable one.

This instability may be remedied if we allow for a more relaxed attitude towards equality. What matters is perhaps not so much that agents are treated equally as that they feel they are. Choose a (reference) bundle $\bar{x} > 0$ in $\mathbb{R}_+^l$. If an allocation $(x_1, \ldots, x_m)$ satisfies $\bar{x} \in \operatorname{bd} P_i(x_i)$ for all i, then it might be considered as equivalent to the egalitarian allocation $(\bar{x}, \ldots, \bar{x})$. This equivalence may be somewhat far-fetched in the general case outlined here, but in the case that preferences are described by utilities, the bundle obtained by each agent is exactly as good as the fixed bundle $\bar{x}$.

Following up on this idea, Pazner and Schmeidler (1978) define an allocation to be *egalitarian equivalent* with reference bundle $\bar{x}$ if

(i) it is feasible, and
(ii) $x \in \text{bd } P_i(x_i)$ for $i = 1, \ldots, m$.

Thus, if $(x_1, \ldots, x_m)$ is egalitarian equivalent, then each consumer is exactly as well off as if he got the bundle $\lambda \bar{x}$. The allocation $(\lambda \bar{x}, \ldots, \lambda \bar{x})$ need not be feasible. It can be shown (cf. Exercise 3) that if λ is chosen to be maximal subject to the conditions (i) and (ii), then the corresponding allocation is Pareto-optimal. Thus, equality considerations are not *per se* incompatible with the efficiency requirements implicit in the concept of Pareto-optimality.

1.1.3 Envy-free and fair allocations

Another way of modeling equality properties less rigid than those of the egalitarian allocation is represented by the concept of fairness. A feasible allocation $x = (x_1, \ldots, x_m)$ is *envy-free* if

$$x_j \notin P_i(x_i), \quad \text{all } i, j \in \{1, \ldots, m\},$$

so that no consumer prefers the bundle being assigned to any other consumer. An allocation is *fair* if it is envy-free *and* Pareto-optimal. Clearly, fair allocations represent a way of making equity considerations compatible with efficiency; no agent would be better off switching a bundle with anybody else, and no efficiency losses are incurred.

It remains, however, to check that the compatibility is a real one, in the sense that fair allocations do exist in economies satisfying standard assumptions. If the economy can be represented by an Edgeworth box, an allocation is envy-free if mirroring the relevant point through the midpoint gives a point on or below the indifference curve, and it seems plausible, at least in this low-dimensional case,

90 Theory of General Economic Equilibrium

that this will happen for some point on the contract curve (see also Exercise 5). On the other hand, this gives little or no intuition for what happens in the general case.

1.1.4 Productive efficiency

Returning to our general discussion of Pareto-optimality, we note that consumers' preferences were crucial in the definition while producers seemed hardly to matter. Turning the problem around, we might consider an alternative optimality concept that focuses on producers and ignores consumers: An allocation $(x_1^0, \ldots, x_m^0, y_1^0, \ldots, y_n^0)$ in the economy

$$\mathcal{E} = \left((X_i, P_i)_{i=1}^m, (Y_j)_{j=1}^n, \omega \right),$$

is said to be *efficient* (in the aggregate) if

(1) $(x_1^0, \ldots, x_m^0, y_1^0, \ldots, y_n^0)$ is feasible,
(2) there is no feasible allocation $(x_1^1, \ldots, x_m^1, y_1^1, \ldots, y_n^1)$ such that

$$\sum_{j=1}^n y_{jk}^1 \geq \sum_{j=1}^n y_{jk}^0,$$

for all k with strict inequality for some k.

Efficiency of allocations bears resemblance to efficiency of individual production y_j in Y_j, cf. Definition 0.1. Indeed, an allocation $(x_1^0, \ldots, x_m^0, y_1^0, \ldots, y_n^0)$ is efficient (in the aggregate) if the production $\sum_{j=1}^n y_j^0$ is efficient in the aggregate production set

$$\sum_{j=1}^n Y_j = \left\{ y \,\middle|\, y = \sum_{j=1}^n y_j, y_j \in Y_j, j = 1, \ldots, n \right\},$$

and it is easily seen that if $(x_1^0, \ldots, x_m^0, y_1^0, \ldots, y_n^0)$ is efficient, then each of the individual productions y^0 must be efficient. The converse does not hold: Production may be efficient at enterprise level without being efficient in the aggregate (why?).

Welfare Theorems and Market Failures 91

There is a close connection between Pareto-optimality and efficiency, actually so close that the latter concept need not be studied separately in the sequel:

Theorem 3.1. *Let $\mathcal{E} = ((X_i, P_i)_{i=1}^m, (Y_j)_{j=1}^n, \omega)$ be an economy where the consumers satisfy Assumptions 0.1 and 0.2. Suppose that there is some consumer, say consumer i, for which*

$$x_i + \left[\mathbb{R}_+^l \setminus \{0\}\right] \subset P_i(x_i), \quad \text{all } x_i \in X_i \tag{3}$$

(strict monotonicity of preferences), and let $(x_1^0, \ldots, x_m^0, y_1^0, \ldots, y_n^0)$ be a Pareto-optimal allocation. Then

$$(x_1^0, \ldots, x_m^0, y_1^0, \ldots, y_n^0)$$

is efficient.

The proof of Theorem 3.1 is straightforward: If the allocations were not efficient, then a larger amount of some commodity could be left for consumption, and any consumer receiving this larger amount would be better off.

Proof of Theorem 3.1. Suppose that $(x_1^0, \ldots, x_m^0, y_1^0, \ldots, y_n^0)$ is not efficient, and let

$$(x_1^1, \ldots, x_m^1, y_1^1, \ldots, y_n^1)$$

be a feasible allocation such that $\sum_{j=1}^n y_{jk}^1 \geq \sum_{j=1}^n y_{jk}^0$ for all k with at least one strict inequality. Suppose w.l.o.g. that consumer 1 satisfies strict monotonicity of preferences.

Since

$$\sum_{i=1}^m x_i^0 + \left(\sum_{j=1}^n y_j^1 - \sum_{j=1}^n y_j^0\right) \leq \sum_{j=1}^n y_j^1 + \omega,$$

the allocation $(x_1^0 + (\sum_{j=1}^n y_j^1 - \sum_{j=1}^n y_j^0), x_2^0, \ldots, x_m^0, y_1^1, \ldots, y_n^1)$ is feasible, and since $x_1^0 + (\sum_{j=1}^n y_j^1 - \sum_{j=1}^n y_j^0) \in P_1(x_1^0)$ by 3, we have a contradiction of Pareto-optimality. Thus,

$$(x_1^0, \ldots, x_m^0, y_1^0, \ldots, y_n^0)$$

is efficient. $\square$

1.2 The fundamental theorems of welfare economics

The notion of Pareto-optimality has a close relation to the equilibria considered in the previous chapters. In a certain sense, it presents in a precise way what is achieved by the famous *invisible hand* of Adam Smith, the particular type of social optimum obtained as a result of the actions driven by self-interest but governed by the laws of the market. In its contemporary version, the theory is presented as the two *main theorems of welfare economics,* connecting Pareto-optimal allocations to those obtained through the market mechanism, where we allow for redistribution of incomes. To treat the latter type of allocations, we introduce the concept of a *market equilibrium*.

Definition 3.2. Let $\mathcal{E} = ((X_i, P_i)_{i=1}^m, (Y_j)_{j=1}^n, \omega)$ be an economy. An array (x^0, y^0, p), where (x^0, y^0) is an allocation and $p \in \mathbb{R}_+^l \setminus \{0\}$ a price, is a *market equilibrium* if

(i) (x^0, y^0) is feasible and $\sum_{i=1}^m x_i^0 - \sum_{j=1}^n y_j^0 = \omega$,
(ii) for each consumer $i = 1, \ldots, m$,

$$P_i(x_i^0) \cap \left\{ x_i \in X_i \,\middle|\, p \cdot x_i \le p \cdot x_i^0 \right\} = \emptyset,$$

(iii) for each producer $j = 1, \ldots, n$, y_j^0 maximizes $p \cdot y_j$ on the set Y_j,

The notion of a market equilibrium is a more general version of the Walras equilibrium considered already, differing only in the budget condition, since the consumers have no well-defined income in economies with no specification of ownership. Alternatively, the market equilibrium can be seen as emerging if the total income in the society can be redistributed among the consumers.

The main theorems of welfare economics state that the two classes of allocations, Pareto-optimal and achieved in market equilibria, coincide under suitable assumptions.

Box 3. The first welfare theorem in the Edgeworth box: Suppose that a certain Pareto-optimal allocation, corresponding to the point A in Fig. 3, has been singled out for actual implementation. In an economy with only two consumers, this could be done in a most unsophisticated way by delivering the relevant amounts of the two commodities to consumers. However, the task would become too complex when m and l are large.

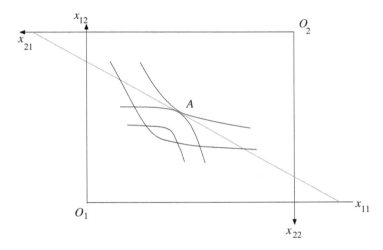

Fig. 3. Decentralizing in the Edgeworth box.

Draw the common tangent line to the indifference curves of the two consumers. If we choose a price system p that is normal to this tangent and assign to the consumer 1 an income I_1 exactly sufficient to buy the bundle represented by A at these prices, then the budget set of consumer 1 will be the area below the price line through A in Fig. 3, and — what is important — the bundle emerges as the one which is maximal for the consumer's preferences in this budget set.

(Continued)

Box 3. *(Continued)*

Similarly for consumer 2: For the same price p normal to the common tangent, we assign an income to the consumer such that he can buy the bundle represented by A as viewed from O_2. Again, this bundle is maximal for the preferences of consumer 2 in the budget set.

We have achieved that the originally given Pareto-optimal allocation will emerge as (the allocation corresponding to) a market equilibrium for a suitable choice of price system p and incomes I_1, I_2. We have thus *decentralized* the decisions in the economy using the market.

1.2.1 The first main theorem: Equilibrium allocations are Pareto-optimal

Our first result in this direction — in the literature traditionally called the first fundamental theorem of welfare economics — says that *allocations belonging to market equilibria are Pareto-optimal*.

Theorem 3.2. *Let $\mathcal{E} = ((X_i, P_i)_{i=1}^m, (Y_j)_{j=1}^n, \omega)$ be an economy where consumers have monotonic preferences, i.e. $x_i + \mathbb{R}_{++}^l \subset P_i(x_i)$ for all $x_i \in X_i$, $i = 1, \ldots, m$. If (x^0, y^0, p) is a market equilibrium, then (x^0, y^0) is Pareto-optimal.*

Proof. Suppose to the contrary that (x^0, y^0) is *not* Pareto-optimal. Then there is another feasible allocation $(x^1, y^1) = (x_1^1, \ldots, x_m^1, y_1^1, \ldots, y_n^1)$, such that $x_i^1 \in \mathrm{cl}P_i(x_i^0)$ for all consumers i and $x_{i_0}^1 \in P_{i^0}(x_{i_0})$ for some i_0. We may assume that $x_i^1 \in P_i(x_i^0)$ for all i, since otherwise, by openness of $P_{i_0}(x_{i_0})$, we can assume that $x_{i_0}^1$ belongs to $\mathbb{R}_{++}^l$, and removing a small enough amount of all commodities from $x_{i_0}^1$ and distributing it among the other consumers, we infer that everybody is better off.

Using (ii) of Definition 3.2, we get that

$$p \cdot x_1^1 > p \cdot x_i^0, \quad \text{all } i,$$

and by (iii) of Definition 3.2, we have that

$$p \cdot y_j^0 \geq p \cdot y_j^1, \quad \text{all } j.$$

Combining the inequalities, we get that

$$p \cdot \left(\sum_{i=1}^{m} x_i^1 - \sum_{j=1}^{n} y_j^1 \right) > p \cdot \left(\sum_{i=1}^{m} x_i^0 - \sum_{j=1}^{n} y_j^0 \right) = p \cdot \omega, \qquad (4)$$

where the last equality follows from (i) of Definition 3.2. Since (x^1, y^1) is feasible, we have that the left-hand side in (4) is $\leq p \cdot \omega$, a contradiction. We conclude that (x^0, y^0) is Pareto-optimal.

Theorem 3.2 shows that allocations coming from the market equilibria are Pareto-optimal, but since there are all sorts of Pareto-optimal allocations, this is not quite satisfactory. However, the converse is obtained from the second main theorem of welfare economics which is as follows:

1.2.2 The second fundamental theorem of welfare economics

The geometric arguments of the previous section suggest that a general result may be established, saying that *Pareto-optimal allocations can be obtained as* (allocations in) *market equilibria* with suitably chosen prices. Moreover, by now we suspect that the argument must involve separation of convex sets. The following will vindicate these beliefs.

Theorem 3.3 ('2nd fundamental theorem of welfare economics').
Let $\mathcal{E} = ((X_i, P_i)_{i=1}^{m}, (Y_j)_{j=1}^{n}, \omega)$ be an economy where consumers have convex and monotonic preferences, and producers have convex production sets.

If (x, y) is a Pareto-optimal allocation in $\mathcal{E}$ with $x_i \in \text{int } X_i, i = 1, \ldots, m$, then there exists a price system $p \in \mathbb{R}_+^l, p \neq 0$, such that (x, y, p) is a market equilibrium.

Proof. Define the set

$$Z = \sum_{i=1}^{m} \text{cl} \, P_i(x_i) - \sum_{j=1}^{n} Y_j - \{\omega\}.$$

Since all the sets $\text{cl} \, P_i(x_i)$ and Y_j are convex by our assumptions, so

> **Box 4. The need for convexity assumptions in the second theorem of welfare economics:** Figure 4 gives an example of an economy where the decentralization argument fails.
>
>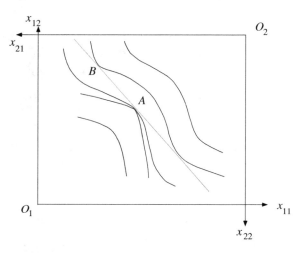
>
> Fig. 4. Non-convexities in the Edgewoth box.
>
> The common tangent at A defines the budget sets for consumers 1 and 2, and the latter will certainly choose A as the solution to his/her maximization problem. However, consumer 1 will choose another bundle, namely the one corresponding to B, since this takes him/her to an indifference curve which lies above the one through A. Consequently, the decisions of the two consumers are inconsistent. The problem with the present example is the preferences of consumer 1 fail to be well-behaved. In general, the decentralization arguments will be seen to depend crucially on the convexity assumptions on consumers and producers.

is Z as a sum of these sets (and a singleton). If $Z \cap \mathbb{R}^l_{--} \neq \emptyset$, then there is $u \in \mathbb{R}^l_{--}$ which can be written as

$$u = \sum_{i=1}^{m} x'_i - \sum_{j=1}^{n} y_j - \omega,$$

with $x'_i \in \mathrm{cl}\, P_i(x_i), i = 1, \ldots, m, y_j \in Y_j, j = 1, \ldots, n$. Define the allocation (x^1, y) by

$$x^1_i = x'_i - \frac{1}{m}, \quad i = 1, \ldots, m.$$

Then

$$\sum_{i=1}^m x^1_i = \sum_{i=1}^m x'_i - u = \sum_{j=1}^n y_j + \omega,$$

so (x^1, y) is feasible, and $x^1_i \in P_i(x_i)$ for each i by monotonicity. It follows that (x^1, y) is a Pareto improvement of (x, y), a contradiction showing that $Z \cap \mathbb{R}^l_{--} = \emptyset$.

By separation of convex sets (cf. Theorem 3.5 in Section 4), there exists $p \in \mathbb{R}^l, p \neq 0$, such that

$$\begin{aligned} p \cdot z &\geq 0 \quad z \in Z, \\ p \cdot z &< 0 \quad z \in \mathbb{R}^l_{--}, \end{aligned} \tag{5}$$

from which we get that $p \in \mathbb{R}^l_+ \backslash \{0\}$.

We check that (x, y, p) is a market equilibrium. The feasibility condition (i) of Definition 3.2 is fulfiled trivially. From (5), we have that $\sum_{i=1}^m x_i - \sum_{j=1}^n y_j - \omega$ minimizes $p \cdot z$ on Z. Using the definition of Z, we see that x_i minimizes $p \cdot x'_i$ on $\mathrm{cl}\, P_i(x_i)$, for $i = 1, \ldots, m$, and that y_j maximizes $p \cdot (-y'_j)$ on $-Y_j$, which is property (iii) in Definition 3.2. Finally, since $x_i \in \mathrm{int}\, X_i$ minimizes expenditure at the prices p, we get from Lemma 1.2 that $P_i(x_i) \cap \{x'_i \mid p \cdot x'_i \leq p \cdot x_i\} = \emptyset$, which is property (ii), so (x, y, p) is indeed a market equilibrium. □

Remark 3.1. We have shown that for any Pareto-optimal allocation (x, y), there is a price system $p \in \mathbb{R}^l_+ \backslash \{0\}$ which

(i) supports each preferred set $P_i(x)$, $i = 1, \ldots, m$ in the sense that $p \cdot x_i \leq p \cdot x'_i$, all $x'_i \in P_i(x_i)$,
(ii) supports each production set Y_j, $j = 1, \ldots, n$ in the sense that $p \cdot y_j \geq p \cdot y'_j$, all $y'_j \in Y_j$.

98   *Theory of General Economic Equilibrium*

This property of *common support* is often referred to as equality of marginal rates of substitution (in consumption as well as in production) for all agents.

2. Pareto-optimality and Social Optimum

As we have seen, the class of Pareto-optimal allocations in a given economy $\mathcal{E} = ((X_i, P_i)_{i=1}^m, (Y_j)_{j=1}^n, \omega)$ is quite large, so that Pareto-optimality cannot stand alone as a criterion for collective or social desirability. Having established the equivalence of Pareto-optimal allocations and those obtained from a market equilibrium, these considerations pertain to the market mechanism as well: Very different types of allocation, when considered from the point of view of distributional justice or other desiderata, can be the equilibrium result of market forces, and if these viewpoints matter, some further selection must take place.

In the classical approach to this problem, it is assumed that social desirability of allocations can be expressed in terms of a *social utility function:* For this, we shall assume that consumer preferences P_i are described by utility functions $u_i, i = 1, \ldots, m$, and in this case, a social choice function $S : \mathbb{R}^m \to \mathbb{R}$ assigns to each allocation $(x, y) \in F(\mathcal{E})$ its social utility $S(u_1(x_1), \ldots, u_m(x_m))$.

There are several reasons why we are not happy with this concept of the social utility function, even if we accept that consumers have ordered preferences. It is known from the theory of social choice that aggregating individual preferences typically has undesired side effects, so that the function cannot be considered as logically prescribed by some reasonable principles. But for the moment, we shall accept that a social utility function has been selected and proceed from this point.

An allocation (x^0, y^0) is said to be a *social optimum* (relative to S) if it maximizes $S(u_1(x_1), \ldots, u_m(x_m))$ over all feasible allocations $(x, y) \in F(\mathcal{E})$. The following is a straightforward consequence of the definitions and Theorem 3.3 and is stated without proof.

Lemma 3.1. *Assume that the social utility function S is increasing in each argument. Then a social optimum (x^0, y^0) relative to S is Pareto-optimal, and there is a price system $p \in \mathbb{R}^l_+ \setminus \{0\}$ such that (x^0, y^0, p) is a market equilibrium.*

The availability of a common support in the social optimum is useful if one considers possible displacements of the allocation, possibly brought about by changes in the production sets of the economy. Since we have already accepted a social utility, we may as well assume that S and the utility functions u_i are differentiable and that each Y_j is represented by a differentiable concave indicator f_j (such that $f_j(y_j) \leq 0$ if and only if $y \in Y_j$).

Assume that the allocation is changed by a small amount dx_{ih} for each consumer i and commodity h, giving rise to a change dx_i in the bundle of consumer i, $i = 1, \ldots, m$. These changes may occur as a consequence of changes in production possibilities, the details of which need not bother us here. If we accept the social utility function S as measuring a society's welfare, then the change in consumption implies a change in social welfare of size

$$dS = \sum_{i=1}^{m} S_i' \, du_i = \sum_{i=1}^{m} \sum_{h=1}^{l} S_i' u_{ik}' \, dx_{ik},$$

where all partial derivatives are assessed at (x^0, y^0). Since the latter is a social optimum, we must have

$$S_i' u_{ih}' = S_k' u_{kh}'$$

for all i and k in $\{1, \ldots, m\}$, and using that p is a common support, so that it has the same direction as the gradient $\nabla u_i(x_i^0)$ of u_i at x_i^0, meaning that $u_{ih}' = \lambda_i p_h$, $h = 1, \ldots, l$, for some constant λ_i, $i = 1, \ldots, m$, we obtain that

$$dS = K \sum_{i=1}^{m} p \cdot dx_{ik}, \tag{6}$$

where $K = S_1' \lambda_1 = \cdots = S_m' \lambda_m$.

100 *Theory of General Economic Equilibrium*

The expression in (6) is useful in several contexts, and we consider some of these in what follows.

2.1 *The foundations of cost–benefit analysis*

In many cases where an investment project must be assessed, one encounters the problem that much of the output created is not directly disposed of in a market, rather it contributes to the final allocation of ordinary commodities, which in their turn have a market price. This happens, for example, when the project has to do with infrastructure or with facilitating the development of new industries. Even though output cannot be assessed directly, an assessment of the desirability for society can be carried out using (6): If the market value of the change in consumption achieved by the project is positive, that is, if

$$\sum_{i=1}^{m} p \cdot \mathrm{d}x_i > 0,$$

then the project is advantageous for the society, whereas it is disadvantageous if this change is negative. Thus, an assessment of the value to society of the project can be carried out even though the individual utilities *and* the social utility function are unknown, since the sign of the expression in (6) depends only on observed quantities and prices.

This being said, it should of course be admitted that the argumentation presupposes that the initial allocation constitutes a social optimum, something which would be difficult to verify without a knowledge of S and the individual utilities. Also, the reasoning will hold only if the changes are marginal; if not, we run into the problem that the linear approximation in (6) is no longer credible.

While the problems connected with non-marginal changes in allocation need another approach, we may reconsider the above approach slightly to avoid — at least superficially — the claim that the initial situation is a social optimum. Indeed, as long as consumers obtain their bundle x_i^0 in a market at prices p, we know from

Section 2 of the Introduction chapter that p supports the set cl $P_i(x_i^0)$ in x_i^0, and if preferences are described by utility functions, we have that

$$u_i' = \lambda_i p, \quad i = 1, \ldots, m,$$

for $\lambda_i > 0$ a constant depending only on i, which may be interpreted as the marginal utility of the income of consumer i. To get to (6) we need to assume that the *income is distributed optimally among consumers*, which of course remains somewhat unclear without explicit reference to a social utility function, so in this sense, we have not moved very far. On the other hand, one may argue that what is of interest when assessing a project is the net contribution to overall social utility of this project, that is the contribution over and above what could be achieved by redistributing income.

2.2 *National accounting*

Another field of application of the reasoning leading to (6) can be found in the system of national accounts. Finding the GDP of a country essentially consists in collecting data on value added in every productive unit of society and assessing this at the market prices. In our notation, this is not fully captured by the aggregate value of net production

$$p \cdot \sum_{j=1}^{n} y_j^0, \tag{7}$$

since commodities (such as, e.g. labor power) available in the endowment and used in production are counted negatively in (7). Moreover, to obtain that total value produced equals total value of goods available for consumption and investment, we should add the term $p \cdot \omega$ in (7) so as to obtain

$$p \cdot \sum_{i=1}^{m} x_i^0$$

102 *Theory of General Economic Equilibrium*

as the formal counterpart of Gross Domestic Product (GDP) in market prices.

Returning once again to (6), we see that a change in final consumption, evaluated at the prices belonging to the original market equilibrium, corresponds to a change in GDP, and we may conclude that *if* welfare can be measured by the social utility function S, and *if* the initial situation was a social optimum with respect to S, then an increase in GDP implies an increase in the overall social welfare.

Considered in the light of everyday usage of GDP statistics, this result may seem utterly self-evident, but it is worth noting that if any one of the premises is deleted, the conclusion will necessarily no longer be true. In particular, growth rates in GDP should be interpreted at arm's distance: Sectors contributing very little to overall social utility may produce goods which receive high prices in the market and consequently have a large impact on GDP. Also, it should be noted that market equilibria, which presuppose perfect competition, are theoretical constructions and very useful as such, but they do not give a truthful picture of the actual market and non-market allocations, so economic measurement must be handled with some care.

2.3   Second-best allocations

As we have seen above, the presence of a social utility function permits a detailed analysis of several problems where no clear solution can be obtained without it, but this is achieved at the cost of a considerable lack of realism. As an additional example of this, we consider the following situation.

Suppose that in a given economy $\mathcal{E}$ some institutional barriers to individual actions are introduced, so that the original social optimum can no longer be realized. In this case, the best (in terms of the given social utility function) that the society can do is to maximize the social utility under the original constraints *together with* the new constraints arising from the institutions considered. The resulting constrained maximization problem has as its solution a so-called *second-best optimum.* Formally, if the institutional barriers amount to a restriction of the set of feasible allocations in the economy to some set $A \subset (\mathbb{R}^l)^m \times (\mathbb{R}^l)^n$ of permissible allocations, then a second-best optimum maximizes $S(u_1(x_1), \ldots, u_m(x_m))$ over all allocations $(x_1, \ldots, x_m, y_1, \ldots, y_n)$ with $(x_1, \ldots, x_m, y_1, \ldots, y_n) \in A$.

It is obvious that a second-best optimum will achieve a social utility level which is no higher, and in most cases lower, than that of the (unconstrained) social optimum. The second-best optimum may or may not be obtainable in a market equilibrium for suitably defined prices and incomes, depending of course on whether or not the allocation is Pareto-optimal, something which cannot be decided upon in general.

If no social utility function is available, we are left with the much weaker concept of Pareto-optimality for comparison of allocations. In this setting, the relevant concept for considerations of social desirability of allocations under additional constraints is that of a second-best Pareto-optimum. We leave it to the reader to formulate the precise definition of this concept.

Contrasting with the situation where a social utility function was available, we cannot any more assert that barriers lead to allocations which are inferior to the original ('first-best') situation. In the general case, some consumers will be worse off but others may very well be better off than in the initial allocation. Therefore, no general statements can be made to the effect that additional institutional barriers decrease welfare.

Another more traditional, way of telling the same story is the following: We know that in a Pareto-optimal allocation the marginal rates of substitution are identical for all agents. Suppose now that due to institutional barriers Pareto-optimality, or equivalently, universal equality of marginal rates of substitution, cannot be achieved. If in this world of second-best we move from one allocation with a given (less than full) number of equalities in marginal rates of substitution to another one with a higher (but also not full) number of equalities, does this represent an improvement in society's welfare?

Box 5. An example of a second-best optimum: There are two producers and a single consumer in this economy. The utility function of the consumer can be used as a social utility function (single-consumer economies being the only cases where existence of a social utility function is uncontroversial). We consider two allocations, the consumption bundles (which are equal to the total production bundles) which correspond to the points A and B, respectively. It is seen at once that neither A nor B is Pareto-optimal.

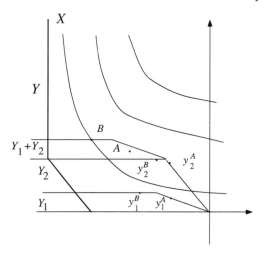

Fig. 5. A case for the general theory of second-best.

In the allocation represented by A, the individual production plans of the two firms are y_1^A and y_2^A, and in allocation B, they are y_1^B and y_2^B. It is seen from Fig. 5 that in A all marginal rates of substitution (=slopes of indifference curve or production functions through the points in question) are different but in B the marginal rates of substitution are the same in the two firms (but still different from that of the consumer). However, it is seen that B is by no means better than A; on the contrary, it is worse. This shows that there is no general connection between the number of equalities in marginal rates of substitution and welfare level of the underlying allocations.

The answer is no, first of all because the concept of society's welfare is ambiguous, but even in cases where it is not, the number of equalities of marginal rates of substitution is quite irrelevant for welfare considerations. For an example, see Box .5.

3. Quantifying the Lack of Efficiency

Efficiency is clearly a fundamental property of allocations, one which in principle should be obtained in any society. In practice, there may, however, be many obstacles to efficient allocation, and in order not to lose sight of the final goal it would be convenient to know how far a given allocation is from being efficient. Clearly, any measure of distance from the set of efficient allocations in an economy depends on the metric used, and since the question arose in the context of economic decision-making, the measure should be economically meaningful.

An intuitive approach toward measuring lack of efficiency is the *coefficient of resource utilization* introduced by Debreu (1951): Let $\mathcal{E} = ((X_i, P_i)_{i=1}^m, \omega)$ be an exchange economy. Choosing an arbitrary feasible allocation $x \in F(\mathcal{E})$, we cannot expect that x is Pareto-optimal, indeed most allocations (at least intuitively speaking, but the statement can easily be made precise, cf. also Chapter 7) in $F(\mathcal{E})$ are not Pareto-optimal. If x is not Pareto-optimal, then the resources of the economy, as specified by ω, could have been reallocated among consumers to achieve another feasible allocation $x' \in F(\mathcal{E})$ with $x'_i \in P_i(x_i)$ for all i. By continuity of preferences, it would even have been possible to allocate some ω' with $\omega'_h < \omega_h$ for $h = 1, \ldots, l$ and achieve an allocation better than x for every consumer i. Achieving inefficient allocations is in this way equivalent to throwing away resources which could otherwise have been used for creating additional consumer satisfaction.

Now a measure of the inefficiency exhibited by the allocation x could be this amount of resources thrown away, only this opens up some ambiguity since the l commodities may be dispensed within different amounts. A natural way of solving this problem would be to use proportional reduction of resources, so that any $\lambda \in [0,1]$ with the property that there is an allocation x' satisfying $\sum_{i=1}^{m} x'_i \leq \sum_{i=1}^{m} \lambda\omega$ and $x'_i \in P_i(x_i)$ represents an admissible reduction of the initial endowment. We may now define the coefficient of resource utilization $\lambda_D(\mathcal{E})$ as

$$\lambda_D(x, \mathcal{E}) = \inf\{\lambda \in [0,1] \mid \exists x' \in F$$
$$((X_i, P_i)_{i=1}^{m}, \lambda\omega) : x'_i \in P_i(x_i), \text{ all } i\}. \tag{8}$$

Box 6. Measures of technical efficiency (1): In applications, efficiency measurement is performed either as *input efficiency* or *output efficiency*.

The Farrell (1957) measure of technical efficiency in input is found as the smallest (proportional reduction) from the given input levels which will yield the given output as follows:

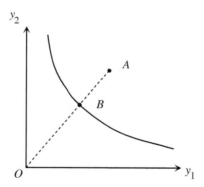

Fig. 6. Farrell measure of input efficiency is $|OB|/|OA|$.

In Fig. 6, the given output level is illustrated by the isoquant II', and actual input is A. The point B represents the maximal proportional reduction possible, and the Farrell measure is then $\dfrac{|OB|}{|OA|}$. In the corresponding measure of output efficiency, the possible output combinations are below the transformation curve TT', and actual output A can be increased proportionally to obtain the point B (Fig. 7).

(Continued)

Box 6. *(Continued)*

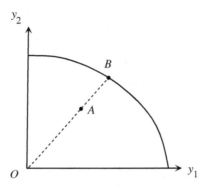

Fig. 7. Farrell measure of output efficiency is $|OA|/|OB|$.

The Farrell measure is now found as the inverse of the maximal proportional increase.

Then $\lambda_D(\mathcal{E})$ is a number between 0 and 1, and if x is Pareto-optimal, then $\lambda_D(x, \mathcal{E}) = 1$. The converse is not true in general, indeed there may be allocations x with $\lambda_D(x, \mathcal{E}) = 1$ which are not Pareto-optimal. This will, however, occur only under particular circumstances. We leave the details to the exercises.

It is easily seen that the coefficient of resource allocation may be defined in the same way as in (8) for economies with production $\mathcal{E} = ((X_i, P_i)_{i=1}^m, (Y_j)_{j=1}^n, \omega)$, namely as

$$\lambda_D((x,y), \mathcal{E}) = \inf\{\lambda \in [0,1] \mid \exists (x', y') \in F((X_i, P_i)_{i=1}^m,$$
$$(Y_j)_{j=1}^n, \lambda\omega) : x_i' \in P_i(x_i), \text{ all } i\}, \qquad (9)$$

where $F((x, y), \mathcal{E})$ is the set of feasible allocations in $\mathcal{E}$,

$$F(\mathcal{E}) = \left\{ (x, y) = (x_1, \ldots, x_m, y_1, \ldots, y_n) \,\middle|\, \sum_{i=1}^m x_i \leq \sum_{j=1}^n y_j + \omega \right\}.$$

As before, Pareto-optimal allocations yield a coefficient value of 1, and although there may be allocations whose coefficient of resource allocation is 1 but which are not Pareto-optimal, we may still consider the value of $\lambda_D((x, y), \mathcal{E})$ as a useful indicator of the degree to which the allocation can be considered as efficient.

The coefficient of resource allocation has found widespread application in productivity analysis. As we have seen, the notion of technical efficiency can be viewed as a special case of Pareto-optimality for an economy $\mathcal{E} = ((X_i, P_i)_{i=1}^m, (Y_j)_{j=1}^n, \omega)$, where consumers are identical with $X_i = \mathbb{R}_+^l$ and P_i given by $P_i(x_i) = \{x_i\} + \left[\mathbb{R}_+^l \setminus \{0\}\right]$ for all $x_i \in \mathbb{R}_+^l$, so that we may assume that there is only one consumer. Writing $Y = \sum_{j=1}^n Y_j$ for the aggregate production set in the economy, we get a special version of the coefficient of resource allocation in this economy, namely

$$\lambda_D((x, y), \mathcal{E}) = \inf \left\{ \lambda \in [0, 1] \,\middle|\, \exists y' \in Y : y' \geq y, x - y' \leq \lambda \omega \right\}.$$

If we interpret the set of feasible aggregate consumption bundles as the set of outputs (in a broad sense) obtained from the economic activity using external input ω, then $\lambda_D((x, y), \mathcal{E})$ determines the largest proportional reduction in inputs possible if the same output should be available.

This interpretation points to applications of the coefficient of resource allocation to measuring output and input productivity for single productive units, cf. Box 6. Since this takes us outside the realm of general equilibrium theory, we shall not pursue this topic any further.

4. Separation Theorems for Convex Sets

Since separation of convex sets has been used in this chapter and will be used repeatedly in the sequel, we state and prove a simple version here.

Theorem 3.4. *Let $C \subset \mathbb{R}^l$ be non-empty, closed and convex, and suppose that $z \notin C$. Then there is $p \in \mathbb{R}^l$, $p \neq 0$, such that $p \cdot z > p \cdot x$ for all $x \in C$.*

Box 7. Data Envelopment Analysis (DEA) is a method for comparing productive efficiency of several units. It consists of two parts:

    (i) measuring technical efficiency of a productive unit with reference to a given set of feasible productions,

    (ii) finding a representation of the set of feasible productions based on the observation of other production plans.

For part (i), one chooses the Farrell measure (cf. Box 6). In part (ii), the production set is reconstructed from the observation of either past productions or productions in other plants. Let these reference observations be $y^1, \ldots, y^r$, and suppose that the unit considered has the production y^0.

Assuming convexity of Y, every convex combination

$$\mu_1 y^1 + \cdots + \mu_r y^r, \quad \mu_i \geq 0, \ i = 1, \ldots, r, \quad \sum_{i=1}^{r} \mu_i = 1, \tag{10}$$

must belong to Y, and we may then find the Farrell index value of y^0 as inverse of the maximal number $\lambda > 0$ such that λy^0 can be written as in (10). This can be reformulated as solving the maximization problem

$$\max \lambda,$$

$$\sum_{i=1}^{r} \mu_i \leq 1, \ \lambda y_h^0 - \sum_{i=1}^{r} \mu_i y_h^i \leq 0, \quad h = 1, \ldots, n, \tag{11}$$

$$\lambda, \mu_1, \ldots, \mu_r \geq 0.$$

It is seen that (11) is a linear programming problem, so that the measuring (Farrell) efficiency relative to the production set spanned by the reference data is a relatively simple procedure.

Proof. Let $x^0 \in C$ be arbitrary. The set $K = \{x \in C | \|x - z\| \leq \|x^0 - z\|\}$ is compact, and $\|x - z\|$ is continuous in x, so there is $\bar{x} \in C$, such that $\|\bar{x} - z\| = \min_{x \in K} \|x^0 - z\|$.

Let $p = \bar{x} - z \neq 0$ and $p \cdot \bar{x} = \alpha$. From the relations

$$0 < \|\bar{x} - z\|^2 = (\bar{x} - z) \cdot (\bar{x} - z) = p \cdot \bar{x} - p \cdot z,$$

we get that $p \cdot z < \alpha$. We show that $p \cdot x \geq \alpha$ for all $x \in C$:

Suppose that there is $x \in C$ with $p \cdot x < \alpha$; define $x^\lambda = \lambda x + (1-\lambda)\bar{x}$ for $\lambda \in [0,1]$. We have that

$$\|\bar{x} - z\|^2 - \|x^\lambda - z\|^2 = \sum_{k=1}^{l}(\bar{x}_k - z_k)^2 - \sum_{k=1}^{l}(\lambda x_k - \lambda \bar{x}_k + \bar{x}_k - z_k)^2$$

$$= -\lambda^2 \sum_{k=1}^{l}(x_k - \bar{x}_k)^2 - 2\lambda \sum_{k=1}^{l}(x_k - \bar{x}_k)(\bar{x}_k - z_k)$$

$$= \lambda(-\lambda\|x - \bar{x}\|^2 - 2p(x - \bar{x})).$$

We have that $p \cdot (\bar{x} - x) > 0$, so that the last expression on the right-hand side is positive whenever $0 < \lambda < 2p \cdot (\bar{x} - x)/\|x - \bar{x}\|^2$. But then, $x^\lambda \in C$ contradicts the definition of $\bar{x}$. □

The situation considered in Theorem 3.4 is illustrated in Fig. 8. In the text, we use a separation theorem for a somewhat different setup.

Theorem 3.5 (Minkowski). *Let $C \subset \mathbb{R}^l$ be a convex set, and suppose that $C \cap \mathbb{R}_{--} = \emptyset$. Then there is $p \in \mathbb{R}^l$, $p \neq 0$, such that $p \cdot x \geq 0$ for all $x \in C$.*

Proof. Choose a sequence $(z_n)_{n=1}^{\infty}$ such that $z_n \in N$ and $z_n \to 0$ (e.g. the sequence with $z_n = -(1/n, \ldots, 1/n)$). For each n, we apply

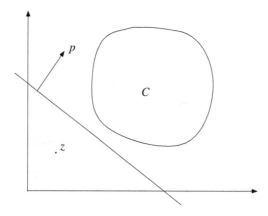

Fig. 8. Separation by a hyperplane of the convex set C and the point.

Theorem 3.4 to get $p_n \in \mathbb{R}^l$, $p_n \neq 0$, such that $p_n \cdot z_n < p_n \cdot x$ for all $x \in \text{cl}C$.

Consider now the sequence $(\bar{p}_n)_{n=1}^{\infty}$, where $\bar{p}_n = p_n/\|p_n\|$. The elements of the sequence belong to the compact set S consisting of all vectors with norm 1. Consequently, there is a subsequence $(\bar{p}_{n_t})_{t=1}^{\infty}$ converging to some $\bar{p} \in S$. Since $\|\bar{p}\| = 1$, we have that $\bar{p} \neq 0$.

Let $x \in C$ be arbitrary. For all t, we have that

$$\bar{p}_{n_t} \cdot z_{n_t} < \bar{p}_{n_t} x,$$

so in the limit we get that

$$0 = \bar{p} \cdot 0 \leq \bar{p} \cdot x.$$

Since $x \in C$ was arbitrary, we must have that $\bar{p} \cdot x \geq 0$ for all $x \in C$. $\square$

5. Exercises

(1) Consider an economy with two goods, one consumer and one producer. The consumer has consumption set $\mathbb{R}^2_+$ and preferences described by the utility function $u(x_1, x_2) = x_1^2 x_2$, and endowment $\omega = (20, 0)$. The producer has the production set $Y = \{(y_1, y_2) \mid y_2 \leq \sqrt{-2y_1}, y_1 \leq 0\}$.

Find all the Pareto-optimal allocations in the economy, and for each such allocation, determine prices at which it can be sustained in a market equilibrium.

(2) Let $\mathcal{E} = ((X_i, P_i)_{i=1}^m, \omega)$ be an economy where each consumer satisfies Assumptions 0.1 and 0.2. Show that if an allocation $(x_1, \ldots, x_m)$ is egalitarian equivalent with maximal scalar multiple of the reference bundle $\bar{x}$, then $(x_1, \ldots, x_m)$ is Pareto-optimal.

(3) Let $\mathcal{E} = (X_i, P_i, \omega_i)_{i=1}^m$ be an economy with private ownership, where consumers satisfy Assumptions 0.1 and 0.2, and let $p \in \mathbb{R}^l_{++}$ be a given price system. Show that there exists a

Pareto-optimal allocation $(x_1, \ldots, x_m)$ which satisfies the budget constraint $p \cdot x_i = p \cdot \omega_i$ for each i.

(4) **Foundation of cost-effectiveness analysis:** Assume that a new medical intervention will change the allocation in the society by the amounts $(dx_1, \ldots, dx_m)$ and will change the health condition of the m individuals by the amounts $(dh_1, \ldots, dh_m)$ (measured by some health indicator). The economy is initially in an equilibrium with price vector p, and both income and health are distributed optimally.

Show that this information is insufficient for determining whether the medical intervention is advantageous for society.

Find conditions such that one medical intervention $(dx_1, \ldots, dx_m, dh_1, \ldots, dh_m)$ can be considered as socially preferable to another intervention $(dx_1', \ldots, dx_m', dh_1', \ldots, dh_m')$ if

$$\frac{\sum_{i=1}^m p \cdot dx_i}{\sum_{i=1}^m dh_i} > \frac{\sum_{i=1}^m p \cdot dx_i'}{\sum_{i=1}^m dh_i'}.$$

(5) **Existence of fair allocations:** Let $\mathcal{E} = ((X_i, P_i)_{i=1}^m, \omega)$ be an exchange economy satisfying Assumptions 0.1 and 0.2. Show that there are fair allocations in $\mathcal{E}$.

(6) **Another interpretation of DEA:** Consider the maximization problem (11) in Box 7. As a linear programming problem, it has a *dual* problem.

Write down this dual problem in variables $v, u_1, \ldots, u_n$.

Show that one may assume $\sum_{h=1}^n u_h = 1$.

Show that u_h, $h = 1, \ldots, r$ can be interpreted as weights or prices of output commodities, chosen in the most favorable way for the unit to be assessed.

(7) Here, is an alternative (to the Farrell measure considered in Box 6) measure of technical efficiency: For L, a given subset of $\mathbb{R}_+^l$, here interpreted as input combinations (with changed sign)

allowing a particular output combination, define the Russell measure

$$\lambda_R(x, L) = \min \left\{ \frac{1}{l} \sum_{h=1}^{l} \lambda_h \middle| (\lambda_h x_h)_{h=1}^{l} \in L, 0 < \lambda_h \leq 1, h = 1, \ldots, l \right\}.$$

Compare the Farrell and the Russell measures and show cases where the two differ.

Chapter 4

Cooperative Game Theory
and Equilibria

1. The Core of an Economy

In Chapter 3, we have considered the use of resources in an economy, taking the viewpoint that all members of society are connected through the processes of production and allocation, so that these processes should be arranged in the best possible way. It seems rather straightforward to extend this point of view so as to also count for reallocation in smaller groups than that for all members of the society. Doing so, we must restrict attention to allocations that are satisfactory for each such group, big or small.

These considerations will lead us to the allocations in the core of the economy, a concept to be discussed at length in this chapter and in Chapter 5. And by introducing the core, we have opened up for the use of concepts and considerations taken from the theory of games, and we shall take a look at several alternative approaches toward singling out allocations with the desirable properties.

1.1 *Definition of the core*

Let $\mathcal{E} = (X_i, P_i, \omega_i)_{i=1}^m$ be a private ownership economy, for the moment without production. Considering allocations which have been achieved in this economy, we imagine that groups of individuals meet and exchange commodities so as to obtain bundles which are better for them than their initial endowments. At a later stage, we shall have more to say about this process of exchange, in particular pairwise exchanges, but for our present considerations, we abstract from the technical details of reallocating commodities within groups and concentrate on the final product of these reallocations. When no more reallocations can take place, the allocation obtained has a property that no group can make its members better off by reallocation of the goods to which it has access. This is the property defining the *core* of the economy.

Let $x = (x_1, \ldots, x_m) \in F(\mathcal{E})$ be a feasible allocation in $\mathcal{E}$. A *coalition* is a non-empty subset $S \subseteq M = \{1, \ldots, m\}$. The set of all coalitions from M is denoted $\mathcal{S}$. The coalition $S \in \mathcal{S}$ can *improve upon* x if there is x'_i, $i \in S$, with $\sum_{i \in S} x'_i = \sum_{i \in S} \omega_i$ such that $x'_i \in P_i(x_i)$ for all $i \in S$. The core of $\mathcal{E}$, written $\mathrm{Core}(\mathcal{E})$, is the set of feasible allocations x in $\mathcal{E}$ which cannot be improved upon by any coalition.

We note that in our definition of the core, what may possibly be reallocated is the initial endowment. This means that if we are thinking of the process of finding the final allocation as a sequence of group reallocations, then none of the reallocations are binding before all groups are satisfied in the sense that they cannot come up with a better reallocation.

1.1.1 *Existence of core allocations*

Having dealt at length with the existence of the Walras equilibria in Chapter 1, it would seem reasonable to be concerned here about the existence of core allocations. It turns out that the existence question can be resolved in a very simple way, using the following straightforward result.

Theorem 4.1. *Let (x, p) be a Walras equilibrium in $\mathcal{E}$. Then $x \in \mathrm{Core}(\mathcal{E})$.*

Proof. Suppose that $x \notin \mathrm{Core}(\mathcal{E})$. Then there is a coalition S which can improve upon x, i.e. there are $x'_i \in X_i$, $i \in S$, such that $\sum_{i \in S} x'_i = \sum_{i \in S} \omega_i$ and $x'_i \in P_i(x_i)$, all $i \in S$. From this set of inequalities, we have

$$p \cdot x_i' > p \cdot \omega_i \quad \text{for all } i \in S.$$

Summing over S, we get $p \cdot \sum_{i \in S} x_i' > p \cdot \sum_{i \in S} \omega_i$, contradicting that $\sum_{i \in S} x_i' = \sum_{i \in S} \omega_i$. Thus, $x \in \text{Core}(\mathcal{E})$. $\qquad\square$

The converse of Theorem 4.1 is not true in general, and we return to this matter shortly. Before doing so, we note that the existence of core allocations might have been established using tools from game theory.

We restrict our attention here to the case where consumer preferences P_i can be described by utility functions u_i. We can then define for each coalition S the set of attainable utility arrays

$$V(S) = \left\{ ((u_i(x_i))_{i \in S} \,\middle|\, x_i \in X_i, \sum_{i \in S} x_i \leq \sum_{i \in S} \omega_i \right\}.$$

The collection of sets $V(S)$ for $S \in \mathcal{S}$ defines a non-transferable utility (NTU) cooperative game, written as (M, V). The core of the game (M, V) is the set of payment vectors $t = (t_1, \ldots, t_n)$ which are Pareto-optimal in $V(M)$ (so that there is no $t' \neq t$ in $V(M)$ with $t_i' \geq t_i$ for all $i \in M$) and which cannot be improved by any coalition S (meaning that there is no $t^S \in V(S)$, such that $t_i^S > t_i$ for all $i \in S$).

The classical result about non-emptiness of the core of an NTU is due to Scarf (1967). For this, we introduce the concept of a balanced family of coalitions: A family $C \subset \mathcal{S}$ of coalitions is *balanced* if there are numbers $\lambda_S \geq 0$ for $S \in C$ (called the balancing weights of the coalitions) such that

$$\sum_{S : i \in S} \lambda_S = 1, \quad \text{each } i \in M. \tag{1}$$

Intuitively, one may think of a balanced family of coalitions as an arrangement by which the consumers are allowed to be part-time members of the coalitions in C. This interpretation may seem far-fetched, and it may be simpler to consider balancedness as just a property guarantees non-emptiness of the core. The game (M, V) is

Box 1. Core allocations which are not Walras allocations: Assuming that $X_i = \mathbb{R}_+^2$ for $i = 1, 2$, we may as usual illustrate feasible allocations in an Edgeworth box.

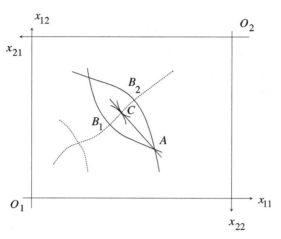

Fig. 1. Core allocations which fail to be Walrasian.

Which allocations belong to the core of $\mathcal{E}$? Since the core is a subset of the set of Pareto-optimal allocations, it must be part of the contract curve. Also, it must be situated in a region between B_1 and B_2 since otherwise one of the consumers could improve by returning to his initial endowment (the point A). Actually, since in the two-person economy all possible coalitions are $\{1\}$, $\{2\}$, and $\{1, 2\}$, there are no further restrictions on the core which therefore is equal to the segment of the contract curve between B_1 and B_2.

The point C in Fig. 1 is an equilibrium allocation and therefore belongs to the core. The point B_1 does *not* belong to a Walras equilibrium, showing that in general the core does not coincide with the set of Walras allocations.

balanced if

$$\bigcap_{S \in C} \widetilde{V}(S) \subseteq V(M),$$

whereby $\widetilde{V}(S) = V(S) \times \mathbb{R}^{M \setminus S}$ is the set of all m-tuples $(t_1, \ldots, t_m) \in \mathbb{R}^m$ such that $(t_i)_{i \in S} V(S)$, for every balanced family C. It was shown by Scarf (1967) that every balanced NTU game has non-empty core.

Theorem 4.2. *Let $\mathcal{E} = (X_i, P_i, \omega_i)_{i=1}^m$ be an exchange economy where consumers satisfy Assumptions 0.1 and 0.2 and preferences are represented by utility functions u_i, $i \in M$, and let (M, V) be the NTU game derived from $\mathcal{E}$. Then (M, V) is balanced.*

Proof. Let C be a balanced family of coalitions with balancing weights $(\lambda_S)_{S \in C}$, and let $(t_1, \ldots, t_m)$ be such that for each $S \in C$, $(t_i)_{i \in S} \in V(S)$, i.e. there are $(x_i^S)_{i \in S}$ with $\sum_{i \in S} x_i^S \leq \sum_{i \in S} \omega_i$ and $u_i(x_i^S) = t_i$ for $i \in S$. For each i, let $x_i = \sum_{S: i \in S} \lambda_S x_i^S$. Then $u_i(x_i) \geq t_i$ by convexity of preferences, and

$$\sum_{i=1}^m x_i = \sum_{i=1}^m \sum_{S: i \in S} \lambda_S x_i^S = \sum_{S \in C} \lambda_S \sum_{i \in S} x_i^S \leq \sum_{S \in C} \lambda_S \sum_{i \in S} \omega_i = \sum_{i=1}^m \sum_{S: i \in S} \lambda_S \omega_i$$

$$= \sum_{i=1}^m \omega_i,$$

so that $(t_1, \ldots, t_m) \in V(M)$. $\qquad\square$

Since (M, V) is balanced, its core is non-empty by the theorem of Scarf, and we have given another proof of the non-emptiness of the core. This more complicated approach may turn out to be useful when we consider economies with production.

1.2 Coalition production economies

So far we have considered only exchange economies, it is now time to introduce production. Judging from previous experience, one would expect this to be a simple matter, but this is not quite the case. In order to define an improvement, we need to specify how the coalition in question is endowed with commodities *and* production possibilities. In our treatment of Walras equilibrium with production, we assumed the existence of a given and fixed number of firms, each characterized by a production set $Y_j \subset \mathbb{R}^l$, and these firms were in their turn owned by the consumers. How the firms came about was not a matter of our concern, at least at that moment.

120   *Theory of General Economic Equilibrium*

When defining an improvement, it was understood that it should be an assignment of bundles to the members of the coalition, which could be achieved by this coalition without recourse to the remaining part of the economy. This means that we must explain how a coalition can produce if left to its own. Formally, we define a *coalition production economy* as an array

$$\mathcal{E} = ((X_i, P_i, \omega_i)_{i=1}^m, \mathcal{Y}),$$

where the first part, $(X_i, P_i, \omega_i)_{i=1}^m$, is the (standard) description of the consumers, whereas the second part is a correspondence $\mathcal{Y} : \mathcal{S} \rightrightarrows \mathcal{P}_0(\mathbb{R}_+^l)$, assigning to each coalition S a production set $\mathcal{Y}(S) \subset \mathbb{R}_+^l$. Thus, if a coalition is formed, it has at its disposal the commodity endowments of its members and a technology, from which it may choose a production plan with which to produce the bundles to be delivered to its consumers. The firms that we have seen in Chapter 2 will then be such that $\sum_{i=1}^m Y_j = \mathcal{Y}(M)$.

We can now define the core of the coalition production economy $\mathcal{E} = ((X_i, P_i, \omega_i)_{i=1}^m, \mathcal{Y})$ as an array $x = (x_1, \ldots, x_m)$ of final consumption bundles such that

(i) x is feasible for M, i.e. $x_i \in X_i$, $i \in M$, and $\sum_{i=1}^m (x_i - \omega) \in \mathcal{Y}(M)$,
(ii) there is no $S \in \mathcal{S}$ and $(x_i')_{i \in S}$ with $\sum_{i \in S}(x_i' - \omega_i) \in \mathcal{Y}(S)$ and $x_i' \in P_i(x_i)$, all $i \in S$.

The question of whether core allocations exist at all is less easily answered for coalition production economies. Indeed, if $(x_1, \ldots, x_m, y, p)$ is a Walras equilibrium in the economy with production $((X_i, P_i, \omega_i)_{i=1}^m, \mathcal{Y}(M))$, then each of the bundles x_i is maximal for P_i in the budget sets $\{x_i' \in X_i \mid p \cdot x_i' \leq p \cdot x_i\}$, and $\sum_{i=1}^m p \cdot (x_i - \omega_i) = \max\{p \cdot y \mid y \in \mathcal{Y}(N)\}$, but this does not preclude that

$$\sum_{i \in S} p \cdot (x_i - \omega_i) < p \cdot y'$$

for some coalition S and $y' \in \mathcal{Y}(S)$, so that the coalition S has an improvement.

Cooperative Game Theory and Equilibria 121

It seems reasonable to demand that each of the production sets $\mathcal{Y}(S)$, for $S \in \mathcal{S}$, satisfies Assumption 0.3. In our present context, we shall, however, need something more, connecting the production possibilities of distinct coalitions. Indeed, if we expect that the final result should be some allocation arranged by the coalition M (of all individuals, also called the *grand* coalition), then the production possibilities of M should not be inferior to those of some subcoalition, that is, $\mathcal{Y}$ should satisfy the condition

$$S_1, S_2 \in \mathcal{S}, S_1 \subseteq S_2 \Rightarrow \mathcal{Y}(S_1) \subseteq \mathcal{Y}(S_2) \text{ (monotonicity).} \qquad (2)$$

Without monotonicity, we might run into existence problems of the type discussed above. However, in order to make sure that core allocations exist, we need something more. In view of our previous considerations, an assumption of balancedness, formulated on the production correspondence $\mathcal{Y}$, appears as a good choice. We shall use it in another version of balancedness (known as *cardinal* balancedness): Let $C \subseteq \mathcal{S}$ be a balanced family of coalitions with weights $(\lambda_S)_{S \in C}$, then

$$\sum_{S \in C} \lambda_S \mathcal{Y}(S) \subseteq \mathcal{Y}(N).$$

We can now prove a counterpart of Theorem 4.2.

Theorem 4.3. *Let $\mathcal{E} = ((X_i, P_i, \omega_i)_{i=1}^m, \mathcal{Y})$ be a coalition production economy, where consumers satisfy Assumptions 0.1 and 0.2 and consumer preferences P_i are represented by utility functions u_i, $i \in M$, where all production sets $\mathcal{Y}(S)$, for $S \in \mathcal{S}$, satisfy Assumption 0.3 and where $\mathcal{Y}$ is cardinally balanced.*

If the NTU game $(M, \widetilde{V})$ is defined by

$$V(S) = \{(t_i)_{i \in S} \mid \exists x_i, i \in S, \sum_{i \in S}(x_i - \omega_i) \in \mathcal{Y}(S), u_i(x_i) = t_i, i \in M\},$$

$$(3)$$

then $(M, \widetilde{V})$ is balanced.

Proof. Choose a balanced family C with weights $(\lambda_S)_{S \in C}$. Suppose that $(t_1, \dots, t_m) \in \bigcap_{S \in C} \widehat{V}(S)$, and for each $S \in C$, choose (in accordance with (3)) coalitional allocations $(x_i^S)_{i \in S}$ with $\sum_{i \in S}(x_i^S - \omega_i) \in \mathcal{Y}(S)$ such that $u_i(x_i^S) = t_i$ for each $i \in S$. By the cardinal balancedness of $\mathcal{Y}$, we have that $\sum_{S \in C} \lambda_S \left(\sum_{i \in S}(x_i^S - \omega_i) \right) \in \mathcal{Y}(M)$.

For each $i \in M$, let $x_i = \sum_{S \in C: i \in S} x_i^S$. By convexity of X_i and of preferences, we have that $u_i(x_i) \geq t_i$ for each $i \in M$. Now,

$$\sum_{S \in C} \lambda_S \sum_{i \in S}(x_i^S - \omega_i) = \sum_{i=1}^{n} \sum_{S \in C: i \in S} \lambda_S(x_i^S - \omega_i) = \sum_{i=1}^{m}(x_i - \omega_i),$$

and it follows that $(t_1, \ldots, t_m) \in V(M)$. $\square$

The result of Theorem 4.3 can be used to show the existence of core allocations in a coalition production economy, considering again that balanced NTU games have non-empty cores. However, this does not provide us with any clue to the relation between the core and allocations achieved using prices and markets. Incidentally, the concept of a Walras equilibrium also needs a reconsideration in the context of coalition production economies, since the shares of the profits from production must also be determined by the equilibrium conditions: When coalitions can form and use their specific technology, the profits for the members of a coalition to be obtained by producing in their own technology should not exceed the share of profits obtained when producing in the grand coalition.

1.3   Core equivalence in replica economies

We have seen that equilibrium allocations belong to the core but that, in general, there may be core allocations which are not obtainable in equilibrium. This, however, is not the complete story, the connection between core and Walrasian allocations is more subtle, and most of the following chapter will deal with exactly this.

As an indication of what happens, consider a k-replication of the given economy, where there are k agents with the same characteristics.

Formally, let $\mathcal{E} = (X_i, P_i, \omega_i)_{i=1}^{m}$ be an exchange economy with private ownership. The k-replica of $\mathcal{E}$, for k a natural number, is the economy $\mathcal{E}^k = (X_{ij}, P_{ij}, \omega_{ij})_{i=1\,j=1}^{m\quad k}$, where for each $j = 1, \ldots, k$,

$$(X_{ij}, P_{ij}, \omega_{ij}) = (X_i, P_i, \omega_i).$$

An equal-treatment allocation in $\mathcal{E}^k$ is an allocation $x^k = (x_{ij})_{i=1\,j=1}^{m\quad k}$ such that

$$x_{ij} = x_i$$

for $j = 1, \ldots, k$. Every allocation $x = (x_1, \ldots, x_m)$ in $\mathcal{E}$ induces an equal-treatment allocation x^k in $\mathcal{E}^k$.

The following result is due to Debreu and Scarf (1963).

Theorem 4.4. *Let $\mathcal{E} = (X_i, P_i, \omega_i)_{i=1}^m$ be an economy such that each consumer i has convex and strongly monotonic preferences (in the sense that $P_i(x_i) + \mathbb{R}_{++}^l \subset P_i(x_i)$ for all $x_i \in X_i$), and $\omega_i \in X_i$. Let x be an allocation in $\mathcal{E}$ such that the equal-treatment allocation x^k belongs to the core of $\mathcal{E}^k$ for every k. Then there exists a price system $p \in \mathbb{R}_+^l, p \neq 0$, such that (x, p) is a Walras equilibrium.*

Proof. For each i, let $z_i = x_i - \omega_i$ be the net trade derived from x_i, and let $\widehat{P}_i(z_i) = \{x_i' \in X_i \mid x_i' + \omega_i \in P_i(x_i)\}$ be the set of preferred net trades.

We claim that the convex hull of $\cup_{i=1}^m \widehat{P}_i(z_i)$ does not intersect the set $\mathbb{R}_{--}^l$ of vectors negative in all coordinates. Indeed, suppose that there were $z_i' \in \widehat{P}_i(z_i)$ and $\lambda_i \in [0,1]$, $i = 1, \ldots, m$, with $\sum_{i=1}^m \lambda_i = 1$, such that $u = \sum_{i=1}^m \lambda_i z_i' \in \mathbb{R}_{--}^l$. Since all the $\widehat{P}_i(z_i)$ are open, we may assume that the numbers λ_i are all rational, and multiplying u by the product $\lambda_1 \lambda_2 \cdots \lambda_m$, we conclude that there are natural numbers $k_1, \ldots, k_m$ such that $\sum_{i=1}^m k_i z_i' \in \mathbb{R}_+^l$. However, if $k = \max_i k_i$, this means that in the k-replica economy $\mathcal{E}^k$, there is a coalition S consisting of k_i consumers of type i, $i = 1, \ldots, m$ such that $\sum_{i \in S} z_i' < \sum_{i \in S} \omega_i$, and $z_i' + \omega_i \in P_i(x_i)$ for all $i \in S$. By monotonicity of preferences, this means that the coalition S can improve the equal-treatment allocation defined by x in $\mathcal{E}^k$, a contradiction proving our claim.

Using separation of convex sets, we see that there is $p \in \mathbb{R}_+^l$ such that $p \cdot z_i' \geq 0$ for $z_i' \in \widehat{P}_i(x_i)$, $i = 1, \ldots, m$. In particular, $p \cdot z_i \geq 0$ for

Box 2. Core allocations in replica economies: Thus, let B_1 represent an equal-treatment allocation x in $\mathcal{E}^k$. Does x belong to the core of $\mathcal{E}^k$? The answer is no, at least for suitably chosen (large) values of k. To see this, we specify an improvement, which will be illustrated in the Edgeworth box introduced in Box 1.

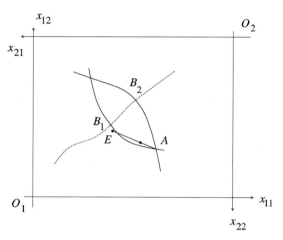

Fig. 2. Some allocations in the core of the original economy cease to be in the core when the economy is replicated.

Consider the point E in the box; from the point of view of an agent of type 2, E represents a bundle which is better than that specified by the allocation x. Now, let S be the coalition consisting of all k agents of type 1 and a single agent of type 2, and let S reallocate the initial endowment of its members so that each agent of type 1 gets $\omega + \frac{1}{k}z$, where z is the vector AE, and the single agent of type 2 gets $\omega_2 - z$. By our construction, this is an improvement if k is large enough (in Fig. 2 for all $k \geq 2$).

What has happened is that although in the allocation B_1 agents of type 2 are relatively well-off, the remaining agents can still offer single agents of type 2 even more in some alternative reallocation, thereby preventing united action by the agents of type 2. We conclude that B_1 does not belong to the core of $\mathcal{E}^k$.

all i, and since $\sum_{i=1}^{m} z_i = 0$, this means that $p \cdot z_i = 0$ for all i. The conclusion of the theorem now follows by standard reasoning. □

Cooperative Game Theory and Equilibria 125

In Chapter 5, we shall address the problem of making the above statement precise and extend it to sequences of economies which become large in a less restricted sense than by replication.

2. Allocations Induced by Other Game Theoretic Solutions

The use of solution concepts taken from the theory of games opens up for a broader study of allocations with desirable properties than those determined by the market. As we saw, the core almost suggests itself as an extension of the considerations about efficiency, now extended to group efficiency as well as overall efficiency. But game theory offers a wide variety of solutions to the problem of selecting allocations in an economy, and it may be useful to step back somewhat and introduce the game theoretic approach in a more systematic way.

2.1 *Cooperative equilibria of strategic form games and allocations*

A *game form* is a triple $G = (M, (\Sigma_i)_{i \in M}, \pi)$, where M is a set of players, Σ_i for each $i \in M$ a non-empty set of strategies, and π, the outcome map, is a function from the set $\Sigma = \prod_{i \in M} \Sigma_i$ of strategy arrays to some outcome space X. A *game* (in normal form) arises from the game form when adding the array $(\widetilde{P}_i)_{i \in M}$, where for each $i \in M$, P_i is a preference on X. A solution to the game is a strategy array $\sigma \in \Sigma$ with specified properties. Solutions of games can be *cooperative* as well as *non-cooperative,* depending on whether the players can coordinate their choices of strategies $\sigma_i \in \Sigma_i$ within coalitions S, which are non-empty subsets of M. We shall have much to say about non-cooperative solutions and their relation to allocation in an economy at a later stage (Chapter 9), at present we are concerned with the cooperative solutions.

If a coalition S can coordinate the members' choices of strategy, then a strategy array considered as a solution should be robust against possible alternative coordinated strategy choices.

A straightforward interpretation of this robustness property leads to the notion of a *strong Nash equilibrium:* The strategy array $\sigma^0 \in \Sigma$ is a strong Nash equilibrium if there is no coalition $S \in \mathcal{S}$ and coordinated strategy choice in S, $(\sigma_i')_{i \in S}$, such that

$$\pi\left((\sigma_i')_{i \in S}, (\sigma_i^0)_{i \in M \setminus S}\right) \in \widetilde{P}_i\left(\pi(\sigma^0)\right), \quad \text{all } i \in S. \tag{4}$$

Thus, in the strong Nash equilibrium, no coalition can achieve something better for all its members using all the options available.

There are many ways in which a given economy $\mathcal{E} = (X_i, P_i, \omega_i)_{i=1}^m$ can give rise to a game. In what follows we consider one such approach, proposed by Kalai *et al.* (1978), in which the allocations arising in strong Nash equilibrium are exactly those in Core($\mathcal{E}$). The set of players is the set M of consumers in $\mathcal{E} = (X_i, P_i, \omega_i)_{i=1}^m$, and the set of strategies available for consumer i is

$$\sum_i = \{ S \in \mathcal{P}_0(I) \mid i \in S \} \times X_i,$$

where $(S_i, x_i) \in \Sigma_i$ is interpreted as a proposal consisting of a coalition containing i and a consumption bundle to i to be obtained if the coalition S is formed to take care of allocation.

To define the preferences $\widetilde{P}_i$ for $i \in M$, we first introduce a notion of consistency of coalitions at strategy arrays σ: A coalition S is consistent at $\sigma = (S_i, x_i)_{i \in M}$ if

(i) $S_i = S$ for each $i \in S$,
(ii) $\sum_{i \in S} x_i \leq \sum_{i \in S} \omega_i$.

Let T be the (possibly empty) set of consumers not belonging to a consistent coalition, then we see that σ gives rise to a partition $\{S^1, \ldots, S^k, T\}$ of M, where $S^1, \ldots, S^k$ are the consistent coalitions. We may then define the outcome function at σ as

$$\pi_i(\sigma) = \begin{cases} x_i & \text{if } i \in S^j, j = 1, \ldots, k, \\ \omega_i & \text{if } i \in T. \end{cases}$$

The preferences $\widetilde{P}_i$ on allocations are those derived from the preferences P_i on bundles.

With this definition of the game form arising from an economy, the following is straightforward.

Theorem 4.5. *Let $\mathcal{E} = (X_i, P_i, \omega_i)_{i=1}^m$ be an economy and $G(\mathcal{E})$ be the game form associated with $\mathcal{E}$. For $x \in F(\mathcal{E})$, a feasible allocation, the following holds:*

(i) *If $x \in Core(\mathcal{E})$ and x is Pareto-optimal, then x is a strong Nash equilibrium outcome of the game $(G(\mathcal{E}), (P_i)_{i \in M})$,*

(ii) *If x is a strong Nash equilibrium outcome of the game $(G(\mathcal{E}), (P_i)_{i \in M})$, then $x \in Core(\mathcal{E})$.*

Proof. (i) Let $x \in Core(\mathcal{E})$. We claim that the strategy array $((M, x_1), \ldots, (M, x_m))$ is a strong Nash equilibrium. Suppose to the contrary that there is a coalition S and a strategy array $(\sigma_i')_{i \in S}$ such that (4) is satisfied. Then $S \neq M$ since otherwise M could improve x. If the first component of σ_i' is not M, all $i \in S$, outcome is ω_i for $i \notin S$, and then $(\pi_i((\sigma_i')_{i \in S}, (\sigma_i^0)_{i \in M \setminus S})_{i \in S})$ is feasible for S and constitutes an improvement, again contradicting that $x \in Core(\mathcal{E})$. If the first component of σ_i' for $i \in S$ is I, then there would be an allocation x' such that $x_i' \in P_i(x_i)$ for $i \in S$ and $x_i' = x_i \in cl\, P_i(x_i)$ for $i \notin S$, contradicting Pareto-optimality of x.

(ii) Suppose that $x = \pi(\sigma)$ for strong Nash equilibrium strategies σ. If the coalition S has an improvement $(x_i')_{i \in S}$, then the strategies $\sigma_i' = (S, x_i')$ for $i \in S$ are such that (4) will be satisfied, a contradiction. It follows that x belongs to $Core(\mathcal{E})$. $\qquad \square$

2.2 *Value allocations*

Having spent some time discussing the core of an economy, it would be natural to consider allocations inspired by other game-theoretic solution concepts, among which the most prominent is the Shapley value, introduced in Shapley (1953). The point of departure is a transferable utility (TU) game (M, v), $\hat{E}$, where M is the set of players and $v : S \to \mathbb{R}$ is the characteristic function of the game, assigning to each coalition S a number interpreted as the total (utility) payment that the coalition can obtain and distribute among its members.

The Shapley value of the game (M, v) assigns to each player i the payoff

$$\phi_i(v) = \sum_{S: i \in S} \frac{(|S| - 1)!(|M| - |S|)!}{|M|!} \left[v(S) - v(S \setminus \{i\}) \right]. \tag{5}$$

The payoff $\phi_i(v)$ can be interpreted as an average of the marginal contribution of player i, the average being taken over all the possible ways in which the grand coalition can be formed by successive addition of a single player.

It can be easily verified that $\sum_{i \in M} \phi_i(v) = v(M)$, an efficiency property of the Shapley value (nothing is left when payoffs are distributed). Also, it is seen to have the so-called dummy property: If player i makes no difference for any coalition, then i gets nothing. Finally, the Shapley value is additive: If (M, v) and (M, w) are TU games (with the same set of players), then

$$\phi_i(v + w) = \phi_i(v) + \phi_i(w),$$

distributing according to the Shapley value in each of the games separately or distributing in the sum game gives the same final payoff to each player.

Extending the Shapley value to the choice of allocation in an economy is not quite straightforward. Even if preferences can be described by utilities, we would need a Shapley value for NTU games, and the resulting allocation would depend on the utility representation chosen. Here, we follow the idea of Shafer (1974) using the expenditure function defined in Introduction chapter.

Let $\mathcal{E} = (X_i, P_i, \omega_i)_{i=1}^m$ be an exchange economy, where we assume that $X_i = \mathbb{R}_+^l$ for each i. For $x_i \in \mathbb{R}_+^l$ and $p \in \Delta$, the expenditure function e of individual i takes the value

$$e(P_i, p, x_i) = \inf \left\{ p \cdot x_i' \,\middle|\, x_i' \in P_i(x_i) \right\}.$$

Cooperative Game Theory and Equilibria   129

For each $p \in \Delta$, we can define a TU game $(M, v(p, \cdot))$ by

$$v(p, S) = \max \left\{ \sum_{i \in S} e(P_i, p, x_i) \,\middle|\, \sum_{i \in S} x_i = \sum_{i \in S} \omega_i \right\} \qquad (6)$$

for $S \in \mathcal{P}_0(S)$, so that $v(p, S)$ is the maximal sum of expenditures for any allocation which the coalition can establish by itself.

For each of the games $(M, v(p, \cdot))$, we can find a Shapley value $\phi(v(p, \cdot))$. The allocation $x^* \in F(\mathcal{E})$ is said to be a *value allocation* if there is a price $p^* \in \Delta$ such that

$$e(P_i, p^*, x_i^*) = \phi_i(v(p^*, \cdot)), \quad \text{each } i \in M. \qquad (7)$$

Thus, each individual gets a bundle which, measured in terms of expenditure, is a Shapley value payoff. We denote the set of value allocations in $\mathcal{E}$ by $\Phi(\mathcal{E})$.

It is easily seen that allocations in $\Phi(\mathcal{E})$ are Pareto-optimal: Let $x^* \in \Phi(\mathcal{E})$ and suppose on the contrary that there is $x^* \in F(\mathcal{E})$ such that $x_i' \in P_i(x_i^*)$ for all $i \in M$. If $p^* \in \Delta$ satisfies (7), then $p^* \cdot x_i' > p^* \cdot x_i^* = e(P_i, p^*, x_i^*)$ for each i, so that

$$p^* \cdot \sum_{i \in M} \omega_i = p \cdot \sum_{i \in M} x_i' > p^* \cdot \sum_{i \in M} x_i^* = p \cdot \sum_{i \in M} \omega_i,$$

a contradiction. The value allocations also have the dummy property: If consumer i owns nothing, then $v(p, S) = v(p, S \backslash \{i\})$ for all p and S, and agent i will get, if not the bundle 0, then a bundle such that everything close to zero is no better.

On the other hand, the value allocations do not inherit the additivity properties of the Shapley value, at least as long as additivity is considered in its most natural version: Value allocations associated with two different assignments of initial endowments to the consumers do not add up to a value allocation in the economy where the consumers are given the sum of the two initial assignments. This is a consequence of using the expenditure function for constructing a TU game. In order to obtain additivity properties, one has to turn to

130   *Theory of General Economic Equilibrium*

Box 3. Value allocations and additivity: An example showing that value allocations may fail to have the additivity property is discussed here. There are two commodities and three consumers. The preferences of the consumers are identical and representable by the utility function $u(x) = \min_{h=1,2} x_h$. We consider three cases of endowments.

In the first case, endowments are $\omega_1 = (1, 1)$, $\omega_2 = (1, 2)$ and $\omega_3 = (1, 0)$. For $p \in \Delta$ with $p_1 \leq 1/2$, we then get that

$$v(p, \{1\}) = v(p, \{2\}) = 1, v(p, \{3\}) = 0, v(\{1, 2\}, p) = 2,$$
$$v(p, \{1, 3\}) = 1, v(p, \{2, 3\}) = 2, v(p, \{1, 2, 3\}) = 3,$$

and the Shapley value of this game can be computed as follows:

$$\phi_1(v(p, \cdot)) = 1, \quad \phi_2(v(p, \cdot)) = \frac{3}{2}, \quad \phi_3(v(p, \cdot)) = \frac{1}{2}.$$

The Shapley value is independent of prices, and a value allocation can be found at any price as $\left((1, 1), \left(\frac{3}{2}, \frac{3}{2}\right), \left(\frac{1}{2}, \frac{1}{2}\right)\right)$.

In the second case, endowments are $\omega_1 = (1, 1)$, $\omega_2 = (1, 0)$ and $\omega_3 = (0, 0)$. We get that for all p, $v(p, S) = 1$ if the coalition S contains consumer 1 and $v(p, S) = 0$ otherwise. The Shapley value $\phi(v(p, \cdot))$ is 1 for consumer 1 and 0 for 2 and 3, and the value allocation is $((1, 1), (0, 1), (0, 0))$.

Now we consider the economy where endowments are the sum of those in the two cases considered, so that $\omega_1 = (2, 2)$, $\omega_2 = (2, 2)$ and $\omega_3 = (1, 0)$. For $p \in \Delta$, we get the characteristic function

$$v(p, \{1\}) = v(p, \{2\}) = 2, v(p, \{3\}) = 0, v(\{1, 2\}, p) = 4,$$
$$v(p, \{1, 3\}) = v(p, \{2, 3\}) = 2, v(p, \{1, 2, 3\}) = 4,$$

with Shapley value

$$\phi_1(v(p, \cdot)) = \phi_2(v(p, \cdot)) = 2, \phi_3(v(p, \cdot)) = 0,$$

and a value allocation with associated price p is $((2, 2), (2, 2), (1, 0))$.

It is seen that the bundle obtained by consumer 3 is better (in terms of his/her preferences as well as in terms of the price-independent Shapley value) in the first situation than in the third one, so additivity in endowments is not obtained.

Cooperative Game Theory and Equilibria 131

modified versions of exchange economies for which the TU games are obtained in a more direct way, as explained, e.g. in Aumann and Shapley (1974).

2.3 The bargaining set of an economy

In the intuitive argument for selecting the core allocation, it would be mentioned that a coalition can choose to discontinue any relationship with its complement and allocate its own resources between its members. Real-life bargaining might not take this form: instead, it would be a process of proposing some other allocations or rejecting them by reference to further improvements. This type of bargaining is reflected in the *bargaining set*, which was proposed for cooperative games by Aumann and Maschler (1964). In the context of exchange economies, we use the version introduced in Mas-Colell (1989).

We consider an exchange economy $\mathcal{E} = (X_i, P_i, \omega_i)_{i=1}^m$, where for simplicity $X_i = \mathbb{R}_+^l$ for all i, and a feasible allocation $x \in F(\mathcal{E})$. The pair (S, y), where $S \in \mathcal{P}_0(I)$ and $y = (y_1, \ldots, y_m)$ is an allocation, is an *objection against* x if

(i) $\sum_{i \in S} y_i \leq \sum_{i \in S} \omega_i$,
(ii) $y_i \in P_i(x_i)$ for all $i \in S$.

An objection, thus, is equivalent to what we have called an improvement when dealing with the core, and the use of this redundant piece of terminology is explained in what follows: If (S, y) is an objection against x, then the pair (T, z) with $T \in \mathcal{P}_0(I)$ and z an allocation is a *counterobjection* to (S, y) if

(i) $\sum_{i \in T} z_i \leq \sum_{i \in T} \omega_i$,
(ii) $z_i \in P_i(y_i)$ if $i \in S \cap T$, $z_i \in P_i(x_i)$ if $i \in T \backslash S$.

An objection (S, y) is *justified* if there is no counterobjection to it.

The *bargaining set* of $\mathcal{E} = (X_i, P_i, \omega_i)_{i=1}^m$, denoted $\mathcal{B}(\mathcal{E})$, can now be defined as the set of feasible allocations against which there are no justified objections.

132   *Theory of General Economic Equilibrium*

Since there are no objections against an allocation belonging to Core($\mathcal{E}$), we get immediately that the core of $\mathcal{E}$ is contained in the bargaining set, and Theorem 4.1 can then be used to give the following theorem.

Theorem 4.6. *Let $\mathcal{E} = (X_i, P_i, \omega_i)_{i=1}^{m}$ be an exchange economy, and let $W_0(\mathcal{E})$ be the set of allocations belonging to Walras equilibria in $\mathcal{E}$. Then*

$$W_0(\mathcal{E}) \subset Core(\mathcal{E}) \subset \mathcal{B}(\mathcal{E}). \tag{8}$$

All the inclusions in (8) can be strict, as shown by the examples in Box 4.

It should be noted, from Theorem 4.6 as well as from the definition, that the bargaining set tends to be rather large. Thus, if an

---

Box 4. An allocation in the bargaining set but not in the core: In this example, we consider an economy $(\mathbb{R}_+^3, P, \omega_i)_{i=1}^{3}$, where all consumers have the same preferences P described by the utility function $u(x) = \min_i x_i$, and where the endowments are given as follows:

$$\omega_1 = \left(\frac{5}{6}, \frac{1}{2}, 0\right), \quad \omega_2 = \left(0, \frac{1}{2}, \frac{5}{6}\right), \quad \omega_3 = \left(\frac{5}{6}, \frac{1}{2}, \frac{1}{6}\right).$$

Thus, the total endowments are $\left(\frac{10}{6}, \frac{3}{2}, 1\right)$.

We consider the allocation x with $x_i = \left(\frac{1}{3}, \frac{1}{3}, \frac{1}{3}\right)$ for $i = 1, 2$. This allocation is Pareto-optimal: All consumers want the commodities in the fixed proportion 1:1:1, and given these preferences, the total endowments cannot be reallocated in a better way. It is *not* in the core, since the coalition consisting of consumers 1 and 2 has resources $\left(\frac{5}{6}, 1, \frac{5}{6}\right)$ which can be distributed between them in a way that both get more than x, for example $\left(\frac{5}{12}, \frac{5}{12}, \frac{5}{12}\right)$ to both. It is seen that the coalition consisting of consumers 2 and 3 can also originate objections against x, but that the coalition of 1 and 3 cannot.

The allocation x belongs to the bargaining set: If $(\{1,2\}, y)$ involves the coalition $(1, 2)$, then the bundle of consumer i must satisfy $u(y_2) < \frac{5}{12}$, and then $(\{2, 3\}, z)$ with $u(y_2) < z_{2i} < \frac{5}{12}$ is a counterobjection against $(\{1, 2\}, y)$. A similar argument applies to an objection initiated by $\{2, 3\}$.

allocation has two objections (S, x) and (T, x) with $S \cap T = \emptyset$, then neither of these objections are justified, making it rather easy to belong to the bargaining set. The bargaining set will emerge Chapter 5, where we look at economies with many agents and show that in such economies, the allocations in the bargaining set are Walrasian. This result will hold also if the bargaining set had been defined in a more restrictive way.

2.4 Nucleolus allocations

Having seen in the previous sections that the solution concepts of the cooperative game theory can be used, often in a very intuitive way, to single out specific allocations in an economy — and taking into account the abundance of such solution concepts — it should come as no surprise that several such allocations have been considered in the literature. In this section, we take a look at *nucleolus allocations* in an exchange economy.

The relevant solution concept is the *nucleolus* of a cooperative TU game (M, v), due to Schmeidler (1969). We begin by introducing the *excess function* of the game: For $z \in \mathbb{R}^M$ an imputation in (M, v), that is a payoff vector with $\sum_{i \in M} z_i = v(M)$, and $S \in \mathcal{S}$ a coalition, the excess of S at z is defined as

$$e(S, z) = v(S) - \sum_{i \in S} z_i, \tag{9}$$

expressing the potential gain to the coalition from working on its own rather than accepting the payoff vector z. Define the map θ to which to each imputation z assigns the array of $|\mathcal{S}|$ excesses $e(S, z)$ written in non-increasing order, so that

$$i < j \Rightarrow e(S_i, z) \geq e(S_j, z).$$

Now the nucleolus of (M, v), written as $\mathcal{N}(v)$, is defined as the lexicographic minimum over all the vectors $\theta(z)$, for z an imputation of (M, v). The nucleolus is single valued and well-defined for all games having a non-empty set of imputations.

Now we return to our standard framework, considering an exchange economy $\mathcal{E} = (X_i, P_i, \omega_i)_{i=1}^m$ with private ownership of resources, where for simplicity we assume that $X_i = \mathbb{R}_+^l$ and $\omega_i \in \mathbb{R}_+^l$ for all i. Following MacLean and Postlewaite (1989), we introduce a notion of excess which does not need preferences described by utility functions, but uses Debreu's coefficient of resource allocation, cf. Chapter 3, Section 3. For $S \in \mathcal{S}$ a coalition, we may define the exchange economy restricted to S as $\mathcal{E}_S = (X_i, P_i, \omega_i)_{i \in S}$. If $x \in F(\mathcal{E})$ is a feasible allocation, then x induces an allocation (not necessarily feasible) in $\mathcal{E}_S$, and we may assess the degree of dissatisfaction with x^S by

$$e_D(S, x) = \frac{1}{\lambda_D(x^S, \mathcal{E}_S)}, \qquad (10)$$

which is the inverse of the proportional change of the endowment vector $\omega^S = \sum_{i \in S} \omega_i$ needed for the reallocation to yield a result as good as x^S.

The excesses constructed in this way may be exploited to define a nucleolus allocation.

Definition 4.1. Let $\mathcal{E} = (X_i, P_i, \omega_i)_{i=1}^m$ be an exchange economy, and for each feasible allocation $x \in F(\mathcal{E})$, let $\theta(x) = (e_D(S_{i_1}, x), \ldots, e_D(S_{i_r}, x))$ be the vector of all $r = |\mathcal{S}|$ excesses as defined in (10), written in non-increasing order.

A nucleolus allocation in $\mathcal{E}$ is an allocation x^0, such that $\theta(x^0)$ is a lexicographic minimum among all vectors $\theta(x)$, $x \in F(\mathcal{E})$.

The nucleolus allocation defined in this way may not be a unique allocation, rather there will be a set of nucleolus allocations in a given economy. We denote this set by $N(\mathcal{E})$.

Theorem 4.7. *If* $\mathrm{Core}(\mathcal{E}) \neq \emptyset$, *then* $N(\mathcal{E}) \subset \mathrm{Core}(\mathcal{E})$.

Proof. If $y \in \mathrm{Core}(\mathcal{E})$ and $S \in \mathcal{S}$ is an arbitrary coalition, then $\lambda_D(y^S, \mathcal{E}) \geq 1$, since otherwise S would have an improvement of y. It follows that the $e_D(S, y) \leq 1$ for all S. Conversely, if $e_D(S, y) \leq 1$ for all S, then $y \in \text{Core}(\mathcal{E})$. Since any $x \in \mathcal{N}(\mathcal{E})$ minimizes the largest of the excesses $e_D(S, x)$ for $S \in \mathcal{S}$, we have that $x \in \text{Core}(\mathcal{E})$. $\qquad \square$

3. Fuzzy Games and Equilibria

The theory of fuzzy sets introduced by Zadeh (1965) has found applications in many different disciplines of science, and although its impact on economics has so far been very small, this may well change in the future. Here, we shall have a brief look at allocations that are obtained using solution concepts of fuzzy cooperative games, cf. Aubin (1981).

A fuzzy subset of a given universe U (which is an ordinary set) is defined by its membership function $\mu : U \to [0, 1]$. Thus, an element a of U can be a member of the fuzzy coalition on a partial basis, expressing the definiteness of its membership. What matters in the theory of fuzzy sets is not so much this membership but the rules by which new fuzzy sets are obtained from the given fuzzy sets. The degree of membership should not be interpreted as a probability, since the rules for computing, e.g. unions or intersections, are very different (Box 5).

We restrict ourselves to fuzzy subsets of I, the set of consumers in the exchange economy $\mathcal{E} = (X_i, P_i, \omega_i)_{i=1}^{m}$. Thus, a fuzzy coalition is a fuzzy subset μ of I. A fuzzy cooperative game is defined by its characteristic function, which in its turn now is defined not only on ordinary coalitions (subsets of I) but also on all fuzzy coalitions.

In the context of choice of allocation in an exchange economy $\mathcal{E} = (X_i, P_i, \omega_i)_{i=1}^{m}$, where an ordinary (in the terminology of fuzzy set theory: crisp) coalition can secure for its members the set $\{(x_i)_{i \in S} \mid \sum_{i \in S} x_i \leq \sum_{i \in S} \omega_i\}$, an extension to fuzzy coalitions μ can be obtained straightforwardly if we assume that a fuzzy coalition μ has as its total endowment the quantity

$$\omega_\mu = \sum_{i=1}^{m} \mu(i)\omega_i.$$

Box 5. General equilibrium with fuzzy preferences: We have introduced fuzziness in the context of game-theoretical solutions to the problem of finding suitable allocations in an economy. It may however be used — and has been used — in many other contexts, for example by extending the standard notions of the general equilibrium model by allowing for fuzziness.

Kajii (1988) considers an exchange economy $\mathcal{E}^f = (X_i, R_i; \omega_i)_{i=1}^m$ with *fuzzy preferences*. For each consumer i, R_i is a fuzzy subset of $X_i \times X_i$, with membership function $R_i : X_i \times X_i \to [0, 1]$, whereby the statement $R_i(x, y) > 0$ is interpreted as the degree to which y is considered better than x by consumer i. We may look at $R_i(x, y)$ as describing the irresoluteness of the consumer in preferring y to x.

If there are markets for all the goods determined by a price system $p \in \Delta$, we can define a Walras equilibrium in $\mathcal{E}$ once we have introduced the notion of an element of the budget set being "better" than another element for consumer i with the fuzzy preference R_i. In accordance with the above interpretation of R_i, we propose the following:

$$y \text{ is better for } i \text{ than } x \Leftrightarrow R_i(x, y) > R_i(y, x). \tag{11}$$

Thus, y is better than x when the consumer is more inclined to prefer y to x than to prefer x to y. With this interpretation, the preference relation has a fuzzy completeness property, since all the elements of X_i can be compared to each other, only the degree of certainty about preferences varies.

Now, a Walras equilibrium in $\mathcal{E}^f$ can be defined as a pair (x, p) such that

(i) x is feasible,

(ii) for each i, if $x_i' \in X_i$ and $R_i(x_i, x_i') > R_i(x_i', x_i)$, then $p \cdot x_i' > p \cdot x_i$.

It is easily seen that defining for each i the correspondence $P_i : X_i \rightrightarrows X_i$ in the obvious way, namely by

$$P_i(x_i) = \{x_i' \in X_i \mid R_i(x_i, x_i') > R_i(x_i', x_i)\},$$

we obtain a standard exchange economy $\mathcal{E} = (X_i, P_i, \omega_i)_{i=1}^m$, and $\mathcal{E}^f$ has exactly the same Walras equilibria as $\mathcal{E}$, so that what has been achieved by introducing fuzziness is only a new and perhaps more appealing interpretation of the preference correspondence P_i.

It can then obtain for its members all μ-allocations $x_\mu = (x_i)_{i \in S_\mu}$, where $S_\mu = \{i \in I \mid \mu(i) > 0\}$ is the support of μ, belonging to the set

$$F_\mu(\mathcal{E}) = \left\{ (x_i)_{i \in S_\mu} \,\middle|\, \sum_{i \in S_\mu} \mu_i x_i \leq \omega_\mu \right\}. \qquad (12)$$

The fuzzy coalition μ can improve upon the feasible allocation $x \in F(\mathcal{E})$ if there is a μ-allocation $(x_i')_{i \in S_\mu}$ in $F_\mu(\mathcal{E})$ such that $x_i' \in P_i(x_i)$ for each $i \in S_\mu$. The *fuzzy core* of $\mathcal{E}$, written $\mathrm{Core}^f(\mathcal{E})$, is the set of feasible allocations which cannot be improved by any coalition (neither fuzzy nor crisp).

It is seen from (12) that fuzzy coalitions may improve upon allocations for which the crisp coalitions have no improvement, suggesting that the inclusion

$$\mathrm{Core}^f(\mathcal{E}) \subseteq \mathrm{Core}(\mathcal{E})$$

may be a proper one. The question of non-emptiness of $\mathrm{Core}^f(\mathcal{E})$ is resolved by Theorem 4.8, which goes a step further, this establishing the equivalence between Walras equilibrium allocations and the fuzzy core.

Theorem 4.8. *Let $\mathcal{E} = (X_i, P_i, \omega_i)_{i=1}^m$ be an economy where consumers have convex and monotonic preferences. Then $x \in F(\mathcal{E})$ belongs to $\mathrm{Core}^f(\mathcal{E})$ if and only if (x, p) is a Walras equilibrium for some $p \in \Delta$.*

Proof. Let (x, p) be a Walras equilibrium. If μ is a fuzzy coalition and $(x_i')_{i \in S_\mu}$ a μ-allocation, then $x_i' \in P_i(x_i)$ for each $i \in S_\mu$ implies that $p \cdot x_i' > p \cdot \omega_i$ for each $i \in S_\mu$, and consequently, $p \cdot \sum_{i=1}^m \mu(i) x_i' > p \cdot \sum_{i=1}^m \mu(i) \omega_i$, so that $(x_i')_{i \in S_\mu}$ cannot be an improvement.

Conversely, suppose that $x \in \mathrm{Core}^f(\mathcal{E})$. We claim that the convex hull of the set

$$\bigcup_{S \in \mathcal{P}_0(I)} \left\{ \sum_{i \in S} (x_i' - \omega_i) \,\middle|\, x_i' \in P_i(x_i) \right\}$$

138   *Theory of General Economic Equilibrium*

does not intersect $\mathbb{R}^l_{--}$. Indeed, suppose that there are coalitions $S_1, \ldots, S_k$ and for $j = 1, \ldots, k$, allocations $(x^j_i)_{i \in S_j}$ exist, such that

$$\sum_{j=1}^{k} \lambda_j \sum_{i \in S_j} (x^j_i - \omega_i) = u, \tag{13}$$

where $x^j_i \in P_i(x_i)$ for each $i \in S_j$, $j = 1, \ldots, k$, $\lambda_j \in [0,1]$, $\sum_{j=1}^{k} \lambda_j = 1$, and $u_h < 0$ for $h = 1, \ldots, l$. For each i, define x'_i by

$$x'_i = \sum_{j:i \in S_j} \frac{\lambda^j}{\sum_{j:i \in S_j} \lambda_j} x^j_i.$$

Then $x'_i \in P_i(x_i)$ by convexity of $P_i(x_i)$. Define the fuzzy coalition μ by $\mu(i) = \sum_{j:i \in S_j} \lambda_j$. Rewriting (13), we get that

$$\sum_{j=1}^{k} \lambda_j \sum_{i \in S_j} (x^j_i - \omega_i) = \sum_{i=1}^{m} \sum_{j:i \in S_j} \lambda_j (x^j_i - \omega_i) = \sum_{i=1}^{m} \mu(i)(x'_i - \omega_i) = 0,$$

so that $(x'_i)_{i \in S_\mu}$ is an improvement upon x, a contradiction.

Using separation of convex sets, we obtain a price $p \in \mathbb{R}^l \setminus \{0\}$ such that $p' \cdot (x'_i - \omega_i) > 0$ whenever $x'_i \in P_i(x_i)$, and we conclude that (x, p) is a Walras equilibrium. $\qquad\square$

4. Exercises

(1) Consider an economy with two goods and three consumers, each having consumption set $\mathbb{R}^2_+$, with preferences described by the utility functions

$$u_1(x_{11}, x_{12}) = x_{11}^2 x_{12}^3, \quad u_2(x_{21}, x_{22}) = \min\left\{\frac{x_{21}}{2}, \frac{x_{22}}{3}\right\},$$

$$u_3(x_{31}, x_{32}) = x_{31} + \sqrt{x_{32}},$$

and endowments $\omega_1 = (2,1)$, $\omega_2 = (2,2)$, $\omega_3 = (2,5)$. Find the core of this economy.

(2) **von Neumann–Morgenstern allocations:** A non-empty set **N** of feasible allocations in the economy $\mathcal{E} = (X_i, P_i, \omega_i)_{i=1}^m$ is stable if

  (a) each feasible allocation x' not in **N** can be improved by an allocation $x \in \mathbf{N}$ via some coalition S (in the sense that $x_i \in P_i(x_i')$, all $i \in S$, and $\sum_{i \in S} x_i \leq \sum_{i \in S} \omega_i$),

  (b) no allocation in **N** can be improved by another allocation in **N**.

  Give the conditions on $\mathcal{E}$ under which there exists at least one stable set of allocation (such a set is called a von Neumann–Morgenstern solution).

(3) Let $\mathcal{E} = (X_i, P_i, \omega_i)_{i=1}^m$ be an economy satisfying Assumptions 0.1 and 0.2, and assume that each preference correspondence P_i has strictly convex values, in the sense that for each $x_i \in X_i$, if $x', x'' \in \mathrm{cl}\, P_i(x_i)$ and $0 < \lambda < 1$, then $\lambda x' + (1 - \lambda)x'' \in P_i(x_i)$.

  Show that the core of $\mathcal{E}^k$ for k sufficiently large will consist only of equal-treatment allocations.

(4) Let $\mathcal{E} = ((X_i, P_i, \omega_i)_{i=1}^m, \mathcal{Y})$ be a coalition production economy, where $\mathcal{Y}$ is cardinally balanced. Define a Walras equilibrium in $\mathcal{E}$ under the constraint that no coalition would improve its members' situation by producing and selling the net output in the market rather than receiving their profit shares from producing in the grand coalition.

  Does the resulting Walras equilibrium allocation belong to the core of the coalition production economy $\mathcal{E}$?

(5) Give an example of an economy with at least two goods and three consumers satisfying Assumptions 0.1 and 0.2 such that it has a value allocation which belongs to the core.

(6) Let $\mathcal{E}$ be an exchange economy satisfying standard assumptions, and suppose that x is an allocation in $F(\mathcal{E})$ with the property that the corresponding equal-treatment allocation belongs to $\mathcal{B}(\mathcal{E})$ for each replica-economy $\mathcal{E}^k$. Check whether x belongs to the core of $\mathcal{E}$.

Chapter 5

Large Economies

1. Economies with Many Agents

The discussion in Chapter 4 of cores and equilibria in exchange economies with many agents of the same type pointed to the need for models of markets with a very large number of participants. In economies with many agents, it will be easier for a group of agents to find the right type of trading partners in the right number so as to achieve an improvement of the initial situation. In the following, we consider several ways of making this intuition work in a formal model.

The first of our several models of large economies is technically undemanding since largeness means that the set of consumers is infinite but otherwise has no particular structure. The notions of core and equilibrium must be reformulated with regard to the new environment, but with the new formulation, core allocations turn out to be obtainable as Walras equilibrium allocations. Unfortunately, the converse cannot be established, and to obtain a neat equivalence result, we turn to another, more standard, model, where the set of agents have the structure of a measure space. A third approach, already suggested by the treatment of replica economies in Chapter 4, is to remain within the framework of finite economies and to consider sequences of such economies containing more and more agents. Also, here an equivalence result can be obtained, even if its formulation is less simple.

142   *Theory of General Economic Equilibrium*

Having dealt with infinitely many agents, it seems natural to turn to cases of infinitely many commodities, which we shall do in Section 2. The largeness of the set of commodities has no intuitive implications for the structure of equilibria, as was the case when moving from finitely to infinitely many agents, and indeed even the existence of Walras equilibria needs to be reconsidered. Adding also an infinity of agents, to obtain what is called a large-square economy, will in many cases upset the equivalence results obtained in the first part of the chapter.

1.1   *Economies with infinitely many agents*

The simplest model of an exchange economy with infinitely many agents exploits some properties of accumulation points of finite sums (cf. Bourbaki, 2004),[1] as presented in Keiding (1976). The usual finite index set of consumers $M = \{1, \ldots, m\}$ is replaced by an infinite set I, and an economy is now an I-indexed family, where I is a set of arbitrary cardinality. We let $\mathcal{F}$ be the family of all finite subsets of I. All the consumption sets of the consumers are assumed to be $\mathbb{R}^l_+$ to avoid complications which are irrelevant in our present context. For $i \in I$, $P_i : \mathbb{R}^l_+ \rightrightarrows \mathbb{R}^l_+$ is the preference correspondence, assumed to have open graph in $\mathbb{R}^l_+ \times \mathbb{R}^l_+$ and to satisfy irreflexivity ($x_i \notin P_i(x_i)$, all $x_i \in \mathbb{R}^l_+$) and strong monotonicity ($x' \geq x$ and $x' \neq x$ implies that $x' \in P_i(x)$). No convexity assumptions on preferences are needed. Finally, the initial endowments ω_i belong to $\mathbb{R}^l_{++}$ for all $i \in I$.

An allocation in this infinite economy is a map from I to $\mathbb{R}^l_+$, written as $x = (x_i)_{i \in I}$. By z we denote the corresponding array $z = x - \omega = (x_i - \omega_i)_{i \in I}$ of net trades. When defining the feasibility of allocations, we need another approach than summation of individual bundles, and we propose instead to consider accumulation points of finite sums, points for which every neighborhood will

---

[1]The idea of using accumulation points of finite sums as the basis for a model of an economy with infinitely many agents is due to Karl Vind, who used it in a never published study of pairwise exchanges.

Large Economies 143

contain sufficiently many finite sums: Let

$$A(z) = \bigcap_{H \in \mathcal{F}, \varepsilon > 0} \left\{ z \in \mathbb{R}^l \,\middle|\, \exists H' \in \mathcal{F}, H \subset H' = \emptyset, \left\| z - \sum_{i \in H'} (z_i) \right\| < \varepsilon \right\}$$
(1)

be the set of points z such that no matter with which finite sum of net trades we begin, we can supplement it to get a larger finite sum of net trades arbitrarily close to z. We consider x as a feasible allocation if $A(z)$ contains 0; intuitively, for any finite subset of agents, there is a larger finite set such that the allocation over this set is very close to being balanced.

We now define Walras equilibria in the standard way, except for the occurrence of an *exceptional* set of consumers, for which the individual optimality conditions need not hold. We shall see that such exceptional sets will occur in all the formalizations of large economies, intuitively they should be considered so small as to be negligible. In our case, the exceptional set is countable, so that for this intuition to hold, we should think of I as uncountable.

Definition 5.1. A pair (x, p), where x is an allocation and $p \in \mathbb{R}^l_+ \setminus \{0\}$ is a price system, is a Walras equilibrium if

(i) x is feasible,
(ii) for all agents in I except possibly a countable subset J_0, x_i is maximal for P_i in the set $\{x'_i \mid p \cdot x'_i \leq p \cdot \omega_i\}$.

In order to define the core of $\mathcal{E}$, we need the notion of an improvement. In view of the preceding remarks, it is not quite enough that a coalition, that is a finite subset of I, can achieve something better, since anyway this coalition may be considered as an exceptional (negligible) set. Therefore, we demand that no matter which finite subset J of I we choose, there should be a coalition $H \in \mathcal{F}$ not intersecting J which can do better using its own resources.

Definition 5.2. An allocation x has an improvement if there is an allocation $(x'_i)_{i \in I}$ such that for each $J \in \mathcal{F}$, there is $H \in \mathcal{F}$ with $J \cap H = \emptyset$ and an array $(x'_i)_{i \in H}$ of bundles with $x'_i \in P_i(x_i)$ and $\sum_{i \in H} (x'_i - \omega_i) = 0$.

144 *Theory of General Economic Equilibrium*

The core is the set of feasible allocations which have no improvement.

The main point of constructing the model is to have a framework where core allocations can be obtained in equilibrium. This will actually be the case, and the approach uses some simple properties of finite sums. Consider a feasible allocation $x = (x_i)_{i \in I}$ with its associated family of net trades $z = (x_i - \omega_i)_{i \in I}$. Let

$$B(z) = \bigcap_{H \in \mathcal{F}, \varepsilon > 0} \left\{ z \in \mathbb{R}^l \,\middle|\, \exists H' \in \mathcal{F}, H' \cap H = \emptyset, \left\| z - \sum_{i \in H'} (z_i) \right\| < \varepsilon \right\}, \tag{2}$$

be the closure of all the points which can then be found as finite sums of x, no matter which finite set is deleted. This set has a very useful additivity property:

Lemma 5.1. $B(z) + B(z) \subset B(z)$. If $0 \in A(z)$, then $B(z) = -B(z)$.

Proof. If $b_1, b_2 \in B$, then for each $H \in \mathcal{F}$ and $\varepsilon > 0$, there is $H_1 \in \mathcal{F}$ with $H \cap H_1 = \emptyset$ and $\left\| b_1 - \sum_{i \in H_1} z_i \right\| < \varepsilon/2$, and there is $H_2 \in \mathcal{F}$ with $(H \cup H_1) \cap H_2 = \emptyset$ and $\left\| b_2 - \sum_{i \in H_2} z_i \right\| < \varepsilon/2$. It follows that $(H_1 \cup H_2) \in \mathcal{F}$, which is disjoint from H, satisfies $\left\| b - \sum_{i \in (H_1 \cup H_2)} z_i \right\| < \varepsilon$.

Next, suppose that $b \in B(z)$. Then for each $H \in \mathcal{F}$ and $\varepsilon > 0$ there is $H' \in \mathcal{F}$ such that $\left\| b - \sum_{i \in H'} z_i \right\| < \varepsilon/2$, and since $0 \in A(z)$, there is H'' with $H' \subset H''$ such that $\left\| \sum_{i \in H''} z_i' < \varepsilon/2 \right\|$, or $\left\| -b - \sum_{i \in (H'' \setminus H')} z_i \right\| < \varepsilon$. It follows that $-b \in B(z)$. $\qquad \square$

We use these properties to show that allocations which have no improvement are Walras equilibrium allocations.

Theorem 5.1. *Let x be an allocation in the core. Then there is a price system $p \in \mathbb{R}_+^l \setminus \{0\}$ such that (x, p) is a Walras equilibrium.*

Large Economies   145

Proof. Let $z = x - \omega$ be the array of net trades belonging to x. Define the set $P(z)$ by

$$P(z) = \bigcap_{H \in \mathcal{F}, \varepsilon > 0} \left\{ y \in \mathbb{R}^l \mid \exists H' \in \mathcal{F}, H' \cap H = \emptyset, x_i' \in P_i(x_i), i \in H' : \right.$$
$$\left. \left\| y - \Sigma_{i \in H'}(x_i' - \omega_i) \right\| < \varepsilon \right\}. \tag{3}$$

Reasoning as in the proof of Lemma 5.1, we get that $P(z) + P(z) \subset P(z)$. By monotonicity, $B(z) + \mathbb{R}_+^l \subset P(z)$.

We claim that $\text{conv} P(z)$, the convex hull of $P(z)$, does not intersect the strictly negative orthant $\mathbb{R}_{--}^l$. Indeed, suppose that $\sum_{i=1}^r \lambda_i p_i \in \mathbb{R}_{--}^l$ for some $\lambda_i \geq 0$ with $\sum_{i=1}^r \lambda_i = 1$. We may assume without loss of generality that all the λ_i are rational numbers, and multiplying all the λ_i by their smallest common denominator, we obtain natural numbers $n_1, \ldots, n_r$ such that

$$\sum_{i=1}^r n_i p_i \in \mathbb{R}_{--}^l.$$

Since $P(z) + P(z) \subset P(z)$, we get that $\text{conv} P(z) \subset P(z)$, so that $P(z)$ is a convex set. If $P(z)$ intersects $\mathbb{R}_{--}^l$, then for each $H \in \mathcal{F}$ there is $H' \in \mathcal{F}$ with $H \cap H' = \emptyset$, and bundles $x_i' \in P_i(x_i)$ for $i \in H'$ with $\sum_{i \in H'}(x_i' - \omega_i) \in \mathbb{R}_{--}^l$, meaning that x has an improvement, a contradiction which proves our claim.

Using Theorem 3.5 on the convex sets $\text{conv}(P(z))$ and $\mathbb{R}_{--}^l$, we get the existence of a price system $p \in \mathbb{R}_+^l, p \neq 0$, such that

$$p \cdot y \geq 0 \quad \text{for } y \in P(z),$$
$$p \cdot y < 0 \quad \text{for } y \in \mathbb{R}_+^l$$

from which we get that $p \in \mathbb{R}_+^l$.

Choose an arbitrary compact set K contained in $\mathbb{R}_+^l$, and let $y \in K$. Then there must be a neighborhood $U(y)$ of y such that

$$\{i \in I \mid z_i \in U(y) \text{ or } U(y) \cap [P_i(x_i) + \{\omega_i\}] \neq \emptyset\},$$

is finite, since otherwise $y \in P(z)$, a contradiction. Selecting a finite subcovering of $\{U(y) \mid y \in K\}$, we obtain that

$$\{i \in I \mid z_i \in K \text{ or } K \cap P_i(x_i) + \{\omega_i\} \neq \emptyset\},$$

is finite as well, and since $\{y \mid p \cdot y < 0\}$ can be written as a countable union of compact sets, we conclude that the set of agents $i \in I$ for which either $p \cdot (x_i - \omega_i) < 0$ or $p \cdot (x'_i - \omega_i) < 0$ for some $x'_i \in P_i(x_i)$ is at most countable. To check that also $p_i \cdot (x_i - \omega_i) > 0$ can occur for at most countably many $i \in I$, we note that otherwise, by Lemma 5.1, we would have more than countably many i with $p \cdot (x_i - \omega_i) < 0$, a contradiction.

For the remaining agents i, we have that $p \cdot (x_i - \omega_i) = 0$ and $p \cdot (x'_i - \omega_i) \geq 0$ for $x'_i \in P_i(x_i)$. Using Theorem 0.2.(iii), we get that x_i is maximal for P_i on $\{x'_i \mid p \cdot x'_i \leq p \cdot \omega_i\}$, so that (x, p) is a Walras equilibrium. $\qquad\square$

The main advantage of the present model is its simplicity, it draws only on elementary mathematics. But this comes at a cost: We cannot establish equivalence of core and equilibrium allocations due to the presence of a countable exceptional set in the definition of an equilibrium, so that equilibrium allocations may not necessarily belong to the core.

The idea of improving allocations through finite coalitions has been reconsidered in subsequent contributions (e.g. Hammond *et al.*, 1989) in a context where additional structure is added to the model.

1.2 Markets with a continuum of traders

The most frequently used model of an exchange economy with many consumers is the one proposed in Aumann (1964). Here, the finite set of consumers indexed by $\{1, \ldots, m\}$ is replaced by the unit interval $[0, 1]$. This interval is endowed with the Lebesgue measure on $\mathbb{R}$, which to each measurable subset S of $[0, 1]$ assigns a number $m(S)$ between 0 and 1, the Lebesgue measure of S (so that, e.g. for an interval $S = [t_1, t_2]$ in $[0, 1]$, $m(S)$ is equal to the length $t_2 - t_1$ of S).

A coalition is then a (Lebesgue) measurable subset of $[0, 1]$. It is null (in the sense of being exceptional) if $m(S) = 0$. The existence of null sets, which will show up as we proceed, captures the idea that the model describes mass phenomena, and what happens to a single specific agent is not interesting in this model. If some condition holds for all agents t except for those in a set of measure 0, it is said to hold for almost all t.

An allocation is now a function $x : [0, 1] \to \mathbb{R}^l_+$ (where for simplicity we have assumed that all consumption sets are the positive orthant in $\mathbb{R}^l$), which is integrable so that the average bundle $\int x(t) \, dt$ is well-defined. There is a particular allocation $\omega : [0, 1] \to \mathbb{R}^l_+$ which defines the initial endowment of the agents. The allocation x is feasible if

$$\int_0^1 x(t) \, dt = \int_0^1 \omega(t) \, dt.$$

For a treatment of core equivalence in this model, we need to formulate preferences, which as usual are given for each agent t by a correspondence $P_t : \mathbb{R}^l_+ \rightrightarrows \mathbb{R}^l_+$, where $x' \in P_t(x)$ indicates that the bundle x' is preferred to the bundle x by agent t. We make the standard assumptions on each P_t, namely monotonicity (in the sense that $y \geq x$ and $y \neq x$ implies that $y \in P_t(x)$, and continuity in the sense of openness of the graph of P_t. However, in the new context of a continuum of traders, we shall need an additional assumption of *measurability:* For each pair x, x' of allocations, the set

$$\{t \in [0, 1] \mid x' \in P_t(x)\}$$

is Lebesgue measurable.

An allocation x can be improved by a coalition S if there is an allocation x' such that

$$\int_S x'(t) \, dt = \int_S \omega(t) \, dt, x'(t) \in P_t(x), \text{ all } t \in S. \tag{4}$$

The core is the set of feasible allocations which cannot be improved by any non-null coalition. A pair (x, p), where x is an allocation and

$p \in \mathbb{R}^l_+ \setminus \{0\}$ is a price system, is an equilibrium if

(i) the allocation is feasible, $\int x(t)\, dt = \int \omega(t)\, dt$,
(ii) for almost all t, $x(t)$ is maximal for P_t in the budget set $\{x \mid p \cdot x \leq p \cdot \omega(t)\}$.

It is easily seen that an equilibrium allocation belongs to the core: Suppose to the contrary that (x, p) is an equilibrium but there is a coalition S with $m(S) > 0$ and an allocation (x'), such that (4) holds. Then $p \cdot x'(t) > p \cdot \omega(t)$ for almost all $t \in S$, and it follows that

$$
p \cdot \int_S x'(t)\, dt > p \cdot \int_S \omega(t)\, dt,
$$

contradicting (4). To prove a converse and thereby establish an equivalence between equilibrium allocations and the core, we must understand that except for a null set of agents, the set of preferred net trades of individuals can be separated from the negative orthant by a linear form, which then can be identified with a price system. We state the equivalence theorem here and postpone its proof to the next subsection.

Theorem 5.2 (Aumann). *The core of the economy coincides with the set of allocations obtainable in equilibrium.*

1.3 Large economies as non-atomic measure spaces

In an economy with the set of agents equal to $[0, 1]$ and with coalitions of all measurable subsets $S \subset [0, 1]$ with $m(S) > 0$, an allocation x, which is integrable as a function from $[0, 1]$ to $\mathbb{R}^l$, gives rise to a vector measure ξ on $[0, 1]$ taking values in $\mathbb{R}^l_+$, with

$$
\xi(S) = \int_S x(t)\, dt,
$$

for S any coalition, so that $\xi(S \cup T) = \xi(S) + \xi(T)$ when $S \cap T = \emptyset$. Similarly, if x' is another allocation such that $x'(t) \in P_t$ is preferred

Large Economies   149

to $x(t)$ for all t in some coalition S, then x' defines another vector measure ξ' with $\xi'(S) = \int_S x'(t)\,dt$ for all coalitions S, which may be considered as being preferred to ξ by S.

The treatment of allocations in economies with an infinity of consumers as vector measures was introduced by Vind [1964]. The main advantage is the possibility if using Lyapunov's theorem, stating that the range of an atomless vector measure is a convex set. For a proof of this important result, see Section 1.5.

We consider an economy with a given set A of consumers, endowed with a σ-algebra $\mathbb{A}$ of coalitions. An allocation is a measure ξ on the measurable space $(A, \mathbb{A})$ with values in $\mathbb{R}_+^l$. An allocation ξ is feasible if $\xi(A) = \omega(A)$. It gives rise to a (signed) vector measure ζ defined by $\zeta(S) = \xi(S) - \omega(S)$ for $S \in \mathbb{A}$.

It turns out to be convenient to formulate preferences directly on net trades rather than on allocations: For each coalition $S \in \mathbb{A}$, there is given a preference correspondence P_S which to each allocation ξ assigns a set $P_S(\xi)$ of net trades η such that the allocation $\eta(S) + \omega(S)$ is preferred by S to ξ, and for convenience, we put $P_S(\xi) = \{0\}$ if $\xi(S) + \omega(S) = 0$. We let $P(\xi) = \cup_{S : \xi(S) + \omega(S) \neq 0} P_S(\xi)$ denote the set of net trades preferred to ξ by some non-null coalition, and we let $\mathcal{P}(\xi)$ denote the range of all the vector measures in $P(\xi)$,

$$\mathcal{P}(\xi) = \{\eta(S) \mid \eta \in P(\xi), S \in \mathbb{A}, \xi(S) + \omega(S) \neq 0\}.$$

Preferences are assumed to be monotonic, in the sense that for each $S \in \mathbb{A}$, $P_S(\xi)$ contains all the net trade measures η for which $\eta(T) \geq \xi(T) - \omega(T)$, all $T \in \mathbb{A}$, $T \subseteq A$. We assume that preferences are independent in the sense that $P_S(\xi)$ depends only on the restriction of ξ to S, for $S \in \mathbb{A}$, this will indeed be the case if the net trade measures have the form $\zeta'(S) = \int_S (x'(t) - \omega(t))\,dt$ for some integrable function x.

An allocation ξ can be improved by a coalition $S \in \mathbb{A}$ if $\omega(S) \neq 0$ and there is a net trade $\eta \in P_S(\xi)$ such that $\eta(S) = 0$. The core of the economy is as usual the set of allocations having no improvements. An equilibrium is a pair (ξ, p), where ξ is an allocation and $p \in \mathbb{R}_+^l \setminus \{0\}$

is a price, such that

(i) ξ is feasible,
(ii) for each $S \in \mathbb{A}$ with $\xi(S) + \omega(S) \neq 0$, $p \cdot \xi(S) = p \cdot \omega(S)$ and $p \cdot \eta(S) \geq p \cdot \omega(S)$ for all $\eta \in P_S(\xi)$.

An *atom* for a vector measure ω is a set $B \in \mathbb{A}$ such that $\omega(B') = \omega(B)$ or $\omega(B') = 0$ for all $B' \in \mathbb{A}$ with $B' \subset B$. We assume that ω is *non-atomic* in the sense that there are no atoms for ω.

Theorem 5.3. *Suppose that the endowment vector ω is non-atomic. Then the core and the set of equilibrium allocations coincide.*

Proof. Suppose that (ξ, p) is an allocation. If ξ can be improved by some coalition S, then there is $\eta \in P_S(\xi)$ such that $\eta(S) = 0$. But then the allocation $\xi' = \eta + \omega$ satisfies $p \cdot \xi'(S) = p \cdot \omega(S)$ contradicting that (ξ, p) is an equilibrium.

If ξ belongs to the core and ω is non-atomic, then ξ is non-atomic as well: If $B \in \mathbb{A}$ is an atom for ξ, then any subset T of S with $\omega(T) > 0$ can obtain a preferred allocation by keeping the initial allocation ω on T. Moreover, the set $\mathcal{P}(\xi)$ is unchanged if only non-atomic allocations preferred to ξ are taken into account: Let $S \in \mathbb{A}$ and $\eta \in P_S(\xi)$ and suppose that A contains an atom for η. Then there is an atom $B \subset A$ with $\xi(B) = \omega(B) = 0$ (this is a consequence of Lyapunov's theorem), and the bundle $\eta(B)$ may be given to the coalitions in $A \backslash B$ defining a non-atomic allocation η' with $\eta' \in S \backslash B(\xi)$ by our monotonicity assumption, and $\eta'(S \backslash B) = \eta(B) \in \mathcal{P}(\xi)$. Proceeding similarly with other atoms for η, if such atoms exist, we construct a non-atomic net trade measure with the same range as η.

We claim that $\mathcal{P}(\xi)$ is a convex subset of $\mathbb{R}_+^l$. Suppose that $y_1 = \eta_1(A_1)$, $y_2 = \eta_2(A_2)$ for non-atomic measures $\eta_1, \eta_2 \in \mathbb{P}(\xi)$, $A_1, A_2 \in \mathbb{A}$, and let $\lambda \in [0,1]$. Using Lyapunov's theorem on η_1 and η_2, we get the existence of $B_1 \subset A_1 \backslash A_2$ with $\eta_1(B_1) = \lambda \eta_1(A_1 \backslash A_2)$, and similarly, $B_2 \subset A_2 \backslash A_1$ with $\eta_2(B_2) = \lambda \eta_2(A_2 \backslash A_1)$. Using Lyapunov's theorem on (η_1, η_2), we find B_3 with $\eta_i = \lambda \eta_i(B_3)$, $i = 1, 2$, and if $B_4 = (A_1 \cap B_2) \backslash B_3$, then $\eta_2(B_4) = (1-\lambda)\eta_2(B_1 \cap B_2)$. Let $B = B_1 \cup \cdots \cup B_4$

and define the measure $\hat{\eta}$ by

$$\hat{\eta}(C) = \eta_1(C \cap (B_1 \cup B_3)) + \eta_2(C \cap (B_2 \cup B_4) + \eta_3(C \backslash B),$$

where η_3 is an arbitrary $\mathbb{R}^l$-valued vector measure on $(A, \mathbb{A})$. Then $\hat{\eta}$ is a non-atomic vector measure, and $\hat{\eta}(B) = \lambda\eta_1(A_1) + (1 - \lambda)\eta_2(A_2)$, proving our claim that $\mathcal{P}$ is indeed convex.

Since ξ belongs to the core, we have that $0 \notin \mathcal{P}(\xi)$, and consequently, there is a price system $p \in \mathbb{R}^l \backslash \{0\}$ such that $p \cdot y > 0$ for all $y \in \mathcal{P}(\xi)$. By monotonicity, we have that $\mathbb{R}^l_+ \backslash \{0\} \subset \mathcal{P}(\xi)$, so that $p \in \mathbb{R}^l_+$. The range of the net trade measure $\xi - \omega$ is convex (by Lyapunov's theorem) and symmetric with respect to 0, and by our monotonicity assumption, it belongs to closure of $\mathcal{P}(\xi)$, so we get that $p \cdot \xi(S) = p \cdot \omega(S)$ for all $S \in \mathbb{A}$, and if $\eta \in P_S(\xi)$, then $p \cdot \eta(S) > p \cdot \omega(S)$. Thus, (ξ, p) is an equilibrium. $\qquad\square$

1.4 Sequences of finite economies

By now we have two types of results about the core of a large economy: (1) Core allocations in infinite economies are Walras allocations and (2) allocations in finite economies which belong to the core of all replica are Walras allocations. From a purely formal point of view, the first version is the more elegant — especially in such models of infinite economies that are more sophisticated than ours and where Walras allocations are also in the core. For these models, the result then is an equivalence of the two types of allocations. However, such formal advantages notwithstanding, the whole idea of an infinite economy is somewhat remote from our economic intuition, and for this reason, the replica-economies may be preferable. Unfortunately, the way in which economies get large — by exact replication of the original economy — is a very particular one, and therefore it would be desirable to extend the Debreu-Scarf result to sequences of economies becoming larger in a less restrictive sense than by replication.

In the following, we consider an arbitrary family $(\mathcal{E}_\alpha)_{\alpha \in \mathcal{A}}$ of finite economies. We shall prove the following result which shows that core allocations in a certain sense become close to Walras allocations when the economy becomes large.

There are several ways in which an allocation x^α in $\mathcal{E}_\alpha$ may fail to be Walrasian at a given price $p \in \Delta$: For each individual i, the value of the bundle may differ from the value of the endowment by more than some $\delta > 0$, or there may be bundles preferred to x_i^α which are more than δ cheaper than x_i^α. We introduce a notion of a δ-exceptional set of agents in $\mathcal{E}_\alpha$ associated with an allocation x^α and a price p, defined as

$$G_\alpha^\delta(x_\alpha, p) = \left\{ i \in \mathcal{E}_\alpha \,\big|\, |p \cdot x_i^\alpha - p \cdot \omega_i| \geq \delta \right\}$$
$$\cup \left\{ i \in \mathcal{E}_\alpha \,\big|\, \exists x_i' \in P_i(x_i^\alpha), \, p \cdot x_i' - p \cdot \omega_i \leq -\delta \right\}. \tag{5}$$

With this notation, we may formulate the main result of this section, stating that core allocations in a sufficiently large economy will be almost Walrasian, meaning that few agents fail to satisfy the conditions of a Walras equilibrium within a small margin.

Now we are ready to show that core allocations become nearly Walrasian when the economy gets large. First, we prove the following fundamental result due to Anderson (1978).

Theorem 5.4. *Let there be an economy where each $X_i = \mathbb{R}_+^l$ and $\omega_i \in X_i$ $i = 1, \ldots, m$.*

If $x \in \text{Core}\,\mathcal{E}$, then there is $p \in \Delta$ such that for any coalition S, if $x_i' \in X_i$, $x_i' \in P_i(x_i)$, for $i \in S$, then

$$p \cdot \sum_{i \in S} (x_i - \omega_i) \geq -lK,$$

where $K = \max_{i,k} \omega_{ik}$.

The proof of Theorem 5.4 uses the Shapley–Folkman theorem on convex hulls of sums of sets. A proof of this theorem is given in the next Section 5.

Proof of Theorem 5.4. Let $x \in \text{Core}(\mathcal{E})$, and for each i, define the set $B_i \subset \mathbb{R}^l$ by

$$B_i = \left\{ z_i \,\big|\, z_i = x_i' - \omega_i, x_i' \in P_i(x_i) \right\} \cup \{0\},$$

and let $B = \sum_{i=1}^m B_i$. Let $z_0 = -lKe$, where $e = (1, \ldots, 1)$ is the diagonal vector in $\mathbb{R}^l$, and let $Z_0 = \{z_0\} + \mathbb{R}^l_{--}$.

We claim that conv $(B) \cap Z_0 = \emptyset$. Indeed, suppose that some $z \in$ conv(B) satisfies $z_h < -lK$ for all h. By the Shapley–Folkman theorem (Theorem 5.6), there are $z_i \in$ convB_i, $i = 1, \ldots, m$, such that $\sum_{i=1}^m z_i = z$ and $z_i \in B_i$ for all except l, of the indices i, say, for $i = 1, \ldots, l$.

For these first l agents, we have $z_i + Ke \geq z_i + \omega_i \geq 0$, so that

$$ z - z_0 = \sum_{i=1}^l (z_i + Ke) + \sum_{i=l+1}^m z_i \geq \sum_{i=l+1}^m z_i, $$

and since $z - z_0 \in \mathbb{R}^l_{--}$, there is a subset S of $\{l+1, \ldots, m\}$ (namely $S = \{i \geq l+1 \mid z_i \neq 0\}$) which can improve upon x, contradicting that $x \in$ Core$(\mathcal{E})$.

Applying Theorem 3.5 to the convex sets conv$B$ and Z_0, we get the existence of $p \in \Delta$ such that

$$ p \cdot z \geq p \cdot z_0, \quad z \in \text{conv}(B), $$
$$ p \cdot z < p \cdot z_0, \quad z \in \text{int} \, Z_0. $$

Using the definition of B_i's, we have the conclusion of the theorem.
$\square$

We apply Theorem 5.4 to obtain the following result.

Theorem 5.5. *Let $(\mathcal{E}_\alpha)_{\alpha \in \mathcal{A}}$ be a family of economies $\mathcal{E}_\alpha = (X_i^\alpha, P_i^\alpha, \omega_i^\alpha)_{i \in M_\alpha}$, with $X_i^\alpha = \mathbb{R}^l_+$ and P_i^α satisfying local non-satiation $(P_i^\alpha(x_i)$ intersects every neighborhood of x_i, all $x_i \in X_i)$ for all α and $i \in M_\alpha$. Assume that the initial endowments are uniformly bounded: there is $K > 0$ such that $\omega_{ih}^\alpha \leq K$ for all α, i, h.*

Then for all $\varepsilon, \delta > 0$, there is a number $n \in \mathbb{N}$ such that if $|M_\alpha| > n$ and x^α is an allocation in Core$(\mathcal{E}_\alpha)$, then there is $p_\alpha \in \Delta$ such that

$$ \frac{\left| G_\alpha^\delta(x^\alpha, p_\alpha) \right|}{|M_\alpha|} < \varepsilon. $$

Proof. Let $\varepsilon, \delta > 0$ be given. For each $\alpha \in \mathcal{A}$, let $p^\alpha \in \Delta$ be a price system for $\mathcal{E}_\alpha$ with the properties asserted by Theorem 5.4. Then

$$ \left| G^\delta(x^\alpha, p^\alpha) \right| (-\delta) \geq \sum_{i \in G^\delta(x^\alpha, p^\alpha)} (x_i^\alpha - \omega_i^\alpha) \geq -lK, $$

so that

$$\frac{\left|G^{\delta}(x^{\alpha}, p^{\alpha})\right|}{|M_{\alpha}|} \leq \frac{lK}{\delta|M_{\alpha}|}$$

which is $\leq \varepsilon$ for large enough $|M_{\alpha}|$.

The assumption of uniform boundedness of endowments has been used in order to state the theorem in its simplest version. To avoid it, we may consider sequences of economies $(\mathcal{E}_k)_{k=1}^{\infty}$ with $\mathcal{E}_k = (X_i^k, P_i^k, \omega_i^k)_{i \in M_k}$ satisfying the following conditions:

(1) $|M_k| \to \infty$,
(2) $\forall \varepsilon > 0 \, \exists K > 0, k' \in \mathbb{N}$, such that for all $k \geq k'$,

$$\frac{|E_k|}{|M_k|} > 1 - \varepsilon,$$

where $E_k = \{i \in M_k \mid \omega_{ih} \leq K \text{ for } h = 1, \dots, l\}$,
(3) $\forall \varepsilon > 0 \, \exists \delta > 0$ such that for all $(A_k)_{k=1}^{\infty}$ with $A_k \subseteq M_k$,

$$\frac{|A_k|}{|M_k|} < \delta \text{ for } k > k' \text{ implies } \frac{\sum_{i \in A_k} \omega_{ih}}{\sum_{i \in M_k} \omega_{ih}} < \varepsilon \text{ for } k > k'$$

$$\text{and } h = 1, \dots, l.$$

(4) $\dfrac{1}{|M_k|} \sum_{i \in M_k} \omega_i \to \omega \in \mathbb{R}_+^l$.

This corresponds closely to the *purely competitive sequences* considered in Hildenbrand (1974). For such sequences, the theorem can be proved with only minor changes. The details are left to the reader.

1.5 *The Shapley–Folkman and Lyapunov theorems*

The famous theorem of Lyapunov (1940) states that the range of an n-dimensional non-atomic vector measure is a closed and convex subset of $\mathbb{R}^n$. An elementary proof of this theorem has been given by Halmos (1948) and a short proof was put forward using functional analysis by Lindenstrauss (1966).

Large Economies 155

In the present note, it is shown that Lyapunov's theorem can be derived from the Shapley–Folkman theorem, which we have already used in Section 1.4. We give a formal proof of this result here.

Theorem 5.6 (Shapley–Folkman). *Let X_i, $i \in M = \{1,\dots,m\}$, be non-empty subsets of $\mathbb{R}^n$. Then for every $z \in \text{conv}\left(\sum_{i=1}^m X_i\right)$, there is a partition $\{M_1, M_2\}$ of M with* card $M \leq n$ *such that*

$$z = \sum_{i \in M_1} z_i + \sum_{i \in M_2} x_i,$$

where $z_i \in \text{conv}X_i$, $x_i \in X_i$.

The theorem first appeared in Starr (1969). The following proof is due to Aubin and Ekeland, see Aubin (1979).

Proof. Let $z \in \text{conv}\left(\sum_{i=1}^m X_i\right)$. Then $z = \sum_{j=1}^k \lambda_j z^j$, where $z^j = \sum_{i=1}^m z_i^j$, $z_i^j \in X_i$, $i = 1,\dots,m$, $j = 1,\dots,k$. W.l.o.g., we may assume that $X_i = \{x_i^1,\dots,x_i^k\}$, $i = 1,\dots,m$. Define

$$P = \left\{ (z_1,\dots,z_m) \in \mathbb{R}^{mn} \,\middle|\, z_i \in \text{conv}X_i, i = 1,\dots,m, \sum_{i=1}^m z_i = z \right\}.$$

Then P is non-empty since $z \in \text{conv}(\sum_{i=1}^m X_i) in \subseteq \sum_{i=1}^m \text{conv}X_i$, and since P is the intersection of a polyhedron with an affine subspace, it has extreme points. Let $\bar{z} = (\bar{z}_1 m \dots, \bar{z}_m)$ be an extreme point of P, and suppose that there are $p > n$ indices (say, the first p), such that $\bar{z}_i \in (\text{conv}X_i)\backslash X_i$, $i = 1,\dots,p$. Then there must be vectors $y_i \in \mathbb{R}^n$, $y_i \neq 0$, $i = 1,\dots,p$, such that $\bar{x}_i + y_i$ and $\bar{z}_i - y_i$ are in $\text{conv}X_i$. The p vectors y_i, $i = 1,\dots,p$, are linearly dependent, so there are μ_i, $i = 1,\dots,p$, not all zero, with $\sum_{i=1}^p \mu_i y_i = 0$. We may assume that $|\mu_i| \leq 1$, $i = 1,\dots,p$.

Now the vectors

$$\bar{z}^1 = (\bar{z}_1 + \mu_1 y_1,\dots,\bar{z}_p + \mu_p y_p, \bar{z}_{p+1},\dots,\bar{z}_m),$$
$$\bar{z}^2 = (\bar{z}_1 - \mu_1 y_1,\dots,\bar{z}_p - \mu_p y_p, \bar{z}_{p+1},\dots,\bar{z}_m),$$

156 *Theory of General Economic Equilibrium*

are in P, and

$$\bar{z} = \frac{1}{2}\bar{z}^1 + \frac{1}{2}\bar{z}^2.$$

But this is a contraction to the fact that $\bar{z}$ is extreme. $\square$

Let $(A, \mathcal{A})$ be a measurable space, and let $\lambda = (\lambda_1, \ldots, \lambda_n)$ be a positive n-dimensional vector measure λ on $\mathcal{A}$. The range of λ is the set

$$\mathrm{Rg}\lambda = \{x = (x_1, \ldots, x_n) \mid \exists E \in \mathcal{A}, \lambda_i(E) = x_i, i = 1, \ldots, n\}.$$

For every $E \in \mathcal{A}$, we denote by λ_E the restriction of λ to $\mathcal{A}_E = \{F \in \mathcal{A} \mid F \subseteq E\}$.

For $x \in \mathbb{R}^n$, let $|x| = \sum_{i=1}^n |x_i|$, and define $|\lambda| : \mathcal{A} \to \mathbb{R}$ by $|\lambda|(E) = |\lambda(E)|$. Since λ is a positive vector measure, $|\lambda|$ is a (positive) measure. An *atom* for λ we understand is a set $E \in \mathcal{A}$ such that $\lambda(E) \neq 0$, and for all $E' \in \mathcal{A}_E$, either $\lambda(E') = 0$ or $\lambda(E') = \lambda(E)$. The vector measure λ is *non-atomic* if there are no atoms for λ. If λ is non-atomic, so is $|\lambda|$.

We shall use the following property of a non-atomic (one-dimensional) measure. A proof can be found, e.g. in Dunford and Schwartz (1957, Lemma 7, p. 308).

Lemma 5.2. *For every set $E \in \mathcal{A}$ and every $\varepsilon > 0$, there is a finite measurable partition $\{E_1, \ldots, E_m\}$ of E such that $|\lambda|(E_i) \leq \varepsilon|\lambda|(E)$, $i = 1, \ldots, m$.*

Now we can prove the convexity part of Lyapunov's theorem.

Theorem 5.7. *Let λ be a non-atomic positive vector measure. Then $\mathrm{Rg}\lambda$ is convex.*

Proof. Let $z \in \mathrm{conv}(\mathrm{Rg}\,\lambda)$. Choose a measurable partition $\{E_1, \ldots, E_m\}$ of A with $|\lambda|(E_i) \leq \frac{1}{2n}|\lambda|(A)$. Then $x \in \mathrm{conv} \sum_{i=1}^m \mathrm{Rg}\,\lambda_{E_i}$, whence, by Theorem 5.6,

$$x = \lambda(F_1) + z_1,$$

where $z_1 \in \sum_{i \in M_1} \text{conv}(\text{Rg}\, \lambda_{E_i})$, $|M_1| \le n$, i.e. $z_1 \in \text{conv}(\text{Rg}\, \lambda_{G_1})$, where $G_1 \in \mathcal{A}$, $G_1 \cap F_1 = \emptyset$ and $|\lambda|(G_1) \le \frac{1}{2}(A)$.

Proceeding as above with $z_1 \in \text{conv}(\text{Rg}\, \lambda_{G_1})$ and a measurable partition $\{E'_1, \dots, E'_m\}$ of G_1 with $|\lambda|(E'_1) \le \frac{1}{2n}|\lambda|(G_1)$, we get that

$$x = \lambda(F_1) + \lambda(F_2) + z_2,$$

where $z_2 \in \text{conv}(\text{Rg}\, \lambda_{G_2})$ for some $G_2 \in \mathcal{A}$ with $G_2 \cap F_2 = \emptyset$, $|\lambda|(G_2) \le \frac{1}{2^2}|\lambda|(A)$, and, in general, for each k,

$$x = \sum_{i=1}^{k} \lambda(F_i) + z_k,$$

$z \in \text{conv}(\text{Rg}\, \lambda_{G_k})$, $G_k \in \mathcal{A}$, $G_k \cap F_k = \emptyset$, and $|\lambda|(G_k) \le \frac{1}{2^k}|\lambda|(A)$. Thus, we get a sequence of disjoint sets (F_i) with $\left| x - \lambda\left(\cup_{i=1}^k F_i\right)\right| \le \frac{1}{2^k}|\lambda|(A)$, so $\lambda\left(\cup_{i=1}^\infty F_i\right) = x$ and $x \in \text{Rg}\, \lambda$. $\qquad\square$

Theorem 5.8. *Let λ be a non-atomic positive vector measure. Then $\text{Rg}\, \lambda$ is closed.*

Proof. By Theorem 5.7 it is enough to show that every extreme point x of the compact set $\text{clRg}\, \lambda$ is actually in $\text{Rg}\, \lambda$.

Let (A_k) be a sequence of sets with $\lambda(A_k) \to x$. Suppose that there is $\varepsilon > 0$ such that for all k we can find $k' > k$ with $|\lambda|(A_{k'} \setminus A_k) \ge \varepsilon$. Then there must be vectors z_k with $|z_k| \ge \varepsilon$, such that $y_k^1 = \lambda(A_k) + z_k$ and $y_k^2 = \lambda(A_{k'} - z_k)$ belong to $\text{Rg}\, \lambda$. Taking a convergent subsequence we find that there are $y^1, y^2 \in \text{clRg}\, \lambda$ with $y^1 = x + z$, $y^2 = x - z$, and $|z| \ge \varepsilon$. But then x is not extreme, a contradiction.

We conclude that for all $\varepsilon > 0$, there is k such that $k' > k$ implies $|\lambda|(A_{k'} \setminus A_k) > \varepsilon$. Taking a subsequence if necessary, we may assume that $|\lambda|(A_{k'} \setminus A_k) < 2^{-k}$ for $k' > k$. Now for all k, we have

$$\left|\lambda(\cup_{i=k}^\infty A_i) - \lambda(A_k)\right| < \sum_{i=1}^{\infty} |\lambda|(A_{i+1} \setminus A_i) < 2^{-(k-1)},$$

158 *Theory of General Economic Equilibrium*

and since

$$\left|\lambda(\cup_{i=k}^{\infty} A_i) - x\right| < \left|\lambda(\cup_{i=k}^{\infty} A_i) - \lambda(A_k)\right| + \left|\lambda(A_k) - x\right|,$$

we have a decreasing sequence (E_k), where $E_k = \cup_{i=1}^{\infty} A_i$, with $\lambda(E_k) \to x$, whence $\lambda(\cap_{k=1}^{\infty} E_k) = x$, so $x \in \mathrm{Rg}\,\lambda$. □

2. Economies with Many Commodities

So far, we have interpreted largeness of an economy as pertaining to its set of agents: If there are many, indeed infinitely many, consumers in an exchange economy, then we are dealing with a large economy. However, there are other ways in which the economy could be large, in particular if there are more than finitely many commodities, and in a way this situation is more directly suggested by economic considerations than the many-agents situation. Already when introducing commodities, we noted that distinguishing commodities according to quality, location and date of delivery or according to an event which triggers the delivery, we are facing a possible infinity of different commodities.

2.1 *Economies with a continuum of commodities*

The economic theory of markets with an infinite number of commodities has been the subject of a considerable literature, starting with Bewley (1972).[2]

There are several good reasons for treating the set of commodities as infinite. For one thing, the number of commodities in real-world economies is very large, and it might well be the case that using a continuum instead of a discrete set would simplify the analysis. Also, it might be argued that what matters is not so much *actual*

---

[2]Other contributions are, e.g. Aliprantis and Brown (1983), Florenzano (1983), Mas-Colell (1986), Yannelis and Zame (1986), Araujo and Monteiro (1989), and more recently Aliprantis *et al.* (2004a, 2004b). Many of the results obtained has appeared also in monographs, e.g. Aliprantis *et al.* (1990) and Florenzano (2003).

commodities traded but rather *potential* commodities which may or may not be selected for trade.

In this new situation, a commodity bundle is no longer an ordinary element in a finite-dimensional vector space, rather we must turn to elements of infinite-dimensional vector spaces. An obvious candidate for such a commodity space is $L^\infty(\mathbb{R}_+)$ (the set of essentially bounded functions on $\mathbb{R}_+$, also written as L^∞). Elements $x = x(\cdot)$ of L^∞ are as previously interpreted as specifications of the consumption of the commodity indicated by the argument s, for each s in the (long) list of possible commodities, which may be differentiated by quality, location, time of delivery and event determining delivery. The consumption set of each agent, consisting of individually feasible commodity bundles, is taken as $X_i = L^\infty(\mathbb{R}_+, \mathbb{R}_+)$, that is all elements $x(\cdot)$ of L^∞ taking non-negative values for all s (except possibly in a subset with Lebesgue measure 0, something which needs not concern us at present). Consumers are supposed to have preferences $P_i : X_i \rightrightarrows X_i$ on their consumption sets and an endowment $\omega_i \in X_i$.

With this setup, we can easily define allocations as arrays $x = (x_1, \ldots, x_m)$ of $(L^\infty)^m$ and the feasibility of allocations by the conditions

$$x_i \in X_i,\ i = 1, \ldots, m, \sum_{i=1}^{m} x_i = \sum_{i=1}^{m} \omega_i. \tag{6}$$

Here, we have considered that L^∞ is a vector space so that addition of commodity bundles make sense. In order to define a Walras equilibrium in this economy, we now need to introduce *prices*. Before doing so, we stop for a moment to recall what has been the role of prices in our models. We have used prices in order to assign values to commodity bundles in a way that agrees adding bundles, indeed with the vector space operations of addition and multiplication by scalars. In our present context, this means that the price system should be a real-valued linear map, a *linear form*, on the commodity space. Moreover, $p(x)$ should be non-negative for all $x \in X_i$. In addition, we would like the linear form to be continuous, so that nearby

160   *Theory of General Economic Equilibrium*

bundles cost more or less the same (given that the commodity space is endowed with a topology, as is the case for L^∞, so that "nearby" bundles make sense).

Summing up, a Walras equilibrium is a pair (x, p) where $x \in (L^\infty)^m$ and p is a linear form on L^∞ with $p(x) \geq 0$ for all non-negative elements of L^∞, satisfying the following conditions:

(i) x is feasible (satisfies (6)),
(ii) for all i, $p(x_i) = p(\omega_i)$, $x'_i \in X_i \cap P_i(x_i)$ implies $p(x'_i) > p(\omega_i)$.

Having introduced the model and defined its equilibria, we should check its consistency showing that equilibria exist. For this, we need as usual some additional assumptions, but the standard conditions do not readily translate to similar conditions in the infinite-dimensional case. Many of the standard properties of finite-dimensional vector spaces cease to be valid in the infinite-dimensional surrounding. In the literature on equilibria with an infinite-dimensional commodity space, certain properties of ordered topological vector spaces have been identified as crucial, among which compactness of order intervals in a suitable topology and existence of interior points in the positive cone are the main properties. We therefore postpone the existence question for a moment, or rather, we discuss in a broader setting with emphasis placed on properties of infinite-dimensional vector spaces which make them usable in models of general equilibrium.

2.2 *Admissible commodity spaces*

In what follows, we identify properties of commodity spaces which are needed in a formal theory of economic equilibrium with infinitely many commodities ("admissible" commodity spaces), in the sense that for every well-behaved economy there is at least one equilibrium. This section presupposes some acquaintance with basic notions from the theory of topological vector spaces.

In what follows, a *commodity space* is an ordered Hausdorff topological vector space $(V, \leq)$, where the order relation $\leq$ is defined by

Large Economies 161

a convex closed cone V_+ such that

(i) $V_+ - V_+ = V$,
(ii) if $v_1, v_2 \in V$ and $nv_1 \leq v_2$ for all $n \in \mathbb{N}$, then $v_1 \leq 0$. For $v', v'' \in V$, the order interval $\{v \in V \mid v' \leq v \leq v''\}$ is written as $[v', v'']$.

Let $(V_+)^*$ be the *polar cone* of V_+, consisting of all linear forms p on V such that $p(v) \geq 0$ for all $v \in V_+$. The set $V^+ = (V_+)^* - (V_+)^*$ is called the *order dual* of V.

An *economy over* V is a finite array $\mathcal{E} = (V_+, P_i, \omega_i)_{i=1}^m$ of consumers, where for each consumer i, the consumption set is V_+, $P_i \subset V_+ \times V_+$ is a preference relation on V_+, and $\omega_i \in V$ is an initial endowment of commodities.

Definition 5.3. An economy $\mathcal{E}$ is well-behaved if

(i) the vector ω_i belongs to V_+ for each i,
(ii) for all i, P_i satisfies

(a) $P_i(x_i)$ is convex, $x_i \notin P_i(x_i)$, each $x_i \in V_+$,
(b) for each $x_i' \in V_+$, the set $P_i^{-1}(x_i') \cap [0, \omega]$, is $\sigma(V, V^+)$-open in $[0, \omega]$,

(iii) there is a consumer $i^0 \in I$ such that for all $x_{i^0} \in V_+$, $P_{i^0}(x_{i^0}) + V_+ \subset P_{i^0}(x_{i^0})$.

The conditions of well-behavedness are largely what should by now be expected, only transferred to the given context of infinite-dimensional commodity spaces. Condition (i) is a weak survival condition on individual endowments; condition (ii.a) states the standard properties of convexity and irreflexivity of preferences, and (ii.b) is a continuity property; it is a rather weak assumption, as the openness of $P_i^{-1}(x_i')$ should hold only relative to the set of vectors effectively attainable by consumer i in $\mathcal{E}$, namely the vectors in the order interval $[0, \omega] = \{x_i \mid 0 \leq x_i \leq \omega\}$. Finally, condition (iii) states that some consumers have monotonic preferences.

Following the by now established tradition, we consider quasi-equilibria in the economies considered: A quasi-equilibrium is a

162   *Theory of General Economic Equilibrium*

pair $(x, p) \in V_+^I \times (V_+)^*$, such that

(i) $\sum_{i=1}^m x_i \le \sum_{i=1}^m \omega_i$ (aggregate feasibility),
(ii) $p(\sum_{i=1}^m \omega_i) > 0$, or there exists $q \in (V_+)^*$ such that for each i, $q(x_i) \le q(\omega_i)$ and $x_i' \in P_i(x_i) \cap \operatorname{Ker} p$ implies $q(x_i') \ge q(\omega_i)$.

Here, $\operatorname{Ker} p = \{v \in V \mid p(v) = 0\}$ is the kernel of the linear form p. We may now formulate our condition for admissibility of commodity spaces.

Definition 5.4. A commodity space V is admissible if every well-behaved economy $\mathcal{E} = (V_+, P_i, \omega_i)_{i=1}^m$ over V has a quasi-equilibrium.

As stated in the definition, a commodity space is admissible if it can be used in modeling a classical economic equilibrium theory in such a way that equilibria always exist.

2.2.1 *Properties of admissible commodity spaces*

In what follows, we exhibit some properties of commodity spaces which are admissible according to Definition 5.4. Together they will amount to a characterization of admissible commodity spaces in terms of topology and order.

Theorem 5.9. *Let V be an admissible commodity space. The order interval $[0, v]$ is compact in the $\sigma(V, V^+)$-topology for each $v \in V_+$.*

In the proof of Theorem 5.9, we shall use the following lemma.

Lemma 5.3. *Let V be an ordered topological vector space, and let $v^0 \in V_+$. Assume that $[0, v^0]$ is not $\sigma(V, V^+)$-compact. Then there is a correspondence $\varphi : [0, v^0] \to [0, v^0]$ with non-empty convex values such that*

(i) *for all $v \in [0, v^0]$, $v \notin \varphi(v)$,*
(ii) *for all $v' \in [0, v^0]$, $\varphi^{-1}(v') = \{v \in [0, v^0] \mid v' \in \varphi(v)\}$ is $\sigma(V, V^+)$-open.*

Proof. The family C of sets

$$\{v \in [0, v^0] \mid p(v) > \alpha\}$$
$$\{v \in [0, v^0] \mid p(v) < \beta\}$$

for $p \in V^+$, $\alpha, \beta \in \mathbb{R}$, is a subbasis for the $\sigma(V, W)$-topology on $[0, v^0]$. From Alexander's subbasis theorem (see, e.g. Kelley, 1975, p. 139), we have that there exists a covering $\mathcal{U}$ of $[0, v^0]$ with sets from C such that no finite subfamily of C covers $[0, v^0]$, as the latter set is not compact. We assume that $\mathcal{U}$ is chosen to be minimal for inclusion among such coverings.

Suppose that $\mathcal{U}$ is countable, i.e.

$$\mathcal{U} = \{U_1, U_2, \ldots\}.$$

For each $v \in [0, v^0]$, let $i(v)$ be the smallest index i such that $v \in U_i$, and define φ by

$$\varphi(v) = \{v' \in [0, v^0] \mid v' \notin U_j, \ j \leq i(v)\}.$$

Since $[0, v^0]\backslash U$ is convex for every $U \in \mathcal{U}$, we have that $\phi(v)$ is a convex set. By construction, $v \notin \varphi(v)$; furthermore, $\varphi(v) \neq \emptyset$ since otherwise $\{U_1, \ldots, U_{i(v)}\}$ would be a finite subcovering. Finally, we have for $v' \in \varphi(v)$ that $v' \in \varphi(v'')$ for all v'' with $v'' \notin U_j$, $j \leq i(v)$, so that $\varphi^{-1}(v')$ is open.

If $\mathcal{U}$ is not countable we choose an arbitrary countable subset $\mathcal{V}$ of $\mathcal{U}$. Then the set

$$K_1 = \{v \in [0, v^0] \mid v \notin U, \text{ all } U \in \mathcal{U}\backslash\mathcal{V}\}$$

is non-empty (since otherwise, $\mathcal{U}$ would not be minimal for inclusion), closed, convex (since $K_1 = \cap_{U \in \mathcal{U}\backslash\mathcal{V}}[[0, v^0]\backslash U]$), and not compact: The covering $\mathcal{V}$ of K_1 has no finite subcovering $\mathcal{V}_1$, since then $(\mathcal{U}\backslash\mathcal{V}) \cup \mathcal{V}_\infty$ would be a proper subcovering of $\mathcal{U}$ contradicting minimality.

From the preceding discussion we note that there exists a correspondence $\tilde{\varphi} : K_1 \to K_1$ with non-empty convex values and having the properties (i) and (ii). Define $\varphi : [0, v^0] \to [0, v^0]$ by $\varphi(v) = \tilde{\varphi}(v)$ if $v \in K_1$ and $\varphi(v) = K_1$ if $v \in [0, v^0]\backslash K_1$, then φ has all the desired properties. $\qquad\square$

Proof of Theorem 5.9. Assume that there is $v^0 \in V_+$ such that $[0, v^0]$ is not $\sigma(V, V^+)$ compact. Consider the economy $\mathcal{E} = ((P_1, \omega_1), (P_2, \omega_2))$

164   *Theory of General Economic Equilibrium*

with two consumers, where

$$\omega_1 = v^0, \; \omega_2 = 2v^0,$$

and where the preference relations are defined as follows.

Choose any linear form $p^1 \in V^+$ and let

$$P_1(x_1) = \{x_1' \in V_+ \mid p^1(x_1') > p^1(x_1)\}, \; x_1 \in V_+.$$

Then $P_1^{-1}(x_1') = \{x_1 \mid p^1(x_1) < p^1(x_1')\}$ is $\sigma(V, V^+)$-open for each $x_1' \in V_+$, and clearly, $P_1(x_1)$ is convex and $x_1 \notin P_1(x_1)$ for all $x_1 \in V_+$. Also, $P_1(x_1) + V_+ \subset P_1(x_1)$ for each $x_1 \in V_+$.

For consumer 2, define the preference correspondence $P_2 : V_+ \to V_+$ by $P_2(x_2) = \varphi(x_2)$ if $x_2 \in [0, v^0]$, and $P_2(x_2) = [0, v^0]$ for $x_2 \notin [0, v^0]$. Then P_2 is irreflexive with non-empty convex values; for $x_2 \in [0, v^0]$, this follows from Lemma 5.3, and for $x_2 \notin [0, v^0]$, it is an easy consequence of the definition. To check the continuity condition, we have that for each $x_2' \in [0, v^0]$ the set $P_2^{-1}(x_2)$ may be written as

$$P_2^{-1}(x_2') = \{x_2 \in [0, \omega_2] \mid x_2' \in \varphi(x_2)\} \cup (V_+ \backslash [0, v^0]),$$

where the first set is $\sigma(V, V^+)$-open relative to $[0, v^0]$ by Lemma 5.3, and the second set is $\sigma(V, V^+)$-open in V_+ since $[0, v^0] = \cup_{p \in (V_+)^*}\{v \mid p(v) \le p(v^0)\}$ is $\sigma(V, V^+)$-closed.

We have shown that $\mathcal{E}$ is a well-behaved economy over V. Assume that (x_1^0, x_2^0) is a feasible allocation in $\mathcal{E}$ and $p^0 \in (V_+)^*$ is a price such that $p^0(\omega_1 + \omega_2) > 0$. If $x_2^0 \notin [0, v^0]$, then $P_2(x_2)$ contains 0 contradicting $p^0(x_2') \ge p^0(\omega_2)$ for all $x_2' \in P_2(x_2^0)$, so $x_2^0 \in [0, \omega_2]$. But then $P_2(x_2^0)$ contains a vector x_2' with $x_2' \le v^0$ so that

$$p^0(x_2') \le p^0(v^0) = \frac{1}{2}p^0(\omega_2) < p^0(\omega_2),$$

where the latter inequality follows since $p^0(\omega_1 + \omega_2) > 0$. Consequently, (x_1^0, x_2^0, p^0) cannot be a quasi-equilibrium, and we have exhibited a well-behaved economy with no quasi-equilibria. We conclude that V is not admissible. $\qquad\square$

Large Economies

The next important property of admissible commodity spaces is the following:

Theorem 5.10. *Let V be a commodity space. If V is admissible, then for each $v^0 \in V_+$, either*

(i) *$p(v^0) > 0$ for all $p \in (V_+)^*$ and the order interval $[-v^0, v^0]$ is absorbing in V (meaning that each $v \in V$ belongs to $\lambda[-v^0, v^0]$ for some $\lambda > 0$), or*

(ii) *there is $p \in (V_+)^0$ with $[-v^0, v^0] \subset \operatorname{Ker} p$, such that $[-v^0, v^0]$ is absorbing in any subspace W of V with $[0, v^0] \subset W \subset \operatorname{Ker} p$ such that $W = (W \cap V_+) - (W \cap V_+)$.*

Condition (i) states that v^0 is an *order unit* in V; the fact that the existence of an order unit is a necessary condition for the existence of equilibria was suggested in Aubin (1979).

Before proving Theorem 5.10, we introduce a lemma.

Lemma 5.4. *Let $v^0 \in V_+$ and $T = \{v \in V_+ \mid \forall \lambda > 0, \lambda v \notin [0, v^0]\}$. Then for any subspace W of V with $v^0 \in W$ such that $W = (W \cap V_+) - (W \cap V_+)$, if $[-v^0, v^0] \cap W$ is not absorbing in W, then $W \cap T \neq \emptyset$.*

Proof. If $[-v^0, v^0] \cap W$ is not absorbing in W, then there is $v \in W$ such that $\lambda v \notin [-v^0, v^0]$ for all $\lambda > 0$. Write $v = v_1 - v_2$ for $v_1, v_2 \in (W \cap V_+)$. If $\lambda_i v_i \leq v^0$ for some $\lambda_i > 0$, $i = 1, 2$, then

$$0 \leq \min\{\lambda_1, \lambda_2\} v_1 \leq v^0,$$

$$-v^0 \leq -\min\{\lambda_1, \lambda_2\} v_2 \leq 0,$$

so that

$$-v^0 \leq \min\{\lambda_1, \lambda_2\}(v_1 - v_2) \leq v_0,$$

a contradiction. We conclude that the set

$$\{v \in (W \cap V_+) \mid \forall \lambda > 0, \lambda v \notin [0, v^0]\} = W \cap T$$

is non-empty.

Proof of Theorem 5.10. Let $v^0 \in V_+$ and define T as in Lemma 5.3. Then T is convex: If $v_1, v_2 \in T$, then $\mu v_1 + (1 - \mu) v_2 \notin T$ for some

$0 \le \mu \le 1$ would imply that $\lambda\mu v_1 + \lambda(1 - \mu)v_2 \in [0, v^0]$ for some $\lambda > 0$; but then, $\lambda\mu v_1 \in [0, v^0] - \{\lambda(1 - \mu)v_2\}$, which together with $v_1 \in V_+$ yields that $\lambda\mu v_1 \in [0, v^0]$, a contradiction. From Theorem 5.9 we have that $[0, v^0]$ is compact in the $\sigma(V, V^+)$-topology, so for each $\lambda > 0$, there is a linear form $\tilde{p} \in V^+$, such that

$$\tilde{p}(\lambda v) > \max\{\tilde{p}(v') \mid v' \in [0, v^0]\}.$$

Define the economy $\mathcal{E}$ over V as $\mathcal{E} = ((P_1, \omega_1), (P_2, \omega_2))$, where $\omega_1 = \omega_2 = v^0$, P_1 is defined as in the proof of Theorem 5.9, and P_2 is given by

$$P_2(x_2) = \begin{cases} T & x_2 \in [0, 2v^0], \\ \emptyset & \text{otherwise.} \end{cases}$$

Then P_2 is irreflexive since $T \cap [0, 2v^0] = \emptyset$ and convex valued, and for each $x_2' \in V_+$, we have either $P_2^{-1}(x_2') = \emptyset$ or $P_2^{-1}(x_2') = [0, 2v^0]$, in both cases a set which is open relative to $[0, 2v^0]$. We conclude that $\mathcal{E}$ is well-behaved.

Let $(x_1^0, x_2^0) \in [0, 2v^0]^2$ and $p^0 \in (V_+)^*$. Suppose first that $p^0(v^0) > 0$. Since $P_2(x_2^0) = T$, so that in particular $P_2(x_2^0)$ contains all non-zero points of some ray in V_+, it follows that there are points $x_2' \in P_2(x_2^0)$ such that $p^0(x_2)$ is arbitrarily close to 0, in particular so that $p^0(x_2') < p^0(v^0)$. Thus, if $p^0(v^0) > 0$ then (x_1^0, x_2^0, p^0) is not a quasi-equilibrium; in particular, if $p(v^0) > 0$ for all $p \in (V_+)^*$, then $\mathcal{E}$ has no quasi-equilibria, a contradiction. We conclude that in this case $[-v^0, v^0]$ must be absorbing in V, which is (a).

Suppose now that $p^0(v^0) = 0$. We have that $\operatorname{Ker} p^0 = (\operatorname{Ker} p^0 \cap V_+) - (\operatorname{Ker} p^0 \cap V_+)$. Indeed, each $v \in V_+$ with $p(v) = 0$ can be written as $v = v_1 - v_2$. If $T \cap \operatorname{Ker} p^0 \ne \emptyset$, we have by the same argument as above that there can be no $q \in (V_+)^*$ with $q(x_2^0) \le q(\omega_2)$ and $q(x_2') \ge q(\omega_2)$ for all $x_2' \in P_2(x_2^0) \cap \operatorname{Ker} p^0$, so that (x_1^0, x_2^0, p^0) cannot be a quasi-equilibrium in $\mathcal{E}$. Since (x_1^0, x_2^0, p^0) was chosen arbitrarily, and V is admissible, we may conclude that $T \cap \operatorname{Ker} p^0 = \emptyset$.

Let W be any subspace of V with $[0, v^0] \subset W \subset \operatorname{Ker} p^0$ with $W = (W \cap V_+) - (W \cap V_+)$. Then $[-v^0, v^0]$ is absorbing in W, since

Large Economies 167

otherwise, $T \cap W \neq \emptyset$, a contradiction. This completes the proof of the theorem. □

Theorems 5.9 and 5.10 together show that if V is admissible, then V must satisfy two sets of conditions. The first of these pertains to compactness of order intervals, while the second one states that order intervals are absorbing. This is in accordance with what has been found in the literature on equilibria with infinite-dimensional commodity spaces, and it indicates that further progress can be achieved only by restricting the notion of well-behavedness or by reconsideration of the equilibrium concept.

2.2.2 *Sufficiency of conditions for admissibility*

In this section, we show that the conditions derived above for a commodity space to be admissible are not only necessary but sufficient as well. This amounts to showing that any well-behaved economy over such a commodity space has a quasi-equilibrium. The method of proof follows the approach of Bewley (1972), namely approximation of the infinite-dimensional commodity space by finite-dimensional commodity spaces.

Theorem 5.11. *Let V be a commodity space satisfying the following conditions: For each $v \in V_+$ with* $\mathrm{span}([0, v]) = V$,

(a) *the order interval $[0, v]$ is compact in the $\sigma(V, V^+)$-topology.*
(b) *the order interval $[-v, v]$ is absorbing.*

If $\mathcal{E} = (V_+, P_i, \omega_i)_{i=1}^m$ is a well-behaved economy over V, then $\mathcal{E}$ has a quasi-equilibrium.

Proof. Let $\mathcal{F}$ be the set of all finite-dimensional subspaces F of V spanned by finite sets of vectors in V_+ containing $\{\omega_1, \ldots, \omega_m\}$. For each $F \in \mathcal{F}$, we have an exchange economy with the finite-dimensional commodity space,

$$\mathcal{E}_F = (F_+, P_i|_F, \omega_i)_{i=1}^m,$$

where $F_+ = F \cap V_+$, and where the preference correspondences $P_i|_F :$ $F_+ \to F_+$, for $i = 1, \ldots, m$ are defined by

$$P_i|_F(v) = P_i(v) \cap F$$

for $v \in F_+$. Clearly, preference correspondences are convex-valued and irreflexive ($v \notin P_i|_F(v)$, all $v \in F_+$) and they satisfy the continuity condition that $(P_i|_F)^{-1}(v) \subset F_+$ is open for each $v \in F_+$ (as the intersection of a $\sigma(V, V^+)$-open set with the finite-dimensional subspace F).

By standard results on equilibria in economies with finite-dimensional commodity spaces, $\mathcal{E}_F$ has a quasi-equilibrium (x^F, p^F), with $x^F = (x_i^F)_{i=1}^m \in V_+^I$ and $p^F \geq 0$ in F^+. Without loss of generality, we may assume that $|p^F(v)| \leq 1$ for all $v \in F$ with $-\omega \leq v \leq \omega$.

By the Hahn–Banach theorem, the linear functional p^F extends to a linear functional on V, likewise denoted by p^F, which belongs to the set

$$B^+ = \{p \in V^+ \mid |p(v)| \leq 1, \ -\omega \leq v \leq \omega\}.$$

By the Banach–Alaoglu theorem (cf., e.g. Rudin, 1991, Theorem 3.15), B^+ is compact in the $\sigma(V^+, V)$-topology, and by condition (a), the set $[0, \omega]$ is compact in the $\sigma(V, V^+)$-topology. It follows that the net $(x^F, p^F)_{F \in \mathcal{F}}$ in $K(\omega)^I \times B^*$ has a subnet (which for ease of notation is assumed to be the net itself) converging to some $(x^0, p^0) \in K(\omega)^I \times B^*$. We claim that (x^0, p^0) is a quasi-equilibrium.

The conditions (i) and (iii) of a quasi-equilibrium are straightforward consequences of our construction, so we need only check (ii). Let $i \in \{1, \ldots, m\}$; suppose that $p^0(x_i^0) > p^0(\omega_i)$, and let $p^0(x_i^0) - p^0(\omega_i) = \varepsilon > 0$. By the convergence properties, for each $\sigma(V^+, V)$-neighborhood U' of p^0, and in particular for the neighborhood

$$U' = \left\{p \in V^+ \,\middle|\, |p(x_i^0) - p^0(x_i^0)| < \frac{\varepsilon}{4}, \ p(\omega_i) - p^0(\omega_i)\Big| < \frac{\varepsilon}{4}\right\},$$

Large Economies 169

the net $(p^F)_{F \in \mathcal{F}}$ is eventually in U. Similarly, for each $\sigma(V, V^+)$-neighborhood U of x_i^0, thus also for the neighborhood

$$U = \left\{ x_i \in V \, \middle| \, |p^0(x_i) - p^0(x_i^0)| < \frac{\varepsilon}{4} \right\},$$

the net (x_i^F) is eventually in U. Since for large enough $F \in \mathcal{F}$ we have $(x_i^F, p^F) \in U \times U'$

$$p^F(x_i^F) \geq p^F(x_i^0) - \frac{\varepsilon}{4} \geq p^0(x_i^0) - \frac{\varepsilon}{2} = p^0(\omega_i) + \frac{\varepsilon}{2} \geq p^F(\omega_i) + \frac{\varepsilon}{4},$$

a contradiction, since (x^F, p^F) is a quasi-equilibrium in $\mathcal{E}^F$. Thus, $p^0(x_i^0) \leq p^0(\omega_i)$ for each i.

Next, let $x_i \in P_i(x_i)$; we must show that $p^0(x_i) \geq p^0(\omega_i)$. As above, assume on the contrary that $p^0(x_i^0) = p(x_i) + \varepsilon$ for some $\varepsilon > 0$, and define the neighborhood

$$G' = \left\{ p \in V^+ \, \middle| \, |p(x_i^0) - p^0(x_i^0)| < \frac{\varepsilon}{4}, \, p(x_i) - p^0(x_i) \, \middle| < \frac{\varepsilon}{4} \right\}$$

of p^0. Then the net (x_i^F, p^F) is eventually in $U \times G'$, so that

$$p^F(x_i) \leq p^0(x_i) + \frac{\varepsilon}{4} = p^0(x_i^0) - \frac{3}{4}\varepsilon \leq p^0(x_i^F) - \frac{\varepsilon}{2} \geq p^F(x_i^F) - \frac{\varepsilon}{4}.$$

By openness (in the $\sigma(V, V^+)$-topology) of $P_i^{-1}(x_i^0)$, we have that $x_i \in P_i(x_i^F)$ for F large enough (and containing x_i^F), and thus, we have a contradiction, since (x^F, p^F) is a quasi-equilibrium in $\mathcal{E}^F$. We conclude that $p^0(x_i) \leq p^0(x_i^0)$, so that property (ii) of the definition of a quasi-equilibrium is satisfied. $\qquad \square$

2.2.3 *Admissible commodity spaces are Kakutani spaces*

In the previous sections, admissible commodity spaces have been characterized by conditions on algebra, order and topology. These conditions were presented in Theorems 5.9 and 5.10; however, it might be argued that they give no precise indication as to the class of topological vector spaces which are interesting in the context of economic equilibrium theory. However, it so happens that they

170　　　*Theory of General Economic Equilibrium*

place V in a rather well-studied family of ordered topological vector spaces, those studied by Kakutani and Krein, cf., e.g. Schaefer (1966) and Aliprantis *et al.* (1990).

From now on, we assume that V is a vector lattice, that is that for each pair (v_1, v_2) of elements of v, there is a unique least upper bound $v_1 \vee v_2$ of v_1 and v_2 in V. This is of course an additional property in relation to admissibility, since no lattice properties were obtained in the characterization discussed earlier.

Following Coppel (1998), we say that a vector lattice V is a *Kakutani space* if it has the following properties

(i) if $x, y \in V$ and $nx \le y$ for all positive integers n, then $x \le 0$,
(ii) there exists $e \in V$ such that, for each $x \in V$, there is a positive integer n for which $-ne \le x \le ne$.

The following is an easy consequence of our results in Section 3.

Lemma 5.5. *Let V be an admissible commodity space which is also a vector lattice. Then V is a Kakutani space.*

Proof. Property (i) discussed precedingly is part of the definition of a commodity space, and (ii) follows from Theorem 5.10. □

The classical result about Kakutani spaces is the Krein–Kakutani theorem, here given in a version adapted to our purpose:

Theorem 5.12. *Let V be an admissible commodity space which is also a vector lattice. Then there is a linear homeomorphism and lattice isomorphism of V into a dense subset of the space $C(\mathbb{K})$ of all continuous real-valued functions on the set $\mathbb{K}$ of extreme points of*

$$\{p \in (V_+)^* \mid p(v^0) = 1\},$$

where $v^0 \in V_+$ is such that $\mathrm{span}([0, v^0]) = V$.

A proof of this result can be found in Coppel (1998, pp. 208–209). The result gives a rather detailed description of admissible commodity spaces; however, its practical applicability remains to be shown.

3. Large-square Economies

In this chapter, we have considered the largeness of economies in two different respects, namely (1) economies with infinitely many agents and (2) economies with infinitely many commodities. Due to the common feature of largeness, one might expect the formal approaches to be closely related; however, they turned out to be very different, as were the kind of results obtained. Economies with many agents turned out to be a convenient tool for the treatment of mass

---

Box 1. A large-square economy without core equivalence. Consider an economy composed by infinitely many copies of the two-agent, two-commodity economy of Box 4.1, so that the commodities of each copy are distinct from those of the other copies. This means that there are countably many agents and that the commodity space is $\mathbb{R}^{(\infty)}$, the vector space of real sequences with finite support. We may choose prices for this economy as sequences $p = (p_n)_{n \in \mathbb{N}}$, a well-defined value $p(x) = \sum_{h:x_h \neq 0} p_h x_h$ for each $x \in \mathbb{R}^{(\infty)}$.

Formally, the set of agents are indexed by $\mathbb{N}$, and for each n, the consumption set X_n of agent n is

$$\left\{ x \in \mathbb{R}^{(\infty)} \middle| x_n, x_{n+1} \geq 0, x_h = 0 \text{ otherwise} \right\}$$

if n is odd, and

$$\left\{ x \in \mathbb{R}^{(\infty)} \middle| x_{n-1}, x_n \geq 0, x_h = 0 \text{ otherwise} \right\}$$

if n is even. Preferences of agents are defined on their commodity sets and endowments also belong to X_n.

The core allocation shown in Box 1 extends trivially to a core allocation in this large-square economy. If this allocation was to be obtained in a Walras equilibrium of the large-square economy, then the restriction to each of the two-agent, two-commodity economies would be a Walras equilibrium as well, a contradiction.

Clearly, replicating the given large-square economy would improve the situation by shrinking the set of core allocations to that of Walras equilibrium allocations. This means that core equivalence obtains, but not in the original large-square economy, rather in another one where the set of agents intuitively speaking is larger than the set of commodities.

phenomena, allowing us to obtain equivalence of core and Walras equilibrium allocations; economies with many commodities did not produce new results; rather to the contrary, they posed a number of specific problems not encountered in the usual economies with a finite number of agents, to be overcome only by approximating by finite economies.

It should be added that until now we have allowed for infinitely many commodities only in economies with a finite number of agents. Combining the two features (1) and (2), so that there are infinitely many agents *and* infinitely many commodities, we get what is called a *large-square* economy by Ostroy (1984). It is easily seen that we cannot hope for a general result about core equivalence in large-square economies, cf. the example in Box 1.

The lack of positive results for general large-square economies does not mean that particular versions cannot be useful, and indeed, we shall meet some of these in future chapters, with overlapping-generations economies as the most prominent case. Actually, core equivalence does hold in large-square economies satisfying specific additional conditions, so this class of economies should in no way be discarded. What is lacking is only *general* results of the type we have been looking for in this chapter.

4. Exercises

(1) Fair net trades, Schmeidler and Vind (1972): Let $(z_1, \ldots, z_m)$ be an array of net trades for each of the consumers in the private ownership economy $\mathcal{E} = (X_i, P_i, \omega_i)_{i=1}^m$. The net trades are *fair* if no individual i is better off replacing z_i by the net trade of another consumer or a group of other consumers.

Show that a Walras equilibrium has fair net trades. Does the converse hold?

Assume now that there are infinitely many consumers in the economy $(\mathbb{R}_+^l, P_i, \omega_i)_{i \in I}$. Show that if the family $(z_i)_{i \in I}$ is fair, then the resulting allocation is a Walras equilibrium (in the sense of Definition 5.1).

Large Economies   173

(2) Exchange equilibria: Let $\mathcal{E} = (X_i, P_i, \omega_i)_{i=1}^m$ be an economy with private ownership. An exchange in $\mathcal{E}$ is a map $\mathfrak{x} : M \times M \to \mathbb{R}^l$ such that $\mathfrak{x}(i, j) = -\mathfrak{x}(j, i)$ for all $i, j \in M = \{1, \ldots, m\}$. If consumer i has a stock of commodities x_i before the exchange $\mathfrak{x}$, then the stock of commodities after the exchange is $x_i + \mathfrak{x}(i, j)$. Show that the allocation $\mathfrak{x}$ obtained after exchanges $\mathfrak{x}^1, \ldots, \mathfrak{x}^r$ satisfies

$$\sum_{i=1}^m x_i = \sum_{i=1}^m \omega_i.$$

A family $\mathfrak{x}^1, \ldots, \mathfrak{x}^r$ of exchanges constitute an *exchange equilibrium* if (1) the resulting allocation is feasible, and (2) for each i, it is impossible to obtain a preferred bundle by canceling some of the transactions $\mathfrak{x}^k(i, j)$ and engaging in new exchanges involving other consumers than those affected by the cancellations.

Let (x, p) be a Walras equilibrium in $\mathcal{E}$. Show that any family of exchanges $\{\mathfrak{x}^1, \ldots, \mathfrak{x}^r\}$ such that $p \cdot \mathfrak{x}^k(i, j) = 0$ for all $i, j \in M \times M$ and $k \in \{1, \ldots, r\}$ is an exchange equilibrium. Show by an example that the converse is not true.

Now turn to the context of an infinite economy $\mathcal{E} = (\mathbb{R}_+^l, P_i, \omega_i)_{i \in I}$ as considered in Section 1.1. Formulate the conditions for a family $\{\mathfrak{x}^1, \ldots, \mathfrak{x}^r\}$ of exchanges $\mathfrak{x} : I \times I \to \mathbb{R}^l$ to be an exchange equilibrium.

Show that if $\{\mathfrak{x}^1, \ldots, \mathfrak{x}^r\}$ is an exchange equilibrium, then there is a price system p such that the resulting allocation x together with p is a Walras equilibrium in the sense of Definition 5.1.

(3) Let $\mathcal{E}$ be an economy with a measure space $(A, \mathbb{A}, \lambda)$ of agents, where λ is a finite measure. Show that if an allocation $\mathbf{x}$ can be improved by some coalition $S \in \mathbb{A}$, then for arbitrarily small $\varepsilon > 0$, it can be improved by a coalition $S' \in \mathbb{A}$ with $\lambda(S') \leq \varepsilon$.

(4) Let $\mathcal{E} = (X_i, P_i, \omega_i)_{i=1}^m$ be an economy with private ownership with $X_i = \mathbb{R}_+^l$ for all i, and let $x \in \mathrm{Core}(\mathcal{E})$ be an allocation in the core of $\mathcal{E}$ (Dierker, 1975).

Let $p \in \Delta$ be a price system, and define for each i the set

$$\sigma_i(p) = \left\{ x_i'' \in \mathrm{cl}, P_i(x_i) \middle| p \cdot x_i'' \geq p \cdot x_i', \text{ all } x_i'' \in \mathrm{cl}\, P_i(x_i) \right\}.$$

174   *Theory of General Economic Equilibrium*

Show that for each $\alpha > 0$, there is $p \in \Delta$ such that

$$\sum_{i \in S} (p \cdot \sigma_i(p) - p \cdot \omega_i) \geq -(l-1)\alpha$$

for any coalition S such that $p \cdot \omega_i \leq \alpha$, all $i \in S$. Give an interpretation of the result.

(5) Check the admissibility of the following topological vector spaces: $C(K)$ for K a compact metric space, $L^p(\mu)$ for the measure space $(A, \mathbb{A}, \mu)$, $p = 1, 2, \ldots$, BV (the set of functions of bounded variation on a compact interval), ba($\mathbb{A}$) (the set of bounded, finitely additive set functions on the measurable space $(A, \mathbb{A})$.

(6) Consider an economy with a continuum of indivisible goods, each of which is available in exactly one unit, such that the set of goods constitute a compact metric space K. Each good is initially owned by a distinct consumer, so that there is also a continuum of consumers, and each consumer wants to consume one unit of some good, so that an allocation in this economy corresponds to a function from K to K. A price system in the economy is a real function p on K.

Give a formal description of a Walras equilibrium in this economy. Suggest a set of assumptions on the basic characteristics of the economy such that Walras equilibria exist.

Chapter 6

Uniqueness and Stability of Equilibria

1. The Uniqueness Problem

Having treated at length the problems of existence of equilibria, it is high time we turned to problems of uniqueness. A theory of market behavior which prescribes a unique outcome is attractive since it gives a definite prescription of what will happen, making it easier to check whether the model is adequate.

We restrict the discussion to economies without production in order to keep the problems as simple as possible. In a certain trivial sense, there is always infinitely many Walras equilibria, since if $(x_1^0, \ldots, x_m^0, p^0)$ is a Walras equilibrium and $\lambda > 0$, then so is $(x_1^0, \ldots, x_m^0, \lambda p^0)$ for any $\lambda > 0$, and this trivial type of multiplicity is of course excluded. Even so, it turns out that there are few general results to be obtained.

1.1 *Gross substitution and uniqueness*

What has been said until now does not imply that uniqueness of Walras equilibria never obtains, only that further restrictions on the class of economies under investigation are needed, and as it happens, the assumptions known to imply uniqueness are stronger than we are usually ready to accept. We give a brief treatment of one such assumption.

Let $\mathcal{E} = (X_i, P_i, \omega_i)_{i=1}^m$ be an economy where each consumer satisfies Assumptions 0.1 and 0.2, and where $\omega_i \in \text{int} X_i$, $i = 1, \ldots, m$.

Box 1. Economies with many equilibria. The Edgeworth box provides a tool for a first investigation of the question of whether a Walras equilibrium, if it exists, will be unique. The answer to this question is no, as can be seen from the example. The initial endowment is A, and there are common tangents to the indifference curves at both B_1 and B_2, corresponding to two Walras equilibria with different bundles to the consumers and prices which are not proportional to each other. Thus, we have an economy with (at least) two different Walras equilibria (Fig. 1).

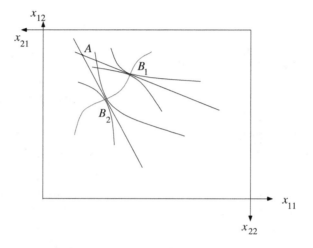

Fig. 1. An exchange economy with at least two Walras equilibria.

In the above example, there are *at least* two Walras equilibria; there might be more than two. Actually, it may happen that the common tangent to the indifference curves at every point between B_1 and B_2 passes through A, so that the economy under consideration has *infinitely many* equilibria.

This latter case is, however, a very special one — moving the point A an arbitrarily small distance away, one may still have more than one, but not infinitely many, equilibria.

Incidentally, the so-called transfer paradox (a consumer can benefit from giving away goods before trade) can occur in this economy. We leave the details to the exercises.

Moreover, we assume that the assumptions in Box 1 are satisfied. Then the excess demand functions ζ_i of each consumer and the

Uniqueness and Stability of Equilibria 177

aggregate excess demand function $\zeta = \sum_{i=1}^{m} \zeta_i$ are defined and differentiable for all $p \in \mathbb{R}^l$ with $p_k > 0$, $k = 1, \ldots, l$ (i.e. on all of $\mathbb{R}^l_{++}$). We say that $\mathcal{E}$ satisfies *gross substitution* if ζ is differentiable in every $p \in \text{int} \mathbb{R}^l_+$, and

$$\frac{\partial \zeta_h(p)}{\partial p_k} > 0 \quad \text{for all } h, k \quad \text{with } h \neq k.$$

The assumption says that all commodities are substitutes in the aggregate ("gross substitutes"): If the price of some commodity k rises, then (excess) demand increases for all other commodities. This implies that the demand for commodity k must decrease: Since ζ is homogeneous of degree zero, cf. (Exercise 3 in the Introduction chapter), it satisfies Euler's equation

$$\sum_{k=1}^{l} \frac{\partial \zeta_h}{\partial p_k} p_k = 0$$

(this follows easily from $\frac{\partial}{\partial \lambda}(\zeta_h(\lambda p)) = 0$ for all $\lambda > 0$); now $\partial \zeta_h / \partial p_k > 0$ and $p_k > 0$ for all k and $h \neq k$ implies that $\partial \zeta_k / \partial p_k < 0$.

There are several reasons why we do not feel comfortable with this assumption. First of all, it violates our principle that assumptions should be made on the individual agents rather than on aggregates where the implications tend to be more difficult to evaluate. Second, it is rather obvious that commodities need not be substitutes, neither for the individual consumer nor in the aggregate. It should be noted that price changes act on individual demand not only through standard substitution as discussed in Chapter 2 but also through the change in income or value of endowments which it affects.

Keeping the above qualifications in mind, we now show that gross substitutability implies uniqueness of Walras equilibria.

Theorem 6.1. *Let $\varepsilon = (X_i, P_i, \omega_i)_{i=1}^{m}$ be an economy satisfying gross substitutability. Then the set of Walras equilibria contains only one element (except for multiplication of the price vector with $\lambda > 0$).*

Proof. If $(x_1^0, \ldots, x_m^0, p^0)$ is a Walras equilibrium, then $p^0 > 0$. Suppose to the contrary that $p_k = 0$ for some k. Then, $p^0 \cdot (x_i^0 + e_k) = p^0 \cdot x_i^0 = p \cdot \omega_i$ for each i, where $e_k = (0, \ldots, 1, \ldots, 0)$ is the kth unit vector, and $u_i(x_i^0 + e_k) > u_i(x_i^0)$ contradicting that $(x_1^0, \ldots, x_m^0, p^0)$ is a Walras equilibrium. $\qquad\square$

Now suppose that $(x_1^0, \ldots, x_m^0, p^0)$ and $(x_1^1, \ldots, x_m^1, p^1)$ are Walras equilibria with non-proportional price vectors. We may assume that $p^1 \geq p^0$ and $p_h^1 = p_h^0$ for a certain commodity h (otherwise, we multiply p^0 with $\min_k(p_k^1/p_k^0)$). Let p_h be constant, while the other prices increase from p^0 to p^1. Then, by the assumption of gross substitutability, ζ_h will increase all the way from p^0 to p^1, contradicting that $\zeta_h(p^0) = 0 = \zeta_h(p^1)$.

1.2 Economies with a finite number of equilibria

While the uniqueness of equilibria cannot be obtained in anything close to generality, one can say something about the other extreme, that of an infinite number of equilibria, a case where it is possible if not to exclude, then at least to relegate to, something very implausible. We shall take a closer look at this, in particular since the tools to be employed, introduced in Debreu (1970), gave rise to a whole new field of differentiable equilibrium theory (Box 2).

In what follows, we are concerned with an exchange economy where the consumption sets are $\mathbb{R}_+^l$ and where preferences can be described by C^2 utility functions (so that the demand functions $\xi_i(p, p \cdot \omega_i)$ are C^1 functions of prices p and initial endowments ω_i). With standard monotonicity assumptions, we have that $\xi_{ih}(p, p \cdot \omega_i)$ tends to $+\infty$ for $p_h \to 0$, so that for a study of equilibria we may restrict attention to prices in the open set $\mathbb{R}_{++}^l$. We shall be interested in all the economies which arise when the initial endowments $(\omega_1, \ldots, \omega_m)$ are sampled from $(\mathbb{R}_{++}^l)^m$. Each such choice of endowments gives rise to an economy $\mathcal{E}[(\omega_1, \ldots, \omega_m)]$ for which, under our standard assumptions, the set $W(\mathcal{E}[(\omega_1, \ldots, \omega_m)])$ of Walras equilibria is non-empty. Its cardinality (the number of different Walras equilibria) may vary from 1 to $+\infty$, but it turns out that we can

> **Box 2. Gross substitution in the Edgeworth box:** The assumption can be illustrated using *offer curves,* which show how the demand of the consumer changes when the budget line revolves around the point given by the initial endowment.
>
> Geometrically, gross substitution at the individual level implies that the offer curve of the consumer must have a particular form; in Fig. 2 it slopes downward for agent 1 and upwards for agent 2.
>
>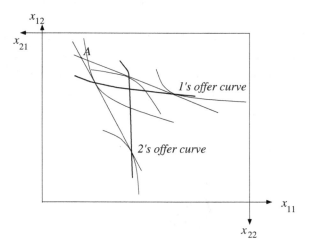
>
> Fig. 2. An economy having the gross substitution property.
>
> It is easily checked that this is not a general shape of offer curves.

exclude the pathological case of infinitely many equilibria, at least if we are willing to discard a small set of "unfortunate" initial distributions of endowments. The key to this is a mathematical result known as Sard's theorem, here given in a simple version adapted to our needs.

Theorem 6.2 (Sard, 1942). *Let $f : U \to \mathbb{R}^n$ be a C^1 map from an open set $U \subset \mathbb{R}^n$ to $\mathbb{R}^n$. Then the set of critical values of f, that is the values $y = f(x)$ for some $x \in U$ such that $Df(a)$ is singular, has Lebesgue measure 0.*

180   *Theory of General Economic Equilibrium*

A simple proof of Sard's theorem can be found in texts on differentiable topology e.g. Milnor and Weaver (1997). It should be noted that the result characterizes the critical *values*, that is the values taken by the functions in points where the Jacobian is singular, not the points themselves (if f is a constant function, then all points in U are critical but there is only one critical value). In our present context, we make use of Sard's theorem and obtain the following result. The proof of Theorem 6.3 follows that given in Debreu (1970).

Theorem 6.3. *Let $\xi_i : \Delta \times \mathbb{R}_{++} \to \mathbb{R}^l$, $i = 1,\ldots,m$, be C^1 demand functions satisfying $\xi_i(p, p \cdot \omega_i) = p \cdot \omega_i$ for all $(p, \omega_i) \in \Delta \times \mathbb{R}^l_{++}$. Then the set of $(\omega_1,\ldots,\omega_m)$ such that $W(\mathcal{E}[(\omega_1,\ldots,\omega_m)])$ has finitely many elements contains an open subset of $\mathbb{R}^l_+$, the complement of which has Lebesgue measure 0.*

Proof. Define the map $F : \mathrm{int}\,\Delta \times \mathbb{R}_{++} \times \left(\mathbb{R}^l_{++}\right)^{m-1} \to \left(\mathbb{R}^l_{++}\right)^m$ by

$$F_i(p, w_1, \omega_2, \ldots, \omega_m)$$

$$= \begin{cases} \xi_1(p, w_1) + \displaystyle\sum_{i=2}^{m} \xi_i(p, p \cdot \omega_i) - \sum_{i=2}^{m} \omega_i & i = 1, \\ \omega_i & i = 2,\ldots, m. \end{cases} \tag{1}$$

We have that $p \cdot F_1(p, w_1, \omega_2, \ldots, \omega_m) = w_1$ for each $(p, w_1, \omega_2, \omega_2, \ldots, \omega_m)$, and p is an equilibrium price if and only if $F(p, w_1, \omega_2, \ldots, \omega_m) = (\omega_1, \ldots, \omega_m)$, so that there is a one-to-one correspondence between $F^{-1}(\omega_1, \ldots, \omega_m)$ and $W(\mathcal{E}[(\omega_1, \ldots, \omega_m)])$. $\qquad\square$

Let C be the set of critical values of F. By Sard's theorem, this set has Lebesgue measure 0. We need some additional properties in order to connect the critical values $(\omega_1, \ldots, \omega_m)$ of F with finiteness of $F(\omega_1, \ldots, \omega_m)$. We begin by showing that $F^{-1}(K)$ for any compact subset K of $\mathbb{R}^{lm}_{++}$. Let $(p^n, w_1^n, \omega_2^n, \ldots, \omega_m^n)_{n=1}^{\infty}$ be a sequence in $F^{-1}(K)$. Let $(\omega_1^n, \ldots, \omega_m^n) = F(p^n, w_1^n, \omega_2, \ldots, \omega_m)$ and consider the sequence $(p^n, w_1^n, \omega_1^n, \omega_2^n, \ldots, \omega_m^n)_{n=1}^{\infty}$. This sequence has a subsequence converging to some $(p^0, w_1^0, \omega_1^0, \omega_2^0, \ldots, \omega_m^n) \in \Delta \times \mathbb{R}_+ \times K$. Here, p^0 must belong to $\mathrm{int}\,\Delta$, since otherwise, $\|\xi_1(p^n, w_1^n)\|$ would tend to infinity, a contradiction since $\xi_i(p^n, w_1^n) \le \sum_{i=1}^n w_i^h$ and the right-hand side is bounded. Since $p^0 \cdot \omega_1^0 = w_1^0$, we also have that $w_1^0 > 0$. It then follows that $(p^n, w_1^n, \omega_2^n, \ldots, \omega_m^n)$ converges to $(p^0, w_1^0, \omega_2^0, \ldots, \omega_m^0)$, so that $F(p^0, w_1^0, \omega_2^0, \ldots, \omega_m^0) = (\omega_1 =, \ldots, \omega_n^0)$, so that the limit belongs to $F^{-1}(K)$.

Now, let $B \subset \text{int}\,\Delta \times \mathbb{R}_{++} \times \left(\mathbb{R}_{++}^l\right)^{m-1}$ be relatively closed, and choose a sequence $(p^n, w_1^n, \omega_2^n, \ldots, \omega_m^n)_{n=1}^\infty$ in $F(B) \cap \mathbb{R}_{++}^{lm}$ converging to some $(\omega_1^0, \ldots, \omega_m^0) \in \mathbb{R}_{++}^{lm}$. The set K consisting of the elements of this sequence and its limit is compact, so that if (p^n, w_1^n) are such that $F_1(p^n, w_1^n, \omega_1^n, \ldots, \omega_m^n) = \omega_1^n$, then the sequence $(p^n, w_1^n, \omega_2^n, \ldots, \omega_m^n)_{n=1}^\infty$ has elements in $F^{-1}(K)$, which is compact by our reasoning above, so a subsequence converges to some $(p^0, w_1^0, \omega_2^0, \ldots, \omega_m^0) \in F^{-1}(K)$. Clearly, $(p^0, w_1^0, \omega_2^0, \ldots, \omega_m^0)$ is in B, and by continuity of F, $(\omega_1^0, \ldots, \omega_m^0) \in F(B)$.

The set of critical points of F, that is the points $(p, w_1, \omega_2, \ldots, \omega_m)$, for which the determinant of the Jacobian matrix, $\det DF(p, w_1, \omega_2, \ldots, \omega_m)$, is equal to zero, is closed, and it follows from Theorem 6.2 that its image, the set of critical values of F, is closed relative to $\mathbb{R}_{++}^l$.

Let $(\omega_1, \ldots, \omega_m)$ be a regular value of F. Then $F^{-1}(\omega_1, \ldots, \omega_m)$ is compact, and for each point $(p, w_1, \omega_2, \ldots, \omega_m)$ in this set, there is an open neighborhood such that DF is regular, so that $(p, w_1, \omega_2, \ldots, \omega_m)$ must be the only element of $F^{-1}(\omega_1, \ldots, \omega_m)$ in this neighborhood. It follows that $F^{-1}(\omega_1, \ldots, \omega_m)$ is a finite set. $\square$

The result reveals that if initial endowments are chosen at random, then the unfortunate case of infinitely many equilibria is an event which has zero probability, so that generally speaking it can be ignored. While the result does not tell us anything about the actual number of equilibria, it ascertains that it makes sense speaking about such a number.

1.3 *Finiteness of equilibria with non-ordered preferences*

Since we have considered consumer preferences in the general form of correspondences attaching a preferred set to each bundle, it seems somewhat unsatisfactory to assume that preferences are represented by utility functions and consequently complete and transitive. In this section, we consider a version where no use is made of demand functions and where consumer preferences are not given by utility functions. For another approach, see, e.g. Bonnisseau (2003).

For simplicity, it is assumed that the consumption set of each consumer is $\mathbb{R}^l_+$. We want to introduce a notion of smoothness for preference correspondences. Let $P_i : \mathbb{R}^l_+ \rightrightarrows \mathbb{R}^l_+$ be the preference correspondence of consumer i, which has the standard properties that for all $x \in \mathbb{R}^l_+$, $P_i(x)$ is non-empty, open, and does not contain x. Then P_i is said to be smooth if there is a C^1 function $g_i : \mathbb{R}^l_+ \to \mathbb{R}^l$, such that for all $x \in \mathbb{R}^l_+$,

$$ P_i(x) \in \{x' \in \mathbb{R}^l_+ \mid g_i(x) \cdot (x' - x) > 0\}. $$

The function g_i is a (generalized) gradient function for P_i. If P_i is representable by a utility function, then g_i is the usual gradient function, $g_i(x) = \nabla u_i(x)$ with $g_{ih}(x) = \frac{\partial g_i}{\partial x_h}(x)$.

We shall need a non-degeneracy assumption on g_i: For any non-trivial coordinate subspace α of $\mathbb{R}^l$ and any $x \in \alpha \cap \mathbb{R}^l_+$, the restriction $g_i(x)\|_\alpha$ of the linear form g_i to the subspace α is non-trivial in the sense that $g_i(x)|_\alpha \neq 0$.

We endow the space $G = C^1\left(\mathbb{R}^l_+, \mathbb{R}^l\right)$ of C^1 functions from $\mathbb{R}^l_+ \to \mathbb{R}^l$ with the Whitney C^1 topology[1]. Now, our space of economies is the set

$$ \mathbb{E} = (G)^m \times \left(\mathbb{R}^l_+\right)^m. $$

An element $(g_1, \ldots, g_m, \omega_1, \ldots, \omega_m)$ of $\mathbb{E}$ is written as (g, ω) with $g \in G^m$, $\omega \in \mathbb{R}^{lm}_+$.

Let $(g, \omega) \in \mathbb{E}$ be an economy. An allocation–price pair in (g, ω) is a point $(x, p) \in \left(\mathbb{R}^l_+\right)^m \times S^{l-1}_+$ with $x = (x_1, \ldots, x_m)$, where $S^{l-1}_+ = \{x \in$

---

[1]Choose a family $(K_n)^\infty_{n=1}$ of compact sets covering $\mathbb{R}^l_+$, then for each $f \in G$, a neighborhood basis at f for the Whitney C^1 topology is given by the sets $\{g \mid \forall x \in K_i, \|g(x) - f(x)\| < \varepsilon, \|Df(x) - Dg(x)\| < \varepsilon\}$, cf., e.g. Hirsch (2012).

$\mathbb{R}_+^l \mid \|p\| = 1 \}$ is the space of prices. For (x, p) to be an equilibrium of (g, ω), we would demand that

(i) $\sum_{i=1}^m x_i = \sum_{i=1}^m \omega_i$,
(ii) for each i, $p \cdot x_i = p \cdot \omega_i$ and $p \cdot x_i' > p \cdot \omega_i$ for all $x_i' \in P_i(x_i)$.

The simplest approach would be to demand that for each i, the gradient $g_i(x_i)$ should be proportional to the price vector, but this condition may not be satisfied if $x_i \in \mathrm{bd}\mathbb{R}_+^l$ but still is maximal for P_i, so we take another, more roundabout, approach.

For this, we shall consider equilibria in coordinate subspaces of $\mathbb{R}^l$. Such equilibria (x, p) occur if x_i belongs to the relative interior of $\mathbb{R}_+^l \cap \alpha_i$, where α_i is a non-trivial coordinate subspace of $\mathbb{R}^l$ (possibly identical to $\mathbb{R}^l$ itself), and $p|_{\alpha_i} = \lambda_i g_i(x_i)|_{\alpha_i}$ for some $\lambda_i > 0$, so that $p \cdot (x_i' - x_i) > 0$ for all $x_i' \in P_i(x_i) \cap \alpha_i$. This does not guarantee that x_i is maximal for P_i in the budget set, so that (x, p) may not be a genuine equilibrium, but since we want a result about finiteness of equilibria, we can accept pairs (x, p) in our extended set of equilibria which are not equilibria in the usual sense.

Definition 6.1. A pair (x, p) is an α-equilibrium in (g, ω) (with $\alpha = (\alpha_1, \ldots, \alpha_m)$, α_i a non-trivial coordinate subspace of $\mathbb{R}^l$) if $x_i \in \alpha_i$ and

(i) $\sum_{i=1}^m x_i = \sum_{i=1}^m \omega_i$,
(ii) for each i, $p \cdot x_i = p \cdot \omega_i$ and $p|_{\alpha_i} = \lambda_i g_i(x_i)|_{\alpha_i}$, $\lambda_i > 0$.

We denote the set of α-equilibria in (g, ω) by $W_\alpha(g, \omega)$. The set $W_{\text{ext}}(g, \omega) = \cup_\alpha W_\alpha(g, \omega)$, where the union is taken over the finite set of all $(\alpha_1, \ldots, \alpha_m)$, is called the set of extended equilibria.

Now we can formulate a result about finiteness of equilibria in this more general setting.

Theorem 6.4. *There is an open, dense subset O of $\mathbb{E}$ such that*

(1) *$W_{\text{ext}}(g, \omega)$ is a finite set for all $(g, \omega) \in O$,*
(2) *The equilibrium correspondence taking each (g, ω) to $W_{\text{ext}}(g, \omega)$ is continuous on O.*

184   *Theory of General Economic Equilibrium*

Proof. Let $(\alpha_1, \ldots, \alpha_m)$ be given. We show that there is $O_\alpha \subset \mathbb{E}$ open and dense such that $W_\alpha(g, \omega)$ is finite and varies continuously for $(g, \omega) \in O_\alpha$.

To this effect, consider the map $\rho^\alpha_{(g,\omega)} : X \to Y$, where

$$X = (\mathbb{R}^l_+ \cap \alpha_1) \times \cdots \times (\mathbb{R}^l_+ \cap \alpha_m) \times S^+$$
$$Y = (S \cap \alpha_1) \times \cdots \times (S \cap \alpha_m) \times S^+ \times \mathbb{R}^{m-1} \times \mathbb{R}^l,$$

defined by

$$\rho^\alpha_{(g,\omega)}(x, p) = \left(\frac{g_1(x_1)}{\|g_1(x_1)\|}\bigg|_{\alpha_1}, \ldots, \frac{g_m(x_m)}{\|g_m(x_m)\|}\bigg|_{\alpha_m}, p, \right.$$
$$\left. p \cdot (x_1 - \omega_1), \ldots, p \cdot (x_{m-1} - \omega_{m-1}), \sum_{i=1}^m (x_i - \omega_i) \right).$$

Clearly, $\rho^\alpha_{(g,\omega)}$ is C^1 for every $(g, \omega) \in \mathbb{E}$.

Define $W \subset Y$ by

$$W = \{ (y_1, \ldots, y_m, p, 0, \ldots, 0, 0) \mid \lambda y_i = p|_{\alpha_i} \text{ for some } \lambda_i > 0,$$
$$i = 1, \ldots, m \}.$$

Then W is a closed submanifold of Y, and $\left(\rho^\alpha_{(g,\omega)} \right)^{-1}(W) = W_\alpha(g, \omega)$. Let O_α be the set of $(g, \omega) \in \mathbb{E}$ for which $\rho^\alpha_{(g,\omega)} \pitchfork W \rho^\alpha_{(g,\omega)}$ is *transversal*[2] to W, written as $\rho^\alpha_{(g,\omega)} \pitchfork W$. We show that O_α is open and dense in $\mathbb{E}$. Openness is obvious. To prove density, fix a point $(\overline{g}, \overline{\omega}) \in \mathbb{E}$ and a neighborhood U of $(\overline{g}, \overline{\omega})$. By the definition of the topology on $\mathbb{E}$, there are m positive C^1 functions $f_1, \ldots, f_m$ and m positive real

---

    [2]Two linear subspaces V_1 and V_2 of a vector space V are *transversal* if $V_1 + V_2 = V$.

    Two submanifolds L and M of a manifold N *intersect transversally*, written $L \pitchfork M$, if the tangent spaces at every $x \in L \cap N$ are transversal, $T_x L + T_x M = T_x N$.

    A differentiable map $f : X \to Y$ is transverse to a submanifold Z of Y, written $f \pitchfork Z$, if at every $x \in f^{-1}(Z)$, the image of $T_x X$ by Df_x intersects $T_{f(x)} Z$ transversally, $Df_x(T_x X) + T_{f(x)} Z = T_{f(x)} Y$.

numbers $\varepsilon_1, \ldots, \varepsilon_m$ such that $(g, \omega) \in U$ whenever

$$\|g_i(x_i) - \overline{g}_i(x_i)\| < f_i(x_i), \|Dg_i(x_i) - D\overline{g}_i(x_i)\| < f_i(x_i), \|\omega_i - \overline{\omega}_i\| < \varepsilon_i$$

for all $x_i \in \mathbb{R}_+^l$, $i = 1, \ldots, m$. Let A be the set of m-tuples of vectors $(a_1, \ldots, a_m)$ with $a_i \in \mathbb{R}^l$, each i. The map $\phi : A \times \left(\mathbb{R}_+^l\right)^m \times X \to Y$ given by

$$\phi(a_1, \ldots, a_m, \omega_1, \ldots, \omega_m, x, p) = \rho_{(h,\omega)}^\alpha(x, p),$$

where $h_i = \overline{g}_i(x_i) + f_i(x_i)a_i$, $i = 1, \ldots, m$, is C^1. Moreover, it is easily seen that $\phi \pitchfork W$. Then, by the transversal density theorem (see, e.g. Abraham and Robbin, 1967, Theorem 19.1), the set of all $(a_1, \ldots, a_m, \omega_1, \ldots, \omega_m)$ for which the map $\rho_{(h,\omega)}^\alpha$ is transversal to W is dense in $A \times \left(\mathbb{R}_+^l\right)$. But this means that there is $(g, \omega) \in U$ with $\rho_{(g,\omega)}^\alpha \pitchfork W$.

Let $(g, \omega) \in O_\alpha$. Then $W_\alpha(g, \omega)$ is a submanifold of X of dimension equal to $\dim X - \operatorname{codim} W = 0$, i.e. a discrete set, and since it is contained in a compact subset of $\left(\mathbb{R}_+^l\right)^m \times S^+$, it is finite. Moreover, an application of the implicit function theorem shows that there is a neighborhood V of (g, ω) and a finite number, say k, of continuous functions γ_i, $i = 1, \ldots, k$, such that $W_\alpha(g', \omega') = \cup_{i=1}^k \gamma_i(g', \omega')$ for $(g', \omega') \in V$.

To complete the proof of the theorem, we note that finiteness of $W_\alpha(g, \omega)$ for all α implies finiteness of $W_{\text{ext}}(g, \omega)$ on $O = \cap_\alpha O_\alpha$, which again is open and dense in $\mathbb{E}$. This is part (1) of the theorem. The continuity part follows similarly except for the set of economies (g, ω), for which $W_\alpha(g, \omega) \cap W_\beta(g, \omega) \neq \emptyset$ for some $\alpha \neq \beta$. As this happens on the complement of an open, dense set of economies, we have proved part (2). $\qquad\square$

As was mentioned earlier, the result about extended equilibria derived earlier can be reformulated as one about usual equilibria. If $(x, p) \in W_{\text{ext}}(g, \omega)$ and $x_i \in \mathbb{R}_{++}^l$, each i, then (x, p) is a usual equilibrium. If $x_i \in \alpha_i$, $i = 1, \ldots, m$, for some m-tuple of non-trivial coordinate subspaces $\alpha = (\alpha_1, \ldots, \alpha_m)$, then (x, p) may or may not be

186 *Theory of General Economic Equilibrium*

a usual equilibrium, but whatever might be the case, the situation remains stable in a neighborhood of $(g, \omega) \in O$, unless we have that $W_\alpha(g, \omega) = W_\beta(g, \omega) \neq \emptyset$ for some $\alpha \neq \beta$. However, this case was already ruled out earlier.

2. Stability

A problem that is somewhat related to the comparative-statics exercises, is stated as follows: Suppose that the economy in question has one or two Walras equilibria: how are these equilibria actually reached? Considerations of this problem have a long tradition in economic theory, having been discussed already in Walras (1874).

2.1 *The tâtonnement process*

Suppose that (in a given economy) a positive price vector $p = (p_1, \ldots, p_l)$ has been chosen. If p is not an equilibrium price, there must be positive excess demand for some commodity k, that is

$$\zeta_k(p) > 0,$$

and since Walras' law holds for any price, irrespective of whether it is an equilibrium or not, we have

$$\zeta_h(p) < 0$$

for some other commodity h.

Since at the price p there is excess demand for some commodities and excess supply of others, a reasonable proposal for what would happen next is that the price would rise for commodities in excess demand and fall for those in excess supply. This gives us an adjustment process for prices, which may be formalized as

$$\frac{\mathrm{d}p_k}{\mathrm{d}t} = a_k \zeta_k(p(t)), \quad k = 1, \ldots, l, \tag{2}$$

where $a_k > 0$ is a constant. Thus, at any time t, the price of commodity k is increased (decreased) if the excess demand is positive (negative), and the constants a_k determine the speed of adjustment.

This process is called a *tâtonnement process*. A somewhat primitive visualization of what is going on is the following: The consumers are gathered in a large room, and a market agent (the celebrated Walrasian "auctioneer") quotes prices for all commodities. Given the quoted prices, consumers calculate their excess demands and submit them to the market agent, who now adjusts the price quotations following the tâtonnement process. This goes on as long as excess demands do not sum to zero. When this happens, the process terminates, the realized prices are equilibrium prices, and trade takes place at these prices.

Note that no trade is allowed *before* equilibrium prices have been arrived at (otherwise, the excess demand function ζ would change during the process). This feature of the tâtonnement process is somewhat unappealing, and it may be preferable to study *non-tâtonnement* processes where such trades are permitted. On the other hand, such processes are considerably more complex.

From a formal point of view, the tâtonnement process is a system of differential equations or a *dynamical system*, and a solution to (2) is a family of functions $p(t)$ describing the adjustment of prices from their initial value $p(0) = (p_1(0), \ldots, p_l(0))$. The central problem is to determine whether the tâtonnement process (2) is *stable*, that is whether for any initial price vector $p(0)$ the solution $p(t)$ will converge to an equilibrium price as t goes to infinity.

We shall have to be more precise on the meaning of "stability" of the price adjustment process (2), or rather, of the equilibria of the process, those p for which the right-hand side of (2) equals zero. We say that an equilibrium (price) is *globally asymptotically stable* if for all choices of initial values $p(0) = (p_1(0), \ldots, p_l(0))$, the prices $p(t)$ will tend to the equilibrium price vector as t goes to infinity.

As was perhaps to be expected in view of the preceding results (global asymptotical), stability cannot be established as a property holding for a broad class of economies. Indeed, it is possible to exhibit excess demand functions such that the tâtonnement process will result in very strange behavior of prices. Only if we are willing to accept additional (and usually quite strong) assumptions on the excess demand can we get positive results.

The property of gross substitution mentioned earlier would also suffice in the present context; here, we shall use another kind of condition on the overall behavior of the economy: We assume that the excess demand satisfies the *strong axiom of revealed preferences*. In the present context, this means that for all price vectors p^1, p^2 we have

$$\left[p^1 \cdot \zeta(p^2) \le 0, \zeta(p^1) \ne \zeta(p^2) \right] \Rightarrow p^2 \cdot \zeta(p^1) > 0$$

(if the net trades performed at the price system p^2 could have been made on the market at prices p^1, and if excess demand differs in the two situations, then it must be the case that the net trades made at prices p^1 are no longer feasible at prices p^2). It should be noted that the strong axiom on the community excess demand is *not* fulfiled as a consequence of the fact that it holds for individual excess demand functions.

Before we state and prove our stability result, we must take care of a detail concerning the normalization of the price vector: If an initial price $p(0)$ has been chosen, and $p(t)$ is given by (2), then from

$$\frac{d}{dt}\left(\sum_{k=1}^{l} \frac{1}{a_k}(p_k(t))^2 \right) = \sum_{k=1}^{l} \frac{2}{a_h}p_k(t)\frac{dp_h(t)}{dt} = 2\sum_{k=1}^{l} p_k(t)\zeta_k(p(t)) = 0$$

where the last equality follows from Walras' law, we get that the quantity

$$\sum_{k=1}^{l} \frac{1}{a_k}(p_k(t))^2$$

must be constant throughout the process. It follows that if we want $p(t)$ to converge to some p^0, then we must at least make sure that this weighted sum of individual prices is correct from the outset, i.e. that $p(0)$ is chosen from

$$\Delta(p^0) = \left\{ p \in \mathbb{R}^l_{++} \middle| \sum_{k=1}^l \frac{1}{a_k}(p_k)^2 = \sum_{k=1}^l \frac{1}{a_k}(p_k^0)^2 \right\}.$$

This is no restriction from the economic point of view; normalizing prices in one way or another has no effect on the choices of the agents in the economy.

Theorem 6.5. *Let $\mathcal{E} = (X_i, P_i, \omega_i)_{i=1}^m$ be an economy where the excess demand function ζ satisfies the strong axiom of revealed preferences, and where there is a unique (except for normalization) equilibrium price p^0. Then p^0 is globally asymptotically stable for the tâtonnement process, in the sense that for every choice of initial price system $p(0) \in \Delta(p^0)$, the solution $p(t)$ satisfies $p(t) \to p^0$ as $t \to \infty$.*

Proof. From standard results on differential equations, we get that (2) has solutions defined on all of int $\mathbb{R}^l_+$. Let $p(0)$ be chosen arbitrarily in $\Delta(p^0)$. We show that $p(t) \to p^0$ for $t \to \infty$.

Define the function $L : \mathbb{R}_+ \to \mathbb{R}_+$ by

$$L(t) = \sum_{k=1}^l \frac{1}{a_k}\left(p_k(t) - p_k^0\right)^2$$

(this is the so-called Lyapunov function for the problem at hand). We have that

$$\frac{dL(t)}{dt} = 2\sum_{k=1}^l \frac{1}{a_k}\left(p_k(t) - p_k^0\right)\frac{dp_k(t)}{dt}$$

$$= 2\sum_{k=1}^l (p_k(t) - p_k^0)\zeta_k(p(t)) = -2p^0 \cdot \zeta(p(t)),$$

where we have used Walras' law $p(t) \cdot \zeta(p(t)) = 0$. Now we have for each t that $p(t) \cdot \zeta(p^0) = 0$ (since p^0 is an equilibrium price), so

$p(t) \neq p^0$ implies that $p^0 \cdot \zeta(p(t)) > 0$ by the strong axiom. It follows that the function L is a decreasing function of t.

Since L is non-negative and decreasing, it must tend to a limit $L^0 \geq 0$. If $L^0 > 0$, then we can extract from the sequence $p(1), p(2), p(3), \ldots$ (whose elements all belong to the compact set $\mathrm{cl}\Delta(p^0)$) a subsequence converging to some $p^* \neq p^0$. For t such that $p(t)$ is close to p^*, we must have that $dL(t)/dt$ is close to $-2p^0 \cdot \zeta(p^*)$, which is negative. But L cannot converge to some $L^0 > 0$ without its derivative converging to zero, and we have a contradiction. We conclude that $L^0 = 0$; by the definition of L, this means that $p(t) \rightarrow p^0$ as $t \rightarrow \infty$. $\qquad\square$

3. Excess Demand Functions

The preceding short accounts of the problems of uniqueness, comparative statics and stability of Walras equilibria have indicated that stronger results in equilibrium theory depend on additional properties of the excess demand function ζ of the economy. One such property was gross substitutability, which would entail uniqueness of equilibria; another one was that ζ satisfies the strong axiom of revealed preferences, in which case price adjustment according to the tâtonnement process is stable. Both these assumptions are rather restrictive compared with the standard assumptions of our theory, and certainly, there are many reasonable situations where they are simply not fulfiled. The question arises, therefore, whether excess demand functions in general have additional properties which we have managed to overlook up to this point, and which would help us in deriving further results.

Clearly, if ζ is an excess demand function for an economy satisfying our standard assumptions, then we know that

  (i) ζ is continuous,
 (ii) ζ is homogeneous of degree zero,
(iii) ζ satisfies Walras' law, i.e. $p \cdot \zeta(p) = 0$ for all p.

It turns out that these are the only properties universally held by all excess demand functions.

Uniqueness and Stability of Equilibria

Theorem 6.6. *Let $f : \Delta \to \mathbb{R}^l$ be a continuous function (where, as usual, $\Delta = \left\{ p \in \mathbb{R}^l \mid \sum_{h=1}^{l} p_h = 1 \right\}$) such that $p \cdot f(p) = 0$ for all $p \in \Delta$, and let $\varepsilon > 0$. Then for all m with $m \geq l$, there exists an economy $\mathcal{E} = (X_i, P_i, \omega_i)_{i=1}^{m}$ with m consumers satisfying Assumptions 0.1 and 0.2 such that*

$$\zeta(p) = f(p)$$

for all $p \in \Delta$ with $p_h > \varepsilon, h = 1, \ldots, l$.

This remarkable result is due to Debreu (1974) who gave a constructive proof, improving a first result due to Sonnenschein (1972). We prove the result under an additional assumption, namely that f is C^2 on all of Δ (meaning that there exists a function f' defined on an open subset of $\left\{ x \in \mathbb{R}^l \mid \sum_{k=1}^{l} x_k = 1 \right\}$ which is C^2 and whose restriction to Δ is f). This proof is due to Mantel (1976).

Proof (for the differentiable case). Choose l linearly independent vectors $\omega_1, \ldots, \omega_l$ in $\text{int}\mathbb{R}^l_+$, and let A be an $l \times l$ matrix with full rank, such that all column sums are equal to 1. We let W be the matrix whose columns are the vectors $\omega_j, j = 1, \ldots, l$, and we set $B = AW^t$ (where t denotes transposition), $C = (W^t)^{-1}$. Finally, we define $X = \{ x \in \mathbb{R}^l \mid x = By \text{ for some } y \in \mathbb{R}^l_+ \}$.

For $i = 1, \ldots, m$, we define the ith indirect utility function $v^i : P \to \mathbb{R}$, where $P = \{ p \in \Delta \mid p_k > \varepsilon, k = 1, \ldots, l \}$, by

$$v^i(p) = \frac{1}{K} c_i \cdot f(p) - a_i \cdot \log(Bp),$$

where log of a vector means the vector of the logs of its coordinates, and c_i is the ith column of C.

By our choice of ω_i, we have $p \cdot \omega_i \geq 0$ for all $p \in P$. Since W is regular, we have $W^t p \neq 0$, and therefore, we must have that $Bp = AW^t p$ is positive in all coordinates. It follows that the second term in the definition of v^i is strictly convex on P. The second-order partial derivatives of f are bounded on P. Consequently, by choosing K to be large enough, we may achieve that v^i is strictly convex on P (since the first term will be insignificant compared with the second).

192 — Theory of General Economic Equilibrium

Let the endowments of agent i be $K\omega_i$, for $i = 1, \ldots, m$. Now the demand function is given by

$$\xi_i(p, p \cdot K\omega_i) = -(Kp \cdot \omega_i)v_p^i,$$

where v_p^i is the gradient of v^i, the vector with coordinates $\partial v^i / \partial p_k$. Using the definition of v^i, we get that

$$\xi_i(p, p \cdot K\omega_i) = Kp \cdot \omega_i \left[-\frac{1}{K} D f_p(p) c_i + B^t \mathrm{diag}(Bp)^{-1} a_i \right],$$

where $\mathrm{diag}(Bp)$ is the diagonal matrix which has the jth coordinate of Bp in position (j, j). It is seen that the set of vectors $B^t \mathrm{diag}(Bp)^{-1} a_i$ for $p \in P$ is compact and contained in the interior of X. Consequently, X may be taken as a consumption set of each consumer for sufficiently large K.

The demand functions ξ_i, $i = 1, \ldots, m$, give the right excess demand function We have

$$
\begin{aligned}
\sum_{i=1}^{l} \xi_i(p, p \cdot K\omega_i) &= -D f_p(p) C W^t p + K B^t \mathrm{diag}(Bp)^{-1} A W^t p \\
&= -D f_p(p) p + K B^t \mathrm{diag}(Bp)^{-1} Bp = f(p) + KBe \\
&= f(p) + kWA^t e \\
&= f(p) + KWe,
\end{aligned}
$$

where we used the relation $-D f_p(p) p + f(p) = 0$ for all p, a consequence of Walras' law, and also that $A^t e = e$ since the column sums of A are all 1 (recall that e is the diagonal vector with all coordinates equal to 1).

It remains only to state the utility functions of each consumer i, namely

$$u_i(x) = \min \{ v_i(p) \mid p \cdot x \le 1, p \in P \},$$

which give rise to preferences P_i satisfying Assumption 0.2. $\qquad\square$

The conclusion to be drawn from Theorem 6.6 is that every function satisfying the rather weak properties (i)–(iii) discussed earlier may occur as excess demand functions of some economy. This implies in its turn that no general results can be hoped for in the problems of stability, uniqueness, etc. Economies may very well have badly behaved excess demand functions.

This should not worry us too much; anyway, we have emphasized throughout that demand functions are an intermediate concept which is not indispensable in equilibrium theory, since all the main results can be obtained — and often easier — without reliance on demand functions.

3.1 *Global Newton methods*

Since no general results can be established for tâtonnement processes, and since we cannot hope for special properties of the excess demand function which would save the day, it seems reasonable to look for other approaches. An obvious candidate for a process which has found widespread use when solving equation systems such as the one concerning us here, finding a zero of the excess demand function

$$\zeta(p) = 0,$$

is the Global Newton approximation, whereby the solution is approached along a path satisfying the differential equation

$$D\zeta(p)\frac{dp}{dt} = -\lambda\zeta(p), \tag{3}$$

where λ is a parameter to be specified.

The inspiration comes from the Newton method for finding a zero of a function $f : \mathbb{R} \to \mathbb{R}$, where the solution is approached by a sequence $(x_n)_{n=0}^{\infty}$ beginning at an arbitrary choice x_0 and satisfying

$$x_{n+1} = x_n - \frac{f(x_n)}{f'(x_n)}, \quad n = 1, 2, \ldots.$$

Convergence of the sequence can be found under rather weak conditions on f; replacing the discrete approximation to zero by a differential path, one gets a generalized Newton of the form

$$\frac{dp}{dt} = -D\zeta(p)^{-1}\zeta(p).$$

The Global Newton method (3) has the advantage that it may work even when $D\zeta(p)$ is singular.

Returning to our case of excess demand, if (3) is rewritten as

$$\frac{d\zeta}{dt} = -\lambda\zeta,$$

it is seen that in this process we attempt to reduce ζ directly rather than working indirectly through the prices. When using it, one needs the Jacobian of ζ in each point in addition to *zeta* itself, and in this sense, it is more demanding than the tâtonnement process.

To see how the Global Newton should be defined and which properties it will have, we follow the approach of Varian (1977): We assume that ζ is defined on all of $\Delta = \{p \in \mathbb{R}_+^l \mid \sum_{h=1}^l p_h = 1\}$. Since Δ is diffeomorphic to the closed unit disk $D^{l-1} = \{x \in \mathbb{R}_+^{l-1} \mid \|x\| \leq 1\}$ (there is a bijective map h from Δ to D^{l-1} such that both h and h^{-1} are differentiable), we may as well consider η as defined on D^{l-1}.

Theorem 6.7. *Assume that the excess demand function ζ is differentiable on D^{l-1} and that for all $x \in$ bd D^{l-1}, there is no $\mu > 0$ such that*

$$\zeta(x) = \mu x. \tag{4}$$

Then for almost all $x \in$ bd D^{l-1}, there is a solution of (3) with initial point x which converges to an equilibrium.

Proof. For the moment, we shall make another assumption, seemingly much stronger than (4), namely that

$$\zeta(x) = -x \text{ for } x \in \text{bd } C^{l-1}. \tag{5}$$

Let E be the equilibria of ζ, the points $x \in D^{l-1}$ such that $\zeta(x) = 0$, and let $M = D^{l-1} \backslash E$. Then the map g from M to the unit sphere S^{l-1}

given by

$$g(x) = \frac{\zeta(x)}{\|\zeta(x)\|}$$

is well-defined. By Theorem 6.2, for almost every $s \in S^{l-1}$, the set $g^{-1}(s)$ is a one-dimensional differentiable manifold, which means that it is a finite union of closed curves and line segments which begin and end at the boundary of M. Each point in bd D^{l-1} belongs to $g^{-1}(e)$ for some e, and using (4), we have that at most one point of the boundary of D^{l-1} can belong to $g^{-1}(e)$. The path ending in bd D^{l-1} must therefore have its other endpoint in M.

Writing this path as $x(t)$, we have that it must satisfy the equation

$$\zeta(x(t)) = e \, \|\zeta(x(t))\| \,,$$

and differentiating with respect to t, we get

$$D\zeta(x(t))\frac{dx}{dt} = e\frac{d}{dt}\|\zeta(x(t))\| = \frac{\zeta(x)}{\|\zeta(x)\|}\frac{d}{dt}\|\zeta(x(t))\| = \zeta(x)\lambda(t), \quad (6)$$

with $\lambda(t) = \dfrac{1}{\|\zeta(x(t)\|}\dfrac{d}{dt}\|\zeta(x(t)\|$. The function $\lambda(t)$ indicates the speed with which the path is traversed, so its magnitude is not important, but the sign is, and if we want to move from the boundary of D^{l-1}, it should correspond to that of $-\det D\zeta(x)$. Then, starting at almost any point of bd D^{l-1} and moving along a solution to (6), we will end at an equilibrium.

To prove the general case, let $D_2^{l-1} = \left\{x \in \mathbb{R}^{l-1} \,|\, \|x\| \le 2\right\}$ be the disk of radius 2, and let $\gamma : D_2^{l-1} \to \mathbb{R}^l$ be defined by

$$\gamma(x) = \begin{cases} -\dfrac{(1 - \|x\|)x}{\|x\|} + (1 - (1 - \|x\|))\zeta\left(\dfrac{x}{\|x\|}\right) & 1 < \|x\| \le 2, \\ \zeta(x) & 0 \le \|x\| \le 1. \end{cases} \quad (7)$$

There are no new equilibria of γ: If $\bar{x} \in D_2^{l-1}\backslash D^{l-1}$ satisfies $\gamma(\bar{x}) = 0$, then from (7) we would get that

$$\zeta\left(\frac{\bar{x}}{\|\bar{x}\|}\right) = \frac{1 - \|\bar{x}\|}{1 - (1 - \|\bar{x}\|)}\frac{\bar{x}}{\|\bar{x}\|},$$

196   *Theory of General Economic Equilibrium*

meaning that there is a point x^0 on the boundary of D^{l-1} such that $\zeta(x^0)$ points radially outward contradicting our assumption.   □

4. Exercises

(1) Is the uniqueness of Walras equilibria obtained in an economy with a non-atomic measure space agents satisfying standard conditions of well-behavedness? Give a proof of uniqueness or a counterexample.

(2) The transfer paradox: Give an example, elaborating on the economy illustrated in Box 1, of an economy where a consumer may obtain a preferred final outcome giving away some amount of all commodities before engaging in the market.

In this example, with two commodities and two consumers, the possibility of one consumer gaining from transferring some of the endowment to the other consumers before trade entails the non-uniqueness of Walras equilibria at some other initial distribution of endowments. Does this hold for arbitrary numbers of commodities and consumers?

(3) Consider an economy with four commodities and two consumers, having preferences described by utility functions (Hart, 1975)

$$u_1(x_1, x_2, x_3, x_4) = 2^{\frac{3}{2}} \sqrt{x_1} + \sqrt{x_2} + 2^{\frac{3}{2}} \sqrt{x_3} + \sqrt{x_4},$$
$$u_2(x_1, x_2, x_3, x_4) = \sqrt{x_1} + 2^{\frac{3}{2}} \sqrt{x_2} + \sqrt{x_3} + 2^{\frac{3}{2}} \sqrt{x_4}.$$

Show that the economy has the gross substitution property.

(4) Show that the tâtonnement process (2) is stable when aggregate excess demand has the gross substitution property.

(5) Let $B = \{(p^1, z^1), \dots, (p^r, z^r)\}$ be a finite set of pairs $(p^k, z^k) \in \Delta \times \mathbb{R}^l$, interpreted as observations of prices and the corresponding vectors of aggregate excess demand. Assume that each of the pairs satisfies Walras' law, so that $p^k \cdot z^k = 0, k = 1, \dots, r$.

Show that there exists an economy $\mathcal{E} = (X_i, P_i, \omega_i)_{i=1}^m$ with $m \leq l + 1$ satisfying Assumptions 0.1 and 0.2 such that for each

k and $i = 1,\ldots,m$, there is x_i^k with $p^k \cdot x_i^k = p^k \cdot \omega_i = \inf\{p^k \cdot x_i' \mid x_i' \in P_i(x_i^k)\}$, such that $\sum_{i=1}^m (x_i^k - \omega_i) = z_i^k$, so that z^k is an aggregate excess demand for $\mathcal{E}$ at p^k, $k = 1,\ldots,r$.

(6) Consider the equation system $g(x) = 0$, where $g : \mathbb{R}^3 \to \mathbb{R}^3$ is given by

$$g_1(x_1, x_2, x_3) = x_1^2 + x_2^2 + x_3^2 - 3,$$

$$g_2(x_1, x_2, x_3) = x_1^2 + x_2^2 - x_3 - 1,$$

$$g_3(x_1, x_2, x_3) = x_1 + x_2 + x_3 - 3.$$

Show that Newton's method will fail if it starts at $x = (1, 0, 1)$.

Chapter 7

The Equilibrium Manifold and Probabilistic Equilibrium Theory

1. The Equilibrium Manifold

In this chapter, we return to the model considered in Chapter 6, Section 1.2, where economies were given by fixed demand functions ξ_i for $i = 1, \ldots, m$, whereas endowments $(\omega_1, \ldots, \omega_m)$ could take arbitrary values in $\mathbb{R}_{++}^{lm}$. We shall be interested in the structure of the set of equilibria when endowments are allowed to vary in all of $\mathbb{R}_{++}^{lm}$ (or in an open subset of $\mathbb{R}_{++}^{lm}$, which for the results to be obtained amounts to the same). This characterization of the set of equilibria, or the *equilibrium manifold,* as it is called in the present context, was introduced by Balasko (1975). The proof that follows is due to Schecter (1979).

1.1 *Definition of the equilibrium manifold*

We assume that the demand functions ξ_i are defined on $\mathring{\Delta} \times \mathbb{R}_{++}$, where $\mathring{\Delta} = \left\{ p \in \mathbb{R}_{++}^l \mid \sum_{h=1}^l p_h = 1 \right\}$ is the interior of Δ, with values in $\mathbb{R}_{++}^l$, that they are C^r for $r \geq 1$ and that they satisfy

$$p \cdot \xi_i(p, w_i) = w_i, \quad \text{all } (p, w_i) \in \mathring{\Delta} \times \mathbb{R}_{++}$$

for $i = 1, \ldots, m$. With the given demand functions, each array $\omega = (\omega_1, \ldots, \omega_m)$ of initial endowments defines an economy; an equilibrium is a pair (p, ω) such that $\sum_{i=1}^m \xi_i(p, p \cdot \omega_i) = \sum_{i=1}^m \omega_i$.

The *equilibrium manifold*[1] E is the subset of $\mathring{\Delta} \times \mathbb{R}^{lm}_{++}$ consisting of all the equilibria (p, ω).

Theorem 7.1. *The equilibrium manifold E is C^r-diffeomorphic to $\mathring{\Delta} \times \mathbb{R}^{lm}_{++}$.*

The proof of Theorem 7.1 relies on a general result about differentiable maps stated as a Lemma 7.1. We give a proof of the lemma in what follows.

Lemma 7.1. *Let A be an open set in $\mathbb{R}^d$, and let U be an open subset of $A \times \mathbb{R}^k$, such that*

(i) $A \times \{0\} \subset U$,
(ii) *for all $x \in A$, the set $(\{a\} \times \mathbb{R}^k) \cap U$ is convex,*
(iii) $\overline{U} \cap (\{a\} \times \mathbb{R}^k) = \overline{U \cap (\{a\} \times \mathbb{R}^k)}$.

Then U is C^∞-diffeomorphic to $A \times \mathbb{R}^k$.

Here and in the sequel, $\overline{B} = \operatorname{cl} B$ is the closure of the set B; we use this notation when in situations where it is more easily readable.

Proof of Theorem 7.1. Define the map $\varphi : \mathring{\Delta} \times \mathbb{R}^{lm}_{++} \to \mathring{\Delta} \times \mathbb{R}^m_{++}$ by

$$\varphi(p, \omega_1, \ldots, \omega_m) = (p, p \cdot \omega_1, \ldots, p \cdot \omega_m).$$

Then φ is differentiable, actually C^∞, and $D\varphi$ has full rank at each $(p, \omega) \in \mathring{\Delta} \times \mathbb{R}^{lm}_{++}$, so that $B = \varphi(\mathring{\Delta} \times \mathbb{R}^{lm}_{++})$ is an open subset of $\mathring{\Delta} \times \mathbb{R}^m_{++}$. Let $F : B \to \mathring{\Delta} \times \mathbb{R}^{lm}_{++}$ be defined by

$$F(p, w_1, \ldots, w_m) = (p, \xi_1(p, p \cdot \omega_1), \ldots, \xi_m(p, p \cdot \omega_m)).$$

Then F is C^r since each of the maps ξ_i is C^r, and $\varphi \circ F = \operatorname{Id}$, the identical mapping on $\mathring{\Delta} \times \mathbb{R}^{lm}_{++}$. Since $D\varphi$ is surjective, we get that DF is injective, meaning that F is an embedding of B in $\mathring{\Delta} \times \mathbb{R}^m_{++}$.

---

[1]To justify the terminology, we should check that E has the structure of a (differentiable) manifold. Since the main theorem states that E looks like $\mathring{\Delta} \times \mathbb{R}^{lm}_{++}$, which has a trivial manifold structure, we skip this step.

The Equilibrium Manifold and Probabilistic Equilibrium Theory   201

Let $\Gamma = \{ (p, w_1, \ldots, w_m, x_1, \ldots, x_m) \mid p \cdot x_i = 0, i = 1, \ldots, m, \sum_{i=1}^{m} x_i = 0 \}$, so that Γ is the product of B and an $(l-1)(m-1)$-dimensional subspace of $\mathbb{R}^{lm}$ and define $g : \Gamma \to \Delta$ by

$$g(p, w_1, \ldots, w_m, x_1, \ldots, x_m) = (p, \xi_1(p, w_1) + x_1, \ldots, \xi_m(p, w_m) + x_m).$$

Then g is again a C^r-embedding, and the restriction of g to $g^{-1}(\mathring{\Delta} \times \mathbb{R}_{++}^{lm})$ is a C^r diffeomorphism between the latter set and E.

Next, we note that Γ is (C^∞)-diffeomorphic to $B \times \mathbb{R}^{(l-1)(m-1)}$: Let

$$x_i(p, y_i) = \left(y_{i1}, \ldots, y_{i,l-1}, -\frac{1}{p_l}(p_1, \ldots, p_{l-1}) \cdot y_i \right)$$

and define $h : B \times \mathbb{R}^{(l-1)(m-1)} \to \Gamma$ by

$$h(p, w_1, \ldots, w_m, y_1, \ldots, y_m)$$

$$= \left(p, w_1, \ldots, w_m, x_1(p, y_1), \ldots, x_{m-1}(p, y_{m-1}), -\sum_{i=1}^{m-1} x_i \right).$$

The set $D = (g \circ h)^{-1}(\mathring{\Delta} \times \mathbb{R}_{++}^{lm})$ is open in $B \times \mathbb{R}^{(l-1)(m-1)}$, and $B \times \{0\}$ is open in D. Moreover, $g \circ h$ defines a C^r-diffeomorphism from D to E.

We check that the conditions of Lemma 7.1 are satisfied with $A = B$ and $k = lm$. Condition (i) holds since $B \times \{0\} \subset D$. For condition (ii), we note that g is affine and h is a convex function for any fixed (p, w), where $w = (w_1, \ldots, w_m)$, and therefore, $D \cap \left[\{(p, w)\} \times \mathbb{R}^{(l-1)(m-1)} \right]$ is convex. It remains to check condition (iii).

The inclusion $\overline{D \cap [\{(p, w)\} \times \mathbb{R}^{(l-1)(m-1)}]} \subset \overline{D} \cap [\{(p, w)\} \times \mathbb{R}^{(l-1)(m-1)}]$ is obvious. To obtain the converse inclusion, let $(p, w, x) \in \overline{D} \cap \left[\{(p, w)\} \times \mathbb{R}^{(l-1)(m-1)} \right]$, and choose the point y in the line segment $[0, x]$ which is closest to x with the property that $(p, x, y) \in \overline{D \cap [\{(p, w)\} \times \mathbb{R}^{(l-1)(m-1)}]}$. Suppose that y belongs to the interior of $[0, x]$, and consider the line segment $[(p, w, y), (p, w, x)]$ in $B \times \mathbb{R}^{(l-1)(m-1)}$. If G is an open, convex neighborhood of p, then $(g \circ h)([(p, w, y), (p, w, x)])$ does not intersect $G \times \mathbb{R}_{++}^{lm}$, but a neighborhood of $(g \circ h)(p, w, 0)$ is in $G \times \mathbb{R}_{++}^{lm}$ and also points arbitrarily close to $(g \circ h)(p, w, x)$ are in $G \times \mathbb{R}^{lm}_{++}$. But $G \times \mathbb{R}^{lm}_{++}$ is convex, and we obtain a contradiction, showing that $y = x$, so that the converse inclusion must hold.

We now apply Lemma 7.1 to obtain that D is C^∞-diffeomorphic to $B \times \mathbb{R}^{(l-1)(m-1)}$. Furthermore, another application of the lemma gives us that B is C^∞-diffeomorphic to $\mathring{\Delta} \times \mathbb{R}^m$. Indeed, $B = \varphi(\mathring{\Delta} \times \mathbb{R}^{ml})$, and since φ is linear for fixed p, B satisfies conditions (i) and (ii) of the lemma. Condition (iii) can be checked following the same line of argument as that given above. Since D is C^r-diffeomorphic to E, we get that E is C^r-diffeomorphic to $\mathring{\Delta} \times \mathbb{R}^{lm}_{++}$. $\qquad\square$

1.1.1 Proof of Lemma 7.1

In what follows, we give a proof of Lemma 7.1 used in the proof of Theorem 7.1: Let $\mathbb{S}^{d-1}$ be the unit sphere in $\mathbb{R}^d$, and let $\alpha(a, y) = \sup\{t \mid (a, ty) \in U\}$, so that $(a, ty) \in U$ whenever $t < \alpha(a, y)$. Using (ii) and (iii) we get that α is a continuous function on $A \times \mathbb{S}^{d-1}$.

We define a strictly increasing sequence $(\alpha_n)_{n=1}^\infty$ of C^∞ functions from $A \times \mathbb{S}^{d-1}$ to $\mathbb{R}_{++}$ converging pointwise to α: Let $\beta_n : A \times \mathbb{S}^{d-1} \to \mathbb{R}$ be given by

$$\beta_n(a, y) = \inf\left\{\frac{2^{n+1} - 1}{2^n}\alpha_n((a, y))\right\}.$$

Then the sequence $(\beta_n)_{n=1}^\infty$ is strictly increasing and converges pointwise to α, but each β_n is not necessarily C^∞. However, it can be approximated so closely by a C^∞ function α_n, for $n = 1, \ldots$, that the other properties remain satisfied.

Define α_0 as the constant function with $\alpha_0(a, y) = 0$ for all (a, y). We construct a map F which is patched together of maps F_n taking the set

$$\{(a, ty) \in U \mid \alpha_n(a, y) \leq t \leq \alpha_{n+1}(a, y)\}$$

surjectively to

$$A \times \{y \in \mathbb{R}^d \mid n \leq \|y\| \leq n + 1\},$$

for $n = 0, 1, \ldots$. To make sure that the maps F_n and F_{n+1} are compatible, we first construct a C^∞ function $\psi : X \to [0, 1]$, with $X = \{(s, t) \in \mathbb{R}^2 \mid s > 0, 0 \le t \le s\}$, such that

(a) $\frac{\partial \psi}{\partial t}(s, t) > 0$, all (s, t),
(b) $\psi(s, t) = t$ in a neighborhood of $X_0 = \{(s, t) \in X \mid s > 0, t = 0\}$,
(c) $\psi(s, t) = 1 - (s - t)$ in a neighborhood of $X_1 = \{(s, t) \in X \mid s > 0, t = 1\}$.

This is done as follows: For $m \in \mathbb{N}$, there are C^∞ functions $\rho_m : [0, 1] \to \mathbb{R}$ with the properties (1) $0 \le \rho_m(x) \le 1 + \frac{1}{n}$ for all $x \in [0, 1]$, (2) $\rho_m(x) = 0$ for all x in a neighborhood U_n of $\{0\} \cup \{1\}$, and (3) $\int_0^1 \rho_m(x)\,dx = 1$. Define $\sigma_m : X \to \mathbb{R}$ by

$$\sigma_m(s, t) = 1 + \frac{1 - s}{s} \rho_m \left(\frac{t}{s} \right).$$

Then $\sigma_m(s, t) = 0$ in a neighborhood of $X_0 \cup X_1$, and t $\sigma_m(s, t) > 0$ for $s \le m$.

Let $X(m) = \{(s, t) \in X \mid s \le m\}$, and define $\psi_m : X(m) \to \mathbb{R}$ by

$$\psi_m(s, t) = \int_0^t \sigma_n(s, u)\,du.$$

Then

$$\psi_m(s, s) = \int_0^s \left[1 + \frac{1 - s}{s} \rho_m \left(\frac{u}{s} \right) \right]$$

$$du = s + (1 - s) \int_0^s \rho_m \left(\frac{u}{s} \right) \frac{1}{s}\,du = s + (1 - s) = 1.$$

We then have that ψ_m satisfies (a) for all $(s, t) \in X(m)$, the image of ψ_m is all of $[0, 1]$, and ψ_m satisfies (c) on $X(m)$, so that it satisfies all the conditions except that it is defined on $X(m)$ instead of X, but using C^∞-partitions of unity, the maps ψ_m can be glued together to a function ψ with the desired properties.

Using now the constructed function ψ, we define F as follows: If $(a, ty) \in U$ with $\alpha_n(a, y) \le t \le \alpha_{n+1}(a, y)$, then

$$F(a, ty) = (a, (n + \psi(\alpha_{n+1}(a, y) - \alpha_n(a, y), t - \alpha_n(a, y)))y).$$

Then F is the identity map on a neighborhood of $A \times \{0\} \subset U$, and the restriction of F to the set $\{(a, ty) \in U \mid \alpha_n(a, y) \le t \le \alpha_{n+1}(a, y)\}$ is a C^∞-diffeomorphism onto $A \times \{y \in \mathbb{R}^d \mid n \le \|y\| \le n + 1\}$, for $n = 0, 1, \ldots$. Finally, since in a neighborhood of each set $\{(a, ty) \in U \mid t = \alpha_n(a, y)\}$, $n = 1, 2, \ldots$, F takes the form

$$(a, ty) \mapsto (a, (n + t - \alpha_n(a, y))y),$$

we obtain that F is a C^∞-diffeomorphism.

1.2 Properties of the equilibrium manifold

Having now established the equilibrium manifold as a differentiable manifold, we take a closer look at its structure.

1.2.1 The natural projection

Since the equilibrium manifold consists of pairs (p, ω) with p an equilibrium price in the economy defined by $\omega = (\omega_1, \ldots, \omega_m)$, we may separate the parameters from the price using the projection π with

$$\pi(p, \omega) = \omega$$

for all $(p, \omega) \in E$. The map π is called the *natural projection* associated with E.

We collect some properties of the natural projection in a lemma.

Lemma 7.2. *Let E be the equilibrium manifold and π its natural projection. Then the following holds:*

(a) *π is smooth,*
(b) *π is proper, i.e. $\pi^{-1}(K)$ is compact for every compact subset K of $\mathbb{R}^{lm}_{++}$.*

The Equilibrium Manifold and Probabilistic Equilibrium Theory   205

Proof. (a) follows from Theorem 7.1 since it can be written as a composition of the diffeomorphism between E and $\mathring{\Delta} \times \mathbb{R}^{lm}_{++}$ and the projection on the second component.

(b) Let K be a compact subset of $\mathbb{R}^{lm}_{++}$, then K is closed and bounded, and there is $\underline{\omega} \in \mathbb{R}^{l}_{++}$ such that $\underline{\omega} \le \omega_i$, $i = 1, \ldots, m$, and $\overline{\omega} \in \mathbb{R}^{l}_{++}$ with $\sum_{i=1}^{m} \omega_i \le \overline{\omega}$ for all $\omega = (\omega_1, \ldots, \omega_m) \in K$. We then have that

$$\sum_{i=1}^{m} \xi_i(p, p \cdot \omega_i) = \sum_{i=1}^{m} \omega_i \le \overline{\omega}$$

for all $(p, \omega) \in \pi^{-1}(K)$.

For each consumer i, the set $B_i = \{x_i \mid u_i(x_i) \ge u_i(\underline{\omega}), x_i \le \overline{\omega}\}$ is compact, and each (p, ω) in $\pi^{-1}(K)$ is such that

$$D^*u_1(x_1) = \cdots = D^*u_m(x_m) = p$$

for some $(x_1, \ldots, x_m) \in B_1 \times \cdots \times B_m$, where $D^*u_i(x_i)$ is the normalized gradient of u_i in the point x_i. Since D^*u_i is continuous, the set $D^*u_i(B_i)$ is compact, and since $\pi^{-1}(K)$ is a closed subset of $D^*u_1(B_1) \times \cdots \times D^*u_m(B_m)$, we get that $\pi^{-1}(K)$ is compact. $\qquad \square$

1.2.2 *Critical and regular equilibria*

Following the ideas laid out in Chapter 6, Section 1.2, we consider *regular values* ω of the natural projection π. An equilibrium $(p, \omega) \in E$ is said to be *critical* if it is a critical point of π, that is if $D\pi$ is singular. Otherwise, the equilibrium is *regular*. Now the set R of regular economies consist of all ω for which (p, ω) is a regular equilibrium. It is easily seen that the set R of regular economies is an open subset of E.

Theorem 7.2. *For each regular economy* $\omega \in R$, $\pi^{-1}(\omega)$, *the set of Walras equilibria in the economy* $\mathcal{E} = (X_i, P_i, \omega_i)_{i=1}^{m}$, *is finite.*

Proof. Using Lemma 7.2, we get that the set of Walras equilibria is compact. Moreover, $\pi^{-1}(\omega)$ is a discrete set, in the sense that each $(p, \omega) \in \pi^{-1}(\omega)$, has an open neighborhood which intersects $\pi^{-1}(\omega)$

only in (p, ω). Indeed, since (p, ω) is a regular point of E, the tangent map $d\pi$ is locally a diffeomorphism around (p, ω), so that for a small enough neighborhood ω can occur as the image only once. Since $\pi^{-1}(\omega)$ is both compact and discrete, it must be a finite set. $\quad\square$

From the proof of the theorem, we obtain that there is a neighborhood of each regular equilibrium where the equilibrium manifold looks like its tangent, so that no new equilibria occur.

Corollary. The number of equilibria is locally constant on the set of regular economies.

1.2.3 *No-trade equilibria*

A particular subset of E is useful in many contexts, namely those (p, ω) for which $\xi(p, p \cdot \omega_i) = \omega_i$ for each i, so that no trade takes place in equilibrium. The set of such no-trade equilibria is denoted T.

In the Edgeworth box, the no-trade equilibria with a given amount of aggregate endowment can be identified with the contract curve. In the general case, each array $(p, w_1, \ldots, w_m) \in \mathring{\Delta} \times \mathbb{R}_{++}^m$ defines a no-trade equilibrium

$$h(p, w_1, \ldots, w_m) = (p, \xi_1(p, w_1), \ldots, \xi_m(p, w_m)).$$

Conversely, the mapping $(p, \omega_1, \ldots, \omega_m) \mapsto (p, p \cdot \omega_1, \ldots, p \cdot \omega_m)$ takes equilibria to arrays $(p, w_1, \ldots, w_m)$ in $\mathring{\Delta} \times \mathbb{R}_{++}^m$, so that the no-trade equilibria can be parametrized by $\mathring{\Delta} \times \mathbb{R}_{++}^l$.

For each price–income combination $(p, w_1, \ldots, w_m)$, the set of corresponding equilibria, that is the set

$$\left\{ (p, \omega_1, \ldots, \omega_m) \,\middle|\, p \cdot \omega_i = w_i, i = 1, \ldots, m, \sum_{i=1}^m \omega_i = \sum_{i=1}^m \xi_i(p, w_i) \right\},$$

is known as the *linear fiber* over $(p, w_1, \ldots, w_m)$. Each of the linear fibers contains exactly one no-trade equilibrium defined by the mapping h.

1.2.4 Path-connectedness of equilibria

Using no-trade equilibria and their linear fibers, one may construct a continuous path from any one equilibrium (p^0, ω^0) in E to any other equilibrium (p^1, ω^1), moving first in the linear fiber to the no-trade equilibrium $(p^0, \xi_1(p^0, p^0 \cdot \omega_1^0), \ldots, \xi_m(p^0, p^0 \cdot \omega_m^0))$, then along T to $(p^1, \xi_1(p^1, p^1 \cdot \omega_1^1), \ldots, \xi_m(p^1, p^1 \cdot \omega_m^1))$, and finally to (p^1, ω^1) along the new linear fiber.

Path-connectedness of the set of equilibria could have been derived directly from the fundamental result in Theorem 7.1, from which one may derive other topological properties as well. One may think of the continuous move from one equilibrium to another as taking place through political intervention consisting in redistribution of endowments, and the property of path-connectedness secures that such an intervention can actually be used to achieve the transformation from any one equilibrium to any other. It should of course be taken into consideration that so far the political intervention is rather abstract, whereas practical intervention would need a particular institutional setup which is not indicated here.

2. Probabilistic Equilibrium Theory

In Section 1 as well as in Chapter 6, we have considered families of economies represented as sets with a particular structure (differentiable manifolds), where it was possible to single out economies with very undesirable properties as exceptional. This approach was initiated by the search for conditions for uniqueness and stability of equilibria, and it only partially succeeded in this direction, excluding particularly worst cases without yielding positive results on wellbehavedness. It might therefore be of interest to consider alternative approaches, searching for results on the typical or average number of equilibria. This will take us to the field of probabilistic equilibrium theory, to be considered briefly in this section.

2.1 The expected number of economic equilibria

One of the classical problems of mathematical economics, considered at length in Chapter 6, is the determinateness of economic equilibria. It is by now well established that economies having a unique equilibrium constitute a very small subset of the economies which are of potential interest, and since the work of Debreu (1970), effort has mainly been directed toward showing that almost all economies have only finitely many equilibria.

In this section, we propose another approach to the question of determinateness, finding the probability distribution of the number of equilibria, given some initial distribution of agents' characteristics. The tools used are based on Edelman and Kostlan (1995). The equilibrium property is transformed to one of geometry, namely to a situation where a vector of parameters characterizing the economic agents and the vector of their excess demand are orthogonal, and the probability of such an event may then be computed from the data of the problem.

2.1.1 Exchange economies and the number of equilibria

For the purpose of our discussion in this section, we consider exchange economies $\mathcal{E} = (X_i, P_i, \omega_i)_{i=1}^m$, where for each i, the consumption set is $X_i = \mathbb{R}_+^l$, the preference correspondence P_i is represented by a utility function $u_i : \mathbb{R}_+^l \to \mathbb{R}$ assumed to be continuous and monotonic (so that $x'_{ih} > x_{ih}$ for $h = 1, \dots, l$ implies $u_i(x'_i) > u_i(x_i)$), and $\omega_i \in \mathbb{R}_{++}^l$ is an initial endowment of consumer i.

Let $\Delta = \{x \in \mathbb{R}_+^l \mid \sum_{h=1}^l x_h = 1\}$ be the standard simplex in $\mathbb{R}^l$. The demand of the consumer i with characteristics (u_i, ω_i) at the price (system) $p \in \operatorname{int} \Delta$ is written as $\xi_{(u_i, \omega_i)}(p)$, and the excess demand as $\zeta_{(u_i, \omega_i)}(p) = \xi_{(u_i, \omega_i)}(p) - \omega_i$. We restrict our attention to cases where the set $\xi_{(u_i, \omega_i)}(p)$ is well-defined, which, as we have seen in the previous chapters, is the case if the consumer (u_i, ω_i) satisfies the following smoothness conditions:

(i) u_i is C^2,
(ii) for each $x_i \in \mathbb{R}_{++}^l$, $Du_i(x_i) \in \mathbb{R}_{++}^l$,

(iii) for each $x_i \in \mathbb{R}_{++}^l$, the restriction of the quadratic form $D^2 u_i(x_i)$ to $\{x_i' \mid Du_i(x_i) \cdot x_i' = 0\}$ is negative definite.

We recall that each of the individual excess demand functions as well as their sum, $\zeta(p) = \sum_{i=1}^m \zeta_{(u_i, \omega_i)}(p)$, satisfy Walras' law,

$$p \cdot \zeta(p) = 0$$

for all $p \in \mathring{\Delta}$.

An equilibrium price (shorthand: an equilibrium) for $\mathcal{E}$ is a price vector p^0 such that

$$\zeta(p^0) = 0. \tag{1}$$

We have seen that in general there may be more than one equilibrium price satisfying (1). It is easy to construct examples of excess demand functions with multiple zeros, and such counterexamples are in no way pathological. As mentioned in Chapter 6, Section 2.2, any continuous function on an open subset of $\mathring{\Delta}$ satisfying Walras' law can be obtained as the aggregate excess demand function of an economy.

In the special case of $l = 2$ (only two commodities), Δ is the unit interval $[0, 1]$, and to find an equilibrium, we need only solve the equation

$$\zeta_{11}(t) + \cdots + \zeta_{m1}(t) = 0, \tag{2}$$

since by Walras' law we will also have that

$$\zeta_{12}(t) + \cdots + \zeta_{m2}(t) = 0.$$

Geometrically, we consider the curve in $\mathbb{R}^2$ given by the rule $t \mapsto z(t) = (\zeta_{11}(t), \dots, \zeta_{m1}(t))$ (excess demand for commodity 1 of each of the consumers). Clearly, t^0 is an equilibrium of this economy if (2) is satisfied or, expressed otherwise,

$$e \cdot z(t) = 0.$$

The equilibrium conditions have a geometric interpretation: The diagonal vector should be orthogonal to the vector $z(t^0)$. We return to the geometric viewpoint in what follows.

210   *Theory of General Economic Equilibrium*

While the uniqueness of equilibria does not hold in general, special properties of the excess demand do lead to uniqueness . As we saw in Chapter 6, one such property is *gross substitution*. However, such assumptions are however restrictive, and they are not easily reduced to properties of the individuals constituting the economy.

2.1.2   *Families of economies and the average number of equilibria*

Following our general approach in this chapter, we consider families of economies, viewed as collections of consumers. Let C^0 be a *finite* set of consumers satisfying the assumptions made earlier (with cardinality $|C^0| = n$). An *economy* is a map a from C^0 to $\mathbb{R}_+$, and a *family of economies* is a pair (C^0, μ), μ is a probability distribution on the set $\mathbb{R}_+^{C^0}$ of economies over the set C^0. The support of μ, written supp μ, is the smallest closed subset of $\mathbb{R}_+^{C^0}$ for which the complement has zero probability.

Intuitively, an economy is a sample with a_i consumers of the type (u_i, ω_i), where in general, a_i may not be an integer. The probability measure μ assigns weights to the economies in the family according to their importance or relative frequency. An *equilibrium* of a is a price vector p^0 such that $\sum_{i=1}^n a_i \zeta_i(p^0) = 0$ and the set of equilibria of a is given as follows:

$$W(a) = \left\{ p^0 \in \Delta \,\middle|\, \sum_{i=1}^n a_i \zeta_i(p^0) = 0 \right\}.$$

The *number of equilibria* is the extended real-valued function $v :$ supp $\mu \to [0, +\infty]$ defined by

$$v(a) = |W(a)|,$$

and the *expected number of equilibria* (of the family (C^0, μ) of economies) is

$$\mathsf{E}\, v = \begin{cases} \int\int v(a)\, d\mu(a) & \text{if } v \text{ is } \mu\text{-integrable}, \\ +\infty & \text{otherwise}. \end{cases}$$

We shall consider assessing $\mathsf{E}\nu$ for general families (C^0, μ). Before we turn to this, we consider some particular families of economies.

(a) Non-atomic economies with finitely many types (e.g., Hildenbrand and Kirman, 1976): The probability distribution μ has support in $\{a \in \mathbb{R}_+^{C^0} \mid \sum_{i=1}^n a_i = 1\}$. In the interpretation, the set of consumers are subsets of $[0, 1]$ of length a_i. If ζ_i is the excess demand of consumer (u_i, ω_i), then aggregate excess demand is $\zeta = \sum_{i=1}^n a_i \zeta_i$. Thus, the family of economies may alternatively be considered as a probability distribution over sums of the individual excess demand functions.

(b) Densities ν with support in $\mathbb{R}_+^{C^0}$: The finite counterpart of (a) is the situation where $\sum_{i=1}^n \zeta_i$ is interpreted as the excess demand of a finite economy with a_i agents of the type (u_i, ω_i) giving rise to excess demand function ζ_i. For this to make sense, we must have that each a_i is a non-negative integer, so that the support of ν is restricted to $\mathbb{Z}_+^{C^0}$.

(c) Homothetic consumers: If ζ_i is an excess demand function of a homothetic consumer, then by the (first) aggregation theorem of Chipman (1974), $a_i \zeta_i$ is the excess demand of the consumer having the same utility and endowment vector $a_i \omega_i$. Therefore, any family of economies over a finite set C^0_{hom} of homothetic consumers — even the family where the probability distribution has full support $\mathbb{R}_+^{C^0}$ — has a straightforward interpretation, since each member of the support is a genuine exchange economy.

2.1.3 The average number of equilibria: The case of two commodities

We begin with the simple case of two commodities. Let (C^0, μ) be a family of economies with $l = 2$. Then excess demand is a function of a single variable $t = p_1$, and by Walras' law, it suffices to consider the excess demand for commodity 1. The key observation in the study of $\mathsf{E}\nu$ is the following simple geometric version of the equilibrium condition, here formulated as a lemma.

Lemma 7.3. *Let $t^0 \in]0,1[$. Then t is an equilibrium for the random economy a if and only if $a = (a_1, \ldots, a_n)$ is orthogonal to $(\zeta_{11}(t), \ldots, \zeta_{n1}(t))$.*

To proceed, we use a geometric argument, given in Edelman and Kostlan (1995): To find the average number of equilibria, we find for each t the set of economies (weighted by its density) for which t is an equilibrium price (using Lemma 1), and then integrate over t. As before, we write the curve in $\mathbb{R}^m$ of excess demands for the first commodity of each of the agents as $z(t) = (\zeta_{11}(t), \ldots, \zeta_{m1}(t))$, and we introduce the normalized version of this curve, $\gamma(t) = z(t)/\|z(t)\|$.

Theorem 7.3. *The expected number of equilibria of the family (C^0, μ) is given by*

$$\mathsf{E}\nu = \int_0^1 \left[\int_{\{\gamma'(t)\}^\perp} |\gamma'(t) \cdot a| \, d\mu(a) \right] dt, \tag{3}$$

where $\{\gamma(t)\}^\perp$ denotes the subspace of $\mathbb{R}^n$ orthogonal to $\gamma(t)$.

Proof. Fix t and choose an orthonormal basis $e_1, \ldots, e_n$ such that $e_1 = \gamma'(t)$ and $e_2 = \gamma'(t)/\|\gamma'(t)\|$. When we move from t to $t + dt$, the hyperplane perpendicular to $\gamma(t)$ will sweep out a subset of $\mathbb{R}^n$. This set is the Cartesian product of a two-dimensional subset of $\mathrm{span}(e_1, e_2)$, having area $(\gamma'(t) \, dt)(|e_2 \cdot a|)$, with subspace spanned by the remaining $n - 2$ basis directions. The volume of the set is therefore

$$\|\gamma'(t)\| \, dt \int_{\mathbb{R}^{n-1}} |e_2 \cdot a| \, d\mu(a),$$

where the domain of integration is the $(n - 1)$-dimensional space $\{\gamma(t)\}^\perp$ orthogonal to e_1. Inserting the above expression, we get

$$\mathsf{E}\nu = \int_0^1 \left(\|\gamma'(t)\| \int_{\{\gamma(t)\}^\perp} \frac{|\gamma't(a) \cdot a|}{\|\gamma'(t)\|} \, d\mu(a) \right)$$

$$dt = \int_0^1 \left[\int_{\{\gamma(t)\}^\perp} |\gamma'(t) \cdot a| \, d\mu(a) \right] dt,$$

which is (3). $\qquad\square$

The Equilibrium Manifold and Probabilistic Equilibrium Theory

---

Box 1. Average number of equilibria in a Cobb–Douglas economy: A simple application of Theorem 7.3 is the following: Consider the family (C^0, μ) where C^0 consists of two consumers having Cobb–Douglas utilities

$$u_i = x_1^{\alpha^i} x_2^{1-\alpha^i}, \ i = 1, 2,$$

with $\alpha^1 < \alpha^2$, and identical endowment $(\omega_1, \omega_2) = (1, 1)$, and where μ is the uniform distribution on $\{a \in \mathbb{R}^2_+ \mid a_1 + a_2 = 1\}$. Economies in this family have a unique equilibrium, so the computation that follows serves mainly as an illustration of the point method.

The excess demand function of (u_i, ω_i) is

$$z_i(t) = \zeta_{i1}(t) = \alpha^i \frac{t\omega_{i1} + (1 - t)\omega_{i2}}{t} - \omega_{i1} = \frac{\alpha^i}{t} - 1,$$

with $z_i'(t) = -\alpha^i t^{-2}$, $i = 1, 2$. It is seen that for fixed t the set of a with $\gamma(t) \cdot a = z(t) \cdot a = 0$ and $a_1 + a_2 = 1$ is empty if $t \notin [\alpha^1, \alpha^2]$ (since in that case both coordinates of $z(t)$ have the same sign), and uniquely determined as the point

$$a(t) = \left(\frac{t - \alpha^1}{\alpha^2 - \alpha^1}, \frac{\alpha^2 - t}{\alpha^2 - \alpha^1} \right)$$

for $t \in [\alpha^2, \alpha^1]$. Consequently, the integral in (3) reduces to

$$\int_{\alpha^1}^{\alpha^2} \frac{1}{\alpha^2 - \alpha^1} \, dt = 1,$$

which gives us the expected result.

The example in Box 1 was sufficiently simple to allow for explicit computation of the expected number of equilibria using the formula of Theorem 7.3. In general, this is not possible, but the result may be used to derive upper bounds on the expected number of equilibria for given families of economies.

Corollary. *Let (C^0, μ) be a family of economies with* $\operatorname{supp} \mu \subset \{a \in \mathbb{R}^{C^0}_+ \mid \sum_{i=1}^{n} a_i = 1\}$, *let $t_{\min}$ and $t_{\max}$ be such that $t_{\min} \leq t \leq t_{\max}$ for any t with $\zeta_i(t) = 0$, $i \in C^0$, and let*

$$M = \max_{t \in [t_{\min}, t_{\max}]} \max_i \frac{|\zeta_i'(t)|}{\sqrt{\zeta_1(t)^2 + \cdots + \zeta_n(t)^2}}.$$

Then $\mathsf{E}v \leq M \sqrt{n}(t_{\max} - t_{\min})$.

214
Theory of General Economic Equilibrium

Proof. Using that

$$\gamma'(t) = \frac{d}{dt}\left(\frac{z(t)}{\|z(t)\|}\right) = \frac{z'(t)}{\|z(t)\|} + \gamma(t)\frac{d}{dt}\|z(t)\|,$$

and exploiting that in (3) we integrate over a such that $\gamma(t) \cdot a = 0$, so that

$$\mathsf{E}\,\nu = \int_{t_{\min}}^{t_{\max}}\left[\int_{\{\gamma(t)\}^{\perp}} \frac{|z'(t) \cdot a|}{\|z(t)\|}\, d\mu(a)\right] dt.$$

Using the Cauchy–Schwarz inequality, we get

$$\mathsf{E}\,\nu \leq \int_{t_{\min}}^{t_{\max}} \int_{\{\gamma(t)\}^{\perp}} \|a\|\frac{\|z'(t)\|}{\|z(t)\|}\, d\mu(a)\, dt$$

$$\leq \int_{t_{\min}}^{t_{\max}} \int_{\{\gamma(t)\}^{\perp}} \|a\|M\sqrt{n}\, d\mu(a)\, dt = M\sqrt{n}(t_{\max} - t_{\min}),$$

which is the assessment that we were looking for. □

2.1.4 *The general case of arbitrary number of commodities*

Assume that the family of economies (C^0, μ) with $C^0 = \{(u_1, \omega_1), \ldots, (u_n, \omega_n)\}$ has l commodities, where $l < n$ for simplicity. We say that a price $p \in \Delta$ is a *commodity 1 equilibrium price* in the economy given by $(a_1, \ldots, a_n)$ if

$$\sum_{i=1}^{n} a_i \zeta_{i1}(p) = 0.$$

For $a \in C^0$, the set of commodity 1 equilibria in a is denoted by $W^1(a)$, its cardinality by $\nu^1(a)$, and the expected number of commodity 1 equilibria in the family (C^0, μ) by $\mathsf{E}\,\nu^1$. We use the notation $z_1(p) = (\zeta_{11}(p), \ldots, \zeta_{n1}(p))$ and $\gamma_1(p) = z_1(p)/\|z_1(p)\|$.

The following is essentially a restatement of Theorem 7.3 to deal with the situation, the new aspect being that the variable p is now $(l-1)$-dimensional.

The Equilibrium Manifold and Probabilistic Equilibrium Theory 215

Lemma 7.4. *The expected number of commodity 1 equilibria of the family* (C^0, μ) *satisfies*

$$\mathsf{E}v^1 \le \int_\Delta \|D_1 z_1(p)\| \cdots \|D_{l-1} z_1(p)\| \left[\int_{\{\gamma_1(p)\}^\perp} |\operatorname{proj}_{Dz_1(p)} a| \, d\mu(a)\right] dp,$$

(4)

where $\operatorname{proj}_{Dz_1(p)}$ *is the projection on the span of the vectors* $D_1 z_1(p), \ldots,$ $D_{l-1} z_1(p)$ *of derivatives of* z_1 *with respect to* $p_1, \ldots, p_{l-1}$.

Proof. The result follows by the same reasoning as in the proof of Theorem 7.3. Changing p by the vector $dp = (dp_1, \ldots, dp_{l-1})$, the point $z_1(p)$ sweeps out an area $Z_1^{(l-1)}(p) \, dp_1 \cdots dp_{l-1}$, where $Z_1^{(l-1)}(p)$ is the $(l-1)$-dimensional volume element at $z_1(p)$, and the volume of the set of weights $a = (a_1, \ldots, a_n)$ which are orthogonal to $x_1(p)$ covered in this movement can be found as $Z_1^{(l-1)}(p) \, dp_1 \cdots dp_{l-1}$ times $|\operatorname{pr}_{Dz_1(p)}(a)|$, integrated over all $a \in \{\gamma_1(p)\}^\perp$. Using the inequality $Z_1^{(l-1)}(p) \le \|D_1 z_1(p)\| \cdots \|D_{l-1} z_1(p)\|$, we get the expression in (4). $\qquad\square$

The bound which can be derived using (4) is rather crude, neglecting the equilibrium conditions in all but one commodity, but it can be used for deriving a bound which exploits the equilibrium property for all commodities. For the evaluation of the expected number of equilibria, the relevant geometric condition is that a is orthogonal to *all* the vectors $z_h(p)$, $h = 1, \ldots, l-1$, where $z_h(p) = (z_{1h}(p), \ldots, z_{l-1 h}(p))$. Therefore, the right-hand side of (4) overstates the expected number of equilibria. We can improve on the bound by restricting integration to the set where the correct geometric condition is satisfied.

Theorem 7.4. *Let* (C^0, μ) *be a family of economies with* l *commodities. Then the expected number of equilibria satisfies*

$$\mathsf{E}v \le \min_{h=1,\ldots,l-1} \int_\Delta |D_1 z_h(p)| \cdots |D_{l-1} z_h(p)| \left[\int_{\Gamma(p)} |\operatorname{proj}_{D\gamma_h(p)} a| \, d\mu(a)\right] dp,$$

where $\Gamma(p) = \{\gamma_1(p), \ldots, \gamma_{l-1}(p)\}^\perp$.

Proof. Apply Lemma 7.4, which holds for arbitrary $h \in \{1, \ldots, l-1\}$, and restrict integration over a to $\Gamma(p)$. $\qquad\square$

We may derive a more usable version of the bound using the same method as in section 2.1.3:

Corollary. Let (C^0, μ) be a family of economies with l commodities, with supp $\mu \subset \{a \in \mathbb{R}_+^{C^0} \mid \sum_{i=1}^{n} a_i = 1\}$. Then there is a compact subset K of intΔ such that $p \notin K$ implies that there is $h \in \{1, \ldots, l\}$ with $\zeta_{ih}(p) > 0$, all i, and if

$$M_K = \min_{h=1,\ldots,l-1} \max_{i=1,\ldots,l-1} \max_{p \in K} \| D_i z_h(p) \|,$$

then $\mathsf{E}\, \nu \leq M^{l-1} m_{l-1}(K)$, where m_{l-1} is $(l-1)$-dimensional Lebesgue measure.

Proof. The existence of a compact set K with the properties stated follows from the monotonicity assumption on the underlying consumers, since for each consumer type i and commodity h there is $p_h^i > 0$ such that $\zeta_{ih}(p) > 0$ whenever $p_h \leq p_h^i$. Clearly, all equilibrium prices for economies in the family must belong to K. The remaining part of the statement now follows from the theorem. $\qquad\square$

While the one-dimensional precise formula for the average number of equilibria can still be put to use in the many-commodity case, it can be seen from the results that this comes at a cost, partly in the form of bounds instead of the exact formula, partly as more complicated expressions. Although these drawbacks make applications less simple, it is still possible to extract useful information on particular families of exchange economies from the above results.

2.2 How likely is factor price equalization?

The factor price equalization (FPE) theorem, which states that under suitable assumptions, factors of production will obtain the same remuneration in countries trading only in final products, was already mentioned in Chapter 2. Subsequent authors have refined

The Equilibrium Manifold and Probabilistic Equilibrium Theory 217

and reformulated it in several ways, in particular Dixit and Norman (1980), who introduced the FPE domain, the set of initial distributions of factors among countries for which international trade equilibria are also equilibria of an integrated world economy with no restriction on trade in factors. The FPE domain lends itself easily to geometric reasoning, and it gives a first picture of the likelihood of factor prices being equalized as well as some intuition as to what governs this likelihood: If the input vectors in the different industries which occur in the integrated equilibrium have very different directions, then the FPE domain will be large relative to the set of all possible distributions among countries, so if each distribution of a fixed-world endowment among countries is considered equally likely, then the probability of FPE is large.

However, restricting attention to the distribution of the given endowments, taking world factor endowment and technology as given, may not be a very useful approach. In our approach that follows, we shall also allow the technologies for producing the commodities to vary. More specifically, we assume that endowments as well as technologies are sampled at random, and we are then interested in assessing the probability of FPE, both for a fixed technology and for the general case. In the simplest case of two countries, two factors, and two traded commodities, known from textbook versions of international trade theory, the resulting assessment of the probability of factor price equalization is more or less as would be expected, showing it to occur with a probability close to one half. The fact that it actually differs from one may come as a surprise, but this has to do with our sampling of technologies. But moving to more than two commodities and factors will make things change rapidly, and the assessments of the relevant probabilities show among other things that the likelihood of FPE decreases toward zero with rising dimension.

This somewhat unexpected result that FPE becomes increasingly unlikely as the number of goods and factors increases is of course in its turn dependent on the particular way of parametrizing the technologies as well as the choice of probability distribution on

Theory of General Economic Equilibrium

parameters. On the other hand, since the assessments are based on considerations of volumes of suitable convex sets, the results are reasonably robust, related as they are to dimensionality, so they seem to catch some basic features of the problem, which may be overlooked when only low-dimensional cases are considered.

2.2.1 The lens condition

In this section, we introduce some of the concepts and the notation to be used in the sequel. Following a geometric approach to FPE domains, Deardorff (1994) formulated the so-called *lens condition*[2] for FPE. We state a version of this lens condition adapted to our purposes.

We begin with the technologies. It turns out to be useful to identify technologies by the support functions of their 1-isoquant. More specifically, let $\Delta_r = \{q \in \mathbb{R}_+^r \mid \sum_{h=1}^r q_h = 1\}$ be the set of normalized prices of r given factors of production, and let $\mathcal{S}_r$ be the set all of concave functions $\sigma : \Delta_r \to \mathbb{R}_+$. Elements of $\mathcal{S}_r$ are interpreted as (downward) support functions of 1-isoquants, so that the upper level set at the output quantity 1 (all the factor combinations yielding at least 1 unit of output) is

$$Y_\sigma = \{z \in \mathbb{R}_+^r \mid q \cdot z \geq \sigma(q) \text{ for all } q \in \Delta_r\},$$

and the production function associated with σ is

$$f_\sigma(z) = \max\{\lambda \mid \lambda^{-1}z \in Y_\sigma\}.$$

Since elements of $\mathcal{S}_r$ are (support functions of) 1-isoquants, we refer to them as *techniques* rather than as production functions.

Let $\phi_\sigma(q) = \{z \in Y_\sigma \mid q \cdot z = \sigma(q)\}$ be the set of cost-minimizing input combinations in the technique σ at the price q. Clearly, for each $q \in \Delta_r$, the set $\phi_\sigma(q)$ is closed and convex, and $\phi_\sigma(q)$ is a singleton if Y_σ is strictly convex.

---

[2]The lens condition has been intensively discussed in the literature and adapted to different purposes, see, e.g. Deardorff (2001), Kemp and Okawa (1998), Qi (2003) and Wong and Yun (2003).

The Equilibrium Manifold and Probabilistic Equilibrium Theory 219

In the following, a *technology* is an array $\underline{\sigma} = (\sigma_1, \ldots, \sigma_n) \in (\mathcal{S}_r)^n$ consisting of n elements of $\mathcal{S}_r$. In the interpretation, $\underline{\sigma}$ specifies the method of producing n distinct commodities using the r factors of production. The following standard property of technologies is a useful consequence of the concavity of the elements of $\mathcal{S}_r$.

Lemma 7.5. *Let $\underline{\sigma} = (\sigma_1, \ldots, \sigma_n) \in (\mathcal{S}_r)^n$ and $\omega \in \mathbb{R}^r_{++}$ be given, and let $y^0 = (y^0_1, \ldots, y^0_n) \in \mathbb{R}^n_{++}$. Then the following are equivalent:*

(i) *y^0 maximizes some increasing and quasi-concave function $U(y)$ over all $y = (y_1, \ldots, y_n)$ for which there are $z_1, \ldots, z_n \in \mathbb{R}^r_+$ with $y_i = f_{\sigma_i}(z_i)$ and $\sum_{i=1}^n z_i = \omega$,*

(ii) *there is $q \in \Delta_r$ such that*

$$\omega \in \sum_{i=1}^n y^0_i \phi_{\sigma_i}(q). \tag{5}$$

Proof. (i)$\Rightarrow$(ii): Let

$$X^0 = \{z \in \mathbb{R}^r_+ \mid \exists (z_1, \ldots, z_n), U(f_{\sigma_1}(z_1), \ldots, f_{\sigma_n}(z_n))$$

$$\geq U(y^0_1, \ldots, y^0_n), \sum_{i=1}^n z_i = z\}.$$

By convexity of production sets and quasi-concavity of U, this set is convex. By our assumptions, it intersects the set $\{z \in \mathbb{R}^r_+ \mid z \leq \omega\}$ only in ω. By separation of convex sets, there is $q \in \Delta_r$, such that $q \cdot x \geq q \cdot \omega$ for all $x \in X^0$. Using the definition of X^0 we see that for each i, $q \cdot z_i \geq q \cdot z^0_i$ for all z_i such that $f_{\sigma_i}(z_i) \geq f_{\sigma_i}(z^0_i)$, where $f_{\sigma_i}(z^0_i) = y^0_i$, each i, and $\sum_{i=1}^n z^0_i = \omega$.

(ii)$\Rightarrow$(i): Write $\omega = z_1 + \cdots + z_n$ with $z_i \in y^0_i \phi_{\sigma_i}(q)$ for each i, and define $p_i = (q \cdot z_i)/y^0_i$. Then by constant returns to scale, we have that $p_i f_{\sigma_i}(z'_i) \leq q \cdot z'_i$ for all $z'_i \in \mathbb{R}^r_+$, and it follows that y^0 maximizes $p \cdot \sum_{i=1}^n f_{\sigma_i}(z'_i)$ over all $(z'_1, \ldots, z'_n) \in (\mathbb{R}^r_+)^n$ with $\sum_{i=1}^n z'_i = \omega$, which is (i). $\qquad\square$

The situation considered in Lemma 7.5 corresponds to what is called "the integrated equilibrium" in the literature. We now proceed to consider the case where total factor endowments ω are distributed among the K countries according to the array $\underline{\omega} = (\omega^1, \ldots, \omega^K) \in (\mathbb{R}_+^r)^K$. The array $(\underline{\sigma}, \underline{\omega})$ is called a K-country world; the associated 1-country world $(\underline{\sigma}, \omega)$, where $\omega = \sum_{k=1}^K \omega^k$, is called the *integrated economy*.

An *equilibrium* in the K-country world $(\underline{\sigma}, \underline{\omega})$ is an array $(p, (y^k, q^k)_{k=1}^K) \in \Delta_n \times (\Delta_r \times \mathbb{R}_+^n)^K$ such that for each k,

$$y^k \text{ maximizes } p \cdot y \text{ over all } y \text{ such that } \omega_k \in \sum_{i=1}^n y_i \phi_{\sigma_i}(q^k). \quad (6)$$

The equilibrium is a *FPE equilibrium* if $q^k = q^l$ for $k, l = 1, \ldots, K$.

In the notation introduced thus far, the "lens condition" for FPE takes the following form:

Theorem 7.5. *Let* $(\underline{\sigma}, \underline{\omega})$ *be a K-country world, and let* $(p, (y^k, q^k)_{k=1}^K)$ *be an equilibrium with* $y^0 = \sum_{k=1}^K y^k$. *Then the following are equivalent:*

(i) $(p, (y^k, q^k)_{k=1}^K)$ *is an FPE equilibrium with* $q^1 = \cdots = q^K = q$,
(ii) *there is* $q \in \Delta_r$ *such that*

$$\omega^k \in \sum_{i=1}^n \text{conv}\left(\{0\}, y_i^0 \phi_{\sigma_i}(q)\right) \quad \text{all } k. \quad (7)$$

Proof. (i)$\Rightarrow$(ii): Since (6) holds with $q^k = q$ for each k, we have that $\omega^k \in \sum_{i=1}^n y_i^k \phi_{\sigma_i}(q)$ for each k, and since $y_i^k \phi_{\sigma_i}(q) \subset \text{conv}\left(\{0\}, y_i^0 \phi_{\sigma_i}(q)\right)$ for each i and k, we have (7).

(ii)$\Rightarrow$(i): By (7), the factor endowment ω_k has a representation

$$\omega_k = z_1^k + \cdots + z_n^k,$$

where $z_i^k \in y_i^k \phi_{\sigma_i}(q)$ for some $y_i^k \in [0, y_i^0]$, each k, and with $y_i^0 = \sum_{k=1}^K y_i^k$ for each i. We then have that

$$\omega = \sum_{k=1}^K \omega^k \in \sum_{k=1}^K y_i \phi_{\sigma_i}(q),$$

The Equilibrium Manifold and Probabilistic Equilibrium Theory   221

which by Lemma 7.5 means that $(p, (y^0, q))$ is an equilibrium in the integrated economy for some $p \in \Delta_n$. It now follows immediately that $(p, (y^k, q)_{k=1}^K)$ is an FPE equilibrium. $\qquad\square$

It should be noted that the present characterization of FPE equilibriums allows for technologies for which isoquants may not be smooth. A prominent such case is that of Leontief technologies: A technique $\sigma \in S_r$ is Leontief if it has the form σ_a with

$$\sigma_a(q) = q \cdot a$$

for some $a \in \mathbb{R}_+^r$. We denote by S_r^L the set of Leontief techniques. A technology $\underline{\sigma}$ is Leontief if it belongs to $\left(S_r^L\right)^n$, that is if $\sigma_i = \sigma_{a_i}$ for each i, and the output y^0 maximizes $p \cdot y$ for some $p \in \Delta_n$ using total resources ω if

$$\sum_{i=1}^n y_i^0 a_i \leq \omega$$

with equality for at least one of the r coordinates. We have that

$$\sum_{i=1}^n y_i^0 a_i = \sum_{i=1}^n (p_i y_i^0) \left[\frac{1}{p_i} a_i\right] = p \cdot y^0 \left(\sum_{i=1}^n \frac{p_i y_i^0}{p \cdot y^0} \left[\frac{1}{p_i} a_i\right]\right),$$

so that the solution is identical to that of finding the maximal output in a one-good economy with the composite good obtained by valuing n-tuples at prices p and using the technique σ defined by

$$\sigma(q) = \min \left\{\frac{1}{p_1} \sigma_{a_1}(q), \ldots, \frac{1}{p_n} \sigma_{a_n}(q)\right\},$$

and in the solution, q supports each of the points $(1/p_i)a_i, i = 1, \ldots, n$.

A similar result holds in the general case, and it provides the link between the classical approach to FPE, finding the (unique) factor prices corresponding to commodity prices p. Except for uniqueness, this carries over to our present setup.

Theorem 7.6. *Let $(\underline{\sigma}, \underline{\omega})$ be a K-country world, and let $(p, (y^k, q^k)_{k=1}^K)$ be an equilibrium with $y^0 = \sum_{k=1}^K y^k$. Then the following are equivalent:*

(i) *$(p, (y^k, q^k)_{k=1}^K)$ is an FPE equilibrium with $q^1 = \cdots = q^K = q$,*

(ii) *there is $q \in \Delta_r$ such that $\sigma_i(q) = p_i$ for each i and*

$$\omega^k \in (p \cdot y^k) \operatorname{conv}\left(\left\{\frac{1}{p_i}\phi_{\sigma_i}(q) \,\middle|\, i = 1, \ldots, n\right\}\right) \tag{8}$$

for each k.

Proof. (i)$\Rightarrow$(ii): Using the properties of the FPE equilibrium, we get that

$$\omega^k \in \sum_{i=1}^{n} p_i y_i^k \frac{1}{p_i}\phi_{\sigma_i}(q) = (p \cdot y^k) \sum_{i=1}^{n} \lambda_i \frac{1}{p_i}\phi_{\sigma_i}(q)$$

with $\lambda_i = p_i y_i^k/(p \cdot y^k)$ for each i, so that (8) holds for each k.

(ii)$\Rightarrow$(i): Suppose that (8) holds for each k. Adding over k we get that

$$\omega = (p \cdot y^0) \sum_{i=1}^{n} \lambda_i \frac{1}{p_i} z_i = \sum_{i=1}^{n} \frac{\lambda_i(p \cdot y^0)}{p_i} z_i$$

with $z_i \in \phi_i(q)$, each i. Using that $q \cdot z_i' = \sigma_i(q) = p_i$, all $z_i' \in \phi_{\sigma_i}(q)$, each i we have that $(q \cdot z_i')/p_i$ is independent of i, so that q must be a support of $\{z_i \mid f_{\sigma_i}(z_i) \geq 1\}$ for each i, and consequently we may assume that $q^1 = \cdots = q^K = q$, which is (i). $\qquad \square$

2.2.2 *Assessing the probability of FPE: The classical case $(r = n)$*

In this section, we consider the probabilistic approach to FPE in its classical version where the number of traded commodities equals the number of factors. We start with the case of only two factors of production, allowing for a graphical representation using the Edgeworth box. This representation also lends some intuition to the probability of FPE, at least in the simple case of independent and uniform distribution of resources, as the relative area of the subset of the box bounded by the rays of factor inputs in the integrated equilibrium.

For simplicity, let the technology be Leontief, with $(\sigma_1, \sigma_2) = (\sigma_{a_1}, \sigma_{a_2})$. If the technology is fixed, and the commodity prices are given, then the factor equalization domain can be represented in the Edgeworth box as the area of the set A between the rays from the

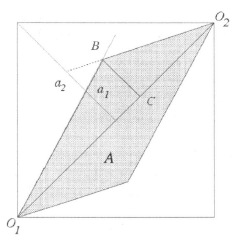

Fig. 1. The factor price equalization domain with Leontief technologies, determined as the parallelogram spanned by the two vectors a_1 and a_2.

origin through a_1 and a_2 (in the coordinate systems with origin O_1 as well as in that with origin O_2), cf. Fig. 1.

We assume that factor endowments in each country are drawn independently according to a probability distribution F with density f. If $\Pi_{\underline{\sigma},p}$ is the probability of FPE, then

$$\Pi_{\underline{\sigma},p} = \int_{(z_1,z_2)\in A} f(z)f(\omega - z)\,dz. \qquad (9)$$

In order to obtain explicit assessments of $\Pi_{\underline{\sigma},p}$ we assume that F is the uniform distribution on $[0,1]^2$ and that $\omega = (1,1)$, so that the expression in (9) equals the area of A. Using the notation as indicated in Fig. 1, we find that the line segment BC has length $2\alpha\beta/(\alpha+\beta)$, so that the probability of FPE is

$$\Pi_{\underline{\sigma},p} = 2\sqrt{2}\,\frac{\alpha\beta}{\alpha+\beta} \qquad (10)$$

in the case illustrated in Fig. 1.

It is evident that the quantity in (10) depends on the technology as expressed here by the two rays $\phi_{\sigma_1}(q), \phi_{\sigma_2}(q)$. Following the probabilistic approach, we would proceed to take expectations over technologies $\underline{\sigma}$ and commodity prices p. This however presupposes a given probability distribution over the set $\left(S_2^L\right)^2$ of pairs of techniques. Such a distribution with reasonable intuitive content does not suggest itself except in the particular case, where all techniques in S_2 are Leontief, and therefore, we restrict our treatment to this case.

Thus, we assume that Leontief technologies are parameterized by pairs (a_1, a_2) of elements of the simplex $\{x \in \mathbb{R}_+^2 \mid x_1 + x_2 = 1\}$ which define the relevant factor proportions in each of the two techniques. We assume that the two techniques are sampled in such a way that the total factor endowment can be exploited efficiently, meaning that $a_{11} \geq 1/2$, $a_{22} \geq 1/2$. The case where one of these inequalities is violated will be considered separately.

In principle, we need not only factor proportions but also the amount of factors needed to produce one unit of commodity. However, the FPE domain will be determined only by factor proportions; the absolute factor productivity will matter only for the commodity prices, and with our parameterization of technologies, we need not take the latter into account. We get the following result in the $2 \times 2 \times 2$ case.

Theorem 7.7. *Consider the family*

$$\left(S_2^L\right)^2 \times \left\{ (\omega_1, \omega_2) \in (\mathbb{R}_+^2)^2 \,\middle|\, \omega_1 + \omega_2 = (1, 1) \right\}$$

of 2-country worlds with two commodities and two factors of production, where techniques and endowments are sampled uniformly given that factors can be used efficiently. Then $\Pi^{2,2,2}$, the probability of FPE, satisfies $\Pi^{2,2,2} < \dfrac{1}{2}$.

Proof. Using the parametrization $u = \| a_1 - (\tfrac{1}{2}, \tfrac{1}{2}) \|$, $\beta = \| a_2 - (\tfrac{1}{2}, \tfrac{1}{2}) \|$, we have the following expression:

$$\Pi^{2,2,2} = 2\sqrt{2} \int_0^{\frac{\sqrt{2}}{2}} \int_0^{\frac{\sqrt{2}}{2}} \frac{\alpha\beta}{\alpha + \beta} \, \mathrm{d}\alpha \, \mathrm{d}\beta.$$

The Equilibrium Manifold and Probabilistic Equilibrium Theory   225

After substitution of y for $\alpha + \beta$, we get that

$$\Pi^{2,2,2} = 2\sqrt{2} \int_0^{\frac{\sqrt{2}}{2}} \int_\beta^{\beta + \frac{\sqrt{2}}{2}} \left(1 - \frac{\beta}{y}\right) dy \, d\beta$$

$$= 2\sqrt{2} \int_0^{\frac{\sqrt{2}}{2}} \left[\frac{\sqrt{2}}{2}\beta - \beta^2 \ln\left(1 + \frac{\sqrt{2}}{2}\frac{1}{\beta}\right)\right] d\beta,$$

and using Taylor expansion of $\ln(1 + \frac{\sqrt{2}}{2}\frac{1}{\beta})$ around $\ln 1 = 0$, we get that

$$\ln\left(1 + \frac{\sqrt{2}}{2}\frac{1}{\beta}\right) < \frac{\sqrt{2}}{2}\frac{1}{\beta} - \frac{1}{4\beta^2},$$

so that

$$\Pi^{2,2,2} < 2\sqrt{2} \int_0^{\frac{\sqrt{2}}{2}} \frac{1}{4} \, d\beta = \frac{1}{2},$$

which gives the assessment of the theorem. $\qquad\square$

The fact that FPE occurs with probability less than $1/2$ may come as a surprise, since the "average" technology would be that where $\alpha = \beta = \frac{\sqrt{2}}{4}$, for which the area of the factor price equalization domain is exactly $1/2$. Needless to say, the result depends on the distribution of techniques; if we had chosen a uniform distribution over the angles between the diagonal and the factor proportion rays, the cases of instances of small FPE domains would have weighted less and the final probability would have been greater. However, the approach taken seems more natural from an economic point of view, using factor bundles rather than factor proportions, and it is much more easy to generalize to more than two factors of production.

For the extension of the results to cases of more than two commodities and factors, we start by considering the case $n = r = 3$ (and, as previously, $K = 2$). With total factor endowment $\omega = e = (1, 1, 1)$, we have that a 2-country world is defined by specifying three Leontief techniques $\sigma_{a_1}, \sigma_{a_2}, \sigma_{a_3}$ with $a_1, a_2, a_3 \in \Delta$. By Theorem 7.6, FPE will obtain whenever (ω_1, ω_2) is such that $\omega_i \in \text{cone}(\{a_1, a_2, a_3\})$

for $i = 1, 2$. Geometrically, this means that ω_1 should belong to the set

$$\text{cone}(\{a_1, a_2, a_3\}) \cap [\{e\} - \text{cone}(\{a_1, a_2, a_3\})]. \tag{11}$$

From this, we obtain a bound for the probability of FPE, $\Pi^{3,3,2}$, namely

$$\Pi^{3,3,2} \leq 2\text{Vol}\left(\text{cone}(\{a_1, a_2, a_3\}) \cap \{z \in \mathbb{R}_+^3 \mid \Sigma_i z_i = 1\}\right), \tag{12}$$

where $\text{Vol}(\cdot)$ denotes the volume in $\mathbb{R}^3$, and this expression may be used to obtain a numerical assessment of the bound. We need a further notion: Let P_3 be the expected value of the area of a triangle spanned by three points in the simplex Δ_3 chosen at random, measured relative to the area of Δ_3.

Theorem 7.8. *Consider the family* $\left(S_3^L\right)^3 \times \{(\omega_i)_{i=1}^3 \in (\mathbb{R}_+^3)^2 \mid \Sigma_i \omega_i = (1, 1, 1)\}$ *of two-country worlds with three commodities and three factors of production, where techniques and endowments are sampled uniformly given that factors can be used efficiently. Then the probability of FPE $\Pi^{3,3,2}$ satisfies*

$$\Pi^{3,3,2} < \frac{\sqrt{3}}{2} P_3 < \frac{\sqrt{3}}{2} \left(\frac{1}{2}\right)^2. \tag{13}$$

Proof. Using (12), we have that the relative area of the FPE at any choice of Leontief technology $(\sigma_{a_1}, \sigma_{a_2}, \sigma_{a_3})$ must be bounded from above by twice the relative area of

$$\text{conv}\left(\left\{0, \frac{\sqrt{3}}{2} a_1, \frac{\sqrt{3}}{2} a_2, \frac{\sqrt{3}}{2} a_3\right\}\right) = \frac{1}{2} \frac{\sqrt{3}}{2} \frac{m(\text{conv}(\{a_1, a_2, a_3\}))}{m(\Delta_2)},$$

where $m(A)$ denotes the Lebesgue measure of the set A. Taking expectations over a_1, a_2, a_3, we get that

$$\Pi^{3,3,2} \leq 2\frac{1}{2}\frac{\sqrt{3}}{2} P_3,$$

The Equilibrium Manifold and Probabilistic Equilibrium Theory   227

and inserting the value of P_3 from Lemma 7.6 (in Section 2.3), we get (13). □

Comparing the expression in (13) to the result obtained in Theorem 7.7, we note that the bound obtained is not exact, being based on the two cones spanned by the techniques from each of the end points of the (three-dimensional) Edgeworth box, which only in exceptional cases is identical to the factor price equalization domain. Even so, it is seen that the probability of FPE is smaller in dimension 3 than in dimension 2. The assessment in dimension 3 can be generalized to higher dimensions with the same line of proof, which is left to the reader. The key ingredient here as above is the assessment of an expected volume of a subsimplex of Δ_n obtained by random selection of its vertices, the quantity P_n considered in Section 2.3.

Theorem 7.9. *Consider the family* $\left(S_r^L\right)^r \times \{(\omega_i)_{i=1}^r \in (\mathbb{R}_+^r)^2 \mid \sum_i \omega_i = e\}$ *of two-country worlds with r commodities and r factors of production, where techniques and endowments are sampled uniformly given that the factors can be used efficiently. Then* $\Pi^{r,r,2}$, *the probability of FPE, satisfies*

$$\Pi^{r,r,2} < \frac{\sqrt{r}}{2} P_r < \frac{\sqrt{r}}{2} \left(\frac{1}{2}\right)^{r-1}. \tag{14}$$

Since the bound in (14) has the magnitude of $\dfrac{\sqrt{r}}{2^r}$, it approaches zero for $r \to \infty$. With a view to the large number of distinct commodities figuring in international trade as well as the large number of factors of production used in real-life, it seems that FPE is a rather unlikely event. Clearly, one should not overdo the importance of results as those obtained here, which pertain to a model of international trade which is anyway lacking in realism. But the result does point to a weakness of the classical theory which may have given too much attention to a phenomenon turning out to be specific for the low-dimensional geometric versions of the model. We return to this point in the concluding remarks.

2.2.3 Assessing the probability of FPE: Unequal numbers of commodities and factors

In the present section, we move beyond the classical case of a equal number of commodities and factors of production, so that we have either (i) $r > n$ or (ii) $r < n$. Case (i) is rather easily resolved: If q is a common factor price vector, then the factor proportion vectors $\phi_{\sigma_1}(q), \ldots, \phi_{\sigma_n}(q)$ span a subspace of $\mathbb{R}_+^r$ of less than full dimension, so that the relative volume of the factor price equalization domain is 0 for all choices of technology. Thus, more factors than commodities means that factor price equalization is a null event independent of technology.

(ii) If $r < n$, the factor endowment of each country must belong to the cone spanned by $\phi_{\sigma_1}(q), \ldots, \phi_{\sigma_n}(q)$. If $K = 2$, we get the so-called "lens condition" illustrated in Fig. 2, where the techniques are defined by a_1 and a_2 as in the case shown in Fig. 1, but where we have an additional technique, for simplicity assumed to belong to the diagonal.

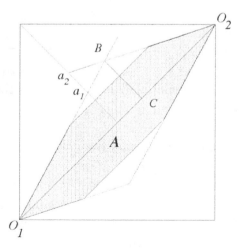

Fig. 2. The lens condition: the factor price equalization domain is determined not only by the two techniques represented by a_1 and a_2, but also by a third technique.

The Equilibrium Manifold and Probabilistic Equilibrium Theory   229

In this case, the FPE domain will depend on commodity prices, or rather, on the amount of each commodity produced in the integrated equilibrium, which shows up in the lengths of the vectors $\phi_{\sigma_1}(q), \ldots, \phi_{\sigma_3}(q)$. This means that for the assessment of the probability of FPE, we need to take the commodity production into account.

We shall not consider the general case but restrict our attention to the case illustrated, where one of the techniques coincides with the diagonal. Then the length of the segment parallel to the diagonal parametrizes the output in the third technique, and given this length, the FPE domain is uniquely determined. Compared to the case of only two techniques, the area of the FPE domain is reduced by a triangle with top in B (and its symmetric counterpart). Letting t be the fraction of the distance from B to C which is not in the FPE domain, we have that the probability of FPE, given that $a_3 = (\frac{1}{2}, \frac{1}{2})$, can be found as

$$
\int_0^{\frac{\sqrt{2}}{2}} \int_0^{\frac{\sqrt{2}}{2}} \left[2\sqrt{2}\frac{\alpha\beta}{\alpha+\beta} - \int_0^1 \sqrt{2}t^2 \frac{2\alpha\beta}{\alpha+\beta} \, dt \right] d\alpha \, d\beta
$$

$$
= \frac{2}{3} \int_0^{\frac{\sqrt{2}}{2}} \int_0^{\frac{\sqrt{2}}{2}} 2\sqrt{2}\frac{\alpha\beta}{\alpha+\beta} \, d\alpha \, d\beta = \frac{1}{3},
$$

showing that the impression obtained from the figure is sustained by the computation. It should of course be stressed that the numerical value of the probability of FPE has been established only for the special case where one of the techniques coincides with the diagonal.

2.3   *Expected volume of a random subsimplex*

In this section, we provide an assessment of P_n, the expected relative volume of a subsimplex of Δ_n the vertices of which are chosen at random. For the one-dimensional simplex $\Delta_2 = [0, 1]$, we have that

$$
P_2 = \int_0^1 |b - a| \, da \, db = \int_0^1 \left[\int_0^b (b - a) \, da + \int_b^1 (a - b) \, da \right] db = \frac{1}{3}.
$$

For the higher dimensional versions, we need some preliminary considerations.

Lemma 7.6. *Let $d \geq 2$ be arbitrary and let $T = \{x \in \mathbb{R}^d_+ \mid \sum_{i=1}^d x_i \leq 1\}$. Consider the subsets $T(t_1, \ldots, t_d) = \mathrm{conv}(\{0, t_1 e_1, \ldots, t_d e_d\})$ of T for $(t_1, \ldots, t_d) \in [0,1]^d$. Then*

$$\int_0^1 \cdots \int_0^1 m_d(T(t_1, \ldots, t_d))\, \mathrm{d}t_1 \ldots \mathrm{d}t_d \leq \frac{1}{2} m_{d-1}(\mathrm{conv}(\{e_1, \ldots, e_d\})),$$

where m_d denotes Lebesgue measure in $\mathbb{R}^d$.

Proof. Since $m(T(t_1, \ldots, t_d)) \leq m(T(t_1, 1, \ldots, 1))$, the assessment follows directly after integrating over t_1. □

As is seen, a much sharper bound could be obtained, depending on the dimension d, but the present crude bound will suffice for our purposes.

Lemma 7.7. *Let $n \geq 3$, and define P_n by*

$$P_n = \int_{a_1, \ldots, a_n \in \Delta_n} \frac{m(\mathrm{conv}(\{a_1, \ldots, a_n\}))}{m(\Delta_n)}\, \mathrm{d}a_1 \cdots \mathrm{d}a_n.$$

Then $P_n \leq \frac{1}{3} 2^{-(n-2)}$.

Proof. By induction in n; for $n = 2$ the result was proved above. Assume that the lemma holds for all $2 \leq k < n$, write $\Delta_n = \mathrm{conv}(\{e_1, \ldots, e_n\})$ and consider an arbitrary subsimplex $D = \mathrm{conv}(\{a_1, \ldots, a_n\})$. Discarding the cases where e_1 belongs to the affine subspace spanned by a facet of D, we may assume that there is some vertex in D, say a_1, such that $D \subset D' = \mathrm{conv}(\{e_1, a_2 \ldots, a_n\})$, and we restrict our attention to subsimplices containing e_1.

Let $\hat{a}_i$, for $i = 2, \ldots, n$, be the intersection of the rays from e_1 through a_i with the facet $\mathrm{conv}(\{e_2, \ldots, e_n\})$. Then using Lemma 7.5 and the fact that relative Lebesgue measure is invariant under linear maps, we get that

$$\frac{m(\mathrm{conv}(\{e_1, a_2, \ldots, a_n\}))}{m(\mathrm{conv}(\{e_1, \hat{a}_2, \ldots, \hat{a}_n\}))} \le \frac{1}{2},$$

and it follows that

$$\frac{m(\mathrm{conv}(\{e_1, a_2, \ldots, a_n\}))}{m(\Delta_n)} \le \frac{1}{2} \frac{m(\mathrm{conv}(\{\hat{a}_2, \ldots, \hat{a}_n\}))}{m(\mathrm{conv}(\{e_2, \ldots, e_n\}))}$$

$$\le \frac{1}{2}\frac{1}{3}2^{-(n-3)} = \frac{1}{3}2^{-(n-2)},$$

where we have used the induction hypothesis. $\qquad \square$

3. Exercises

(1) There are m given consumers (X_i, P_i), , for which $X_i = \mathbb{R}^l_+$ and the preferences P_i are described by Cobb–Douglas utility functions

$$u_i(x_1, \ldots, x_l) = x_1^{\alpha_1} \cdots x_l^{\alpha_l}, \ (\alpha_1, \ldots, \alpha_l) \in \mathbb{R}^l_{++}, \sum_{h=1}^{l} \alpha_h = 1,$$

$i = 1, \ldots, m$. Find the equilibrium manifold for the economy with demand functions ξ_i derived from $(X_i, P_i)_{i=1}^{m}$.

(2) Suppose that m consumers have Leontief preferences, i.e. the preferences can be described by utility functions of the form

$$u_i(x_1, \ldots, x_l) = \min\left\{\frac{x_1}{a_1}, \ldots, \frac{x_l}{a_l}\right\}, (a_1, \ldots, a_l) \in \mathbb{R}^l_{++},$$

$i = 1, \ldots, m$. Find the equilibrium manifold.

(3) Let $(p^0, \omega_1^0, \ldots, \omega_m^0)$ and $(p^1, \omega_1^1, \ldots, \omega_m^1)$ be points on the equilibrium manifold E of Exercise 1. Find a continuous curve in E connecting the two points.

(4) Let $\zeta^\alpha : \mathbb{R}^2_{++} \to \mathbb{R}^2$ be given by

$$\zeta^\alpha_1(p) = \frac{p^\alpha_2}{p_1}, \ \zeta^\alpha_2(p) = -p^{\alpha-1}_2, \ p \in \mathring{\Delta},$$

and let $C^0 = \{\zeta^\alpha \mid \alpha \in [0, 10]\}$. Find an upper bound for the number of equilibria in any of the economies given by the excess demand functions in C^0.

(5) Let $a, a' \in \mathbb{R}^2_+$. A technology, given by its support $\sigma_{\{a,a'\}}$ is the join of the two Leontief technologies σ_a and σ'_a if

$$\sigma_{\{a,a'\}}(q) = \min \{\sigma_a(q), \sigma_{a'}(q)\}$$

for $q \in \Delta = \{q \in \mathbb{R}^2_+ \mid q_1 + q_2 = 1\}$. Let S^* be the set of all such technologies, where $a, a' \in \{x \in \mathbb{R}^2_+ \mid x \le 1\}$, and let $(\sigma_{\{a,a'\}}, \sigma_b, \omega_1) \in S^* \times S^L$ be a pair of technologies.

Find the FPE domain in a two-country world with the given technologies when aggregate factor endowment satisfies $\omega_1 + \omega_2 = (1, 1)$. Assess the probability of FPE for fixed technologies when the all distributions of the total factor endowments among the countries are equally probable.

Chapter 8

General Equilibrium and Imperfect Competition

1. The Subjective Demand Approach

In this section, we present a first attempt, due to Negishi (1960), to treat the monopolistic competitor in a general rather than a partial equilibrium framework. We consider an economy $\mathcal{E} = ((X_i, P_i, \omega_i)_{i=1}^{m}, (Y_j)_{j=1}^{n}, (\theta_{ij})_{i=1j=1}^{m\ n})$ with l commodities, m consumers and n producers. Among the latter, $n' < n$ are competitive in the sense that they maximize profits taking the prices $p_1, \dots, p_l$ of the commodities as given, while the remaining producers entertain some degree of monopoly power.

The monopolistic situation of firms $n' + 1, \dots, n$ will be formalized by their subjective or perceived demand or supply functions. For j a monopolistic firm, let L_j denote the set of commodities for which firm j exercises some monopoly power, which here means that the prices are *not* taken as given. We assume that $L_j \cap L_{j'} = \emptyset$ for $j, j' \in \{n' + 1, \dots, n\}$, $j \neq j'$ so that there is only one firm which is an imperfect competitor with respect to any given commodity. This rules out oligopoly (several firms selling the same commodity in an imperfectly competitive market) as well as bilateral monopoly, where both the buyer and seller of a commodity are imperfect competitors.

Each imperfect competitor $j \in \{n' + 1, \ldots, n\}$ is endowed with a *subjective (inverse) demand function* for each commodity $k \in L_j$,

$$\delta_k^j : Y_j \times \mathbb{R}_+^l \times F(\mathcal{E}) \to \mathbb{R}_+$$

where $F(\mathcal{E})$ is the set of feasible allocations $z = (x_1, \ldots, x_m, y_1, \ldots, y_n)$ in the economy $\mathcal{E}$. To each triple (y'_j, p, z) consisting of a production y'_j for firm j, a price vector p and a feasible allocation z in the economy, the function δ_k^j gives the subjectively perceived price of commodity k at which y'_{jk} can be sold in the market, given the produced and consumed quantities of all other commodities, as represented by z. We assume that δ_k^j is continuous and homogeneous of degree one in p.

Given that the objective of the firm is to maximize profits, a standard argument will give an optimal production choice with a resulting monopoly price p_k. This is not the end of the story, however; if consumers' total demand at the price system $(p_1, \ldots, p_l)$, where some prices have been set by the monopolists and some by the market, is such that the actual demand for some monopolist's commodities differs from what is prescribed by his perceived demand function, then this could not be a situation of equilibrium. Presumably, the subjective demand would have to be revised, but this line of thought is beyond our present model: We are interested only in equilibria, where no such revision is necessary.

More specifically, we define a *monopolistically competitive equilibrium* as an array $(z^0, p) = (x_1^0, \ldots, x_m^0, y_1^0, \ldots, y_n^0, p)$, where

(i) $z^0 \in F(\mathcal{E})$ is a feasible allocation,
(ii) for each consumer i, x_i^0 is maximal for P_i on the budget set
$$\left\{ x_i \in X_i \,\middle|\, p \cdot x_i \le p \cdot \omega_i + \sum_{j=1}^n \theta_{ij} p \cdot y_j^0 \right\},$$
(iii) for each competitive producer $j \in \{1, \ldots, n'\}$, y_j^0 maximizes $p \cdot y_j$ on Y_j,

General Equilibrium and Imperfect Competition

Box 1. Subjective demand: The function δ_k^j giving the monopolist's price depending on the quantity y_k' of the commodity brought to the market (for fixed prices of all other commodities and fixed choices of all other agents, consumers or producers), shown as dd' in the figure, corresponds to standard textbook demand functions (Fig. 1).

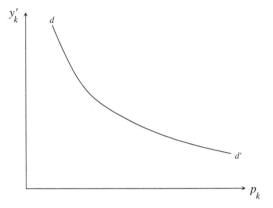

Fig. 1. Subjective inverse demand of a monopolist: Given the production and consumption of all other commodities, the firm expects that quantities sent to the market will be sold at the price given by the function δ.

If this curve represents what the monopolist believes to be the reaction of the market to his choice of quantity to be sold, it is reasonable (even though it does not follow logically from the model) that he will choose a production which maximizes his profits subject to the constraint given by the perceived demand curve.

(iv) for each monopolistic producer $j \in \{n'+1,\ldots,n\}$, y_j^0 maximizes

$$\sum_{k \in L_j} \delta_k^j(y_j^0, p, z^0) y_{jk} + \sum_{k \notin L_j} p_k y_{jk}$$

on Y_j.

Having now specified the model and its equilibria, we must check that equilibria exist. This will establish the internal consistency of

236   *Theory of General Economic Equilibrium*

the model, meaning that the monopolistic behavior of some producers is compatible with the actions of the other producers and
consumers in the economy. We go into some detail about the existence question since it turns out to be somewhat delicate later in this
chapter. Already at this point, we cannot continue with the standard
assumptions, we shall need an assumption on the profit functions
of the imperfect competitors:

Assumption 8.1. For each $j \in \{n' + 1, \dots, n\}$ and $(y_j, p, z) \in Y_j \times \mathbb{R}_+^l \times F(\mathcal{E})$, the set

$$\left\{ y_j' \in Y_j \, \middle| \, \sum_{k \in L_j} \delta_k^j(y_j', p, z) y_{jk}' + \sum_{k \notin L_j} p_k y_{jk}' \geq p \cdot y_j \right\}$$

is convex.

This assumption will be satisfied if the profit function is quasi-
concave. However, it is hard to say whether monopolists tend to
have a subjective demand such that the associated profit function
is quasi-concave. Indeed, Assumption 8.1 is of a type which we
try to avoid, since it is formulated not on the basics of the model
but on some of its derived concepts, in this case the profit function.
However, once we accept the assumption, an existence result is
within reach:

Theorem 8.1. *Let $\mathcal{E} = ((X_i, P_i, \omega_i)_{i=1}^m, (Y_j)_{j=1}^n, (\theta_{ij})_{i=1 \, j=1}^{m \ \ n})$ be a private ownership economy where consumers and producers satisfy the well-behavedness assumptions as stated in Theorem. 2.1, and, in addition, the imperfectly competitive producers satisfy Assumption. 8.1.*

Then there exists a monopolistically competitive equilibrium in $\mathcal{E}$.

There are several ways of proving Theorem. 8.1. In the proof
presented below, we transform the problem to that of finding a
Walras equilibrium in a suitably chosen exchange economy, in order
to apply Theorem. 1.1.

Proof of Theorem 8.1.   First of all, we show that there is
some vector $a \in \mathbb{R}_+^l$, such that for every feasible allocation $z =$

$(x_1, \ldots, x_m, y_1, \ldots, y_n) \in F(\mathcal{E}), y_j \leq a$ for every j, that is to say, such that production in feasible allocations is bounded from above. Suppose not; then there would be a sequence $(y_1^\nu, \ldots, y_n^\nu)_{\nu=1}^\infty$ of productions with, say, $|y_{1k}^\nu| > \nu$ for all ν and $\sum_{j=1}^n y_{jk}^\nu$ bounded (since the productions belong to feasible allocations), and consequently, the sequence $\left(\frac{1}{\nu} y_1^\nu, \ldots, \frac{1}{\nu} y_n^\nu\right)_{\nu=1}^\infty$ would satisfy $\left|\frac{1}{\nu} y_{1k}^\nu\right| > 1$ and $\sum_{j=1}^n \left(\frac{1}{\nu} y_{jk}^\nu\right) \to 0$. Using compactness of the set of feasible allocations, we obtain a contradiction of the irreversibility property of $\sum_{j=1}^n Y_j$.

Next, define an exchange economy with allocation-dependent income functions as follows: There are $m + n$ consumers, the first m being the original consumers of $\mathcal{E}$. For the remaining n consumers, consumption sets are defined as

$$\tilde{X}_j = \{x \in \mathbb{R}^l \mid x \geq -a\}.$$

For each new consumer j, we define preferences by specifying the bundles which are preferred by the consumer to any given bundle; this preferred set will depend on parameters other than the given bundle (prices; other agents' bundles), a feature which is purely technical and which does not interfere with the existence theorem to be applied.

For each j, define the function $\mu_j : \tilde{X}_j \to \mathbb{R}$ by

$$\mu_j(x) = \inf\left\{\frac{1}{\lambda} \middle| \lambda > 0, \lambda(x - a) \in -Y_j - \{a\}\right\}.$$

Then μ_j can be recognized as the Minkowski functional of the convex set $(-Y_j) \cap \tilde{X}_j$ w.r.t. the point a; the number $\mu_j(x)$ can be interpreted as the degree of belonging to $-Y_j$ for the point x. It can be shown rather easily to be continuous and strictly quasi-concave — actually, it is even concave.

Now, for $j \in \{1, \ldots, n'\}$, let $x_j \in \tilde{X}_j$ be given, and define the set of bundles preferred to x_j as

$$P_j(x_j) = \{x_j' \in \tilde{X}_j \mid \mu_j(x_j') > \mu_j(x_j)\}.$$

For $j \in \{n' + 1, \ldots, n\}$, let $x_j \in \tilde{X}_j, p \in \Delta$ and $z \in F(\mathcal{E})$ be given, and define the set of preferred bundles as

$$P_j(x_j; p, z) = \begin{cases} \{x'_j \in \tilde{X}_j \mid \mu_j(x'_j) > \mu_j(x_j)\} & \text{if } \mu_j(x_j) < 1, \\ \{x'_j \in \tilde{X}_j \mid \mu_j(x'_j) > 1, -\pi_j(x'_j, p, z) < p \cdot x_j\} & \text{otherwise,} \end{cases}$$

where $\pi_j(x'_j, p, z)$ is the profit function of "consumer" $j \in \{n'+1, \ldots, n\}$ given by

$$\pi_j(y'_j, p, z) = \sum_{k \in L_j} \delta^j_k(y'_j, p, z) y'_{jk} + \sum_{k \notin L_j} p_k y'_{jk}.$$

For agents $j \in \{1, \ldots, n'\}$, the preferences can be described by a utility function (namely $\mu_j(\cdot)$), whereas for $j \in \{n' + 1, \ldots, n\}$, they cannot. However, in the latter case, whenever x'_j is preferred to x_j given p and z, we can easily check that there are neighborhoods $U_{x'_j}, U_{x_j}, U_p$ and U_z of x'_j, x_j, p and z, respectively, such that every $\tilde{x}'_j$ in $U_{x'_j}$ is preferred to every $\tilde{x}_j$ in U_{x_j} given any $\tilde{p} \in U_p$ and $\tilde{z} \in U_z$. This, together with $x_j \notin P_j(x_j; p, z)$ and convexity of the sets $P_j(x_j; p, z)$ for all x_j, p and z, is all that we need to apply the existence theorem for Walras equilibria in the constructed exchange economy.

It remains now to assign income functions to each consumer. For $i = 1, \ldots, m$, we let the income of consumer i be defined as

$$w_i(p, z) = p \cdot \omega_i + \sum_{j=1}^{n} \theta_{ij} p \cdot y_j,$$

whereas for $j = 1, \ldots, n$, it is defined as

$$w_j(p, z) = - \max\{p \cdot y_j \mid y_j \in Y_j, y_j \leq a\};$$

we leave it to the reader to check that each I_j defined as above is continuous.

Now we may apply Theorem. 1.1 to obtain the existence of an equilibrium $(x_1, \ldots, x_m, x'_1, \ldots, x'_n, p)$ for the exchange economy $\mathcal{E}$. Letting $y_j = -x'_j$ for $j = 1, \ldots, n$, it is easily seen that $(x_1, \ldots, x_m, y_1, \ldots, y_n, p)$ is a monopolistically competitive equilibrium. $\qquad \square$

Theorem. 8.1 establishes that equilibria of the considered type exist, thus opening up for further investigations of the model. However, we shall not proceed to a detailed study of its properties, and this for several reasons.

First of all, few general properties of monopolistically competitive equilibria can be established without a further specification of the model. Thus, equilibria are not in general Pareto-optimal; on the other hand, Pareto-optimal allocations may be obtained as monopolistically competitive equilibria (after a suitable income transfer), but this is a trivial consequence of the fact that a Walras equilibrium is a monopolistically competitive equilibrium with specially constructed (infinitely elastic) subjective demand functions of the type

$$\delta_k^j(y_j, p, z) = p_k$$

where constant, or given, prices are assumed.

Another, more fundamental, reservation against the concept of a monopolistically competitive equilibrium is that there is an equilibrium for any possible subjective demand function, so that the set of price–allocation pairs which may occur as a monopolistically competitive equilibrium for *some* conjectures is very large. This problem arises since conjectures or subjective demand functions are taken as given; they are supposed to be determined outside the model.

There could be several ways of making up for this shortcoming, such as including an activity of information processing and learning by the imperfect competitors, but in the following, we adapt a simpler one, assuming right away that the conjectures formed by imperfect competitors are correct reflections of actual demand conditions. This will take us to models with objective rather than subjective demand.

2. Equilibrium with Objective Demand

In this section, we replace the subjective perceived demand relationships used in the previous model by the true demand of the competitive sector (households and competitive firms): An

240   *Theory of General Economic Equilibrium*

imperfect competitor contemplating a change in output will take into consideration the actual change of prices necessary for market clearing in the new situation.

2.1   *Imperfect equilibrium with quantity-setting firms*

Since we shall need an explicit formulation of market clearing prices as a function of output in the imperfectly competitive sector, the relation between the two sectors will be formalized in a rather primitive way. More specifically, we assume that the goods produced by the firms are handed over physically to the shareholders (in proportion to their number of shares) who then take the goods to the market and perform the trading. Under perfect competition, this arrangement would produce the same result as the usual one, since with prices given it does not matter to the shareholder whether he gets his profit share in value or in actual goods. However, with prices depending on actual supply, it does make a difference. The idea of shareholders being paid off in kind rather than in money may lack realism, but it allows us to formulate the profit maximization conditions of firms in a reasonably simple way.

We consider an economy $\mathcal{E} = ((X_i, P_i, \omega_i)_{i=1}^m, (Y_j)_{j=1}^n, (\theta_{ij})_{i=1\,j=1}^{m\ \ n})$, where, as usual, the share of consumer i in firm j is given by θ_{ij}. However, the economic content of shares differs from that of previous models. We assume that when each firm j has chosen a production plan y_j, then shareholder i's endowment changes from ω_i to $\omega_i + \sum_{j=1}^n \theta_{ij} y_j$. The reestablished endowments are then brought to the market and exchanges take place, resulting in a Walras equilibrium for the exchange economy

$$\mathcal{E}(y_1, \ldots, y_n) = (X_i, P_i, \omega_i(y_1, \ldots, y_n))_{i=1}^m.$$

Since shareholders in our economy must supply the inputs of the firms, feasibility of a production plan y_j demands not only technical feasibility, that is $y_j \in Y_j$ for $j = 1, \ldots, n$, but also availability of inputs in the shareholders' endowments. This, in turn, will in general

General Equilibrium and Imperfect Competition 241

depend on the production plans chosen by the other firms. Let

$$Y_j(y_1,\ldots,y_n) = \left\{ y_j' \,\middle|\, y_j' \in Y_j, \omega_i + \sum_{h\neq j} \theta_{ih} y_h + \theta_{ij} y_j' \in X_i, i = 1,\ldots,m \right\}$$

(1)

be the set of production plans for firm j which are such that each shareholder can survive without trading, given that the other firms have chosen plans y_h, $h \neq j$. For notational convenience, this set is written as $Y_j(y_1,\ldots,y_n)$ even though it does not depend on y_j.

We now turn to the behavior of the firms. For each array of production plans $(y_1,\ldots,y_n)$ which is feasible in the sense that $y_j \in Y_j(y_1,\ldots,y_n)$, $j = 1,\ldots,n$, we let $W[\mathcal{E}(y_1,\ldots,y_n)]$ be the set of Walras equilibria for $\mathcal{E}(y_1,\ldots,y_n)$. We assume now that there is a *price function* π assigning to each array of feasible productions $(y_1,\ldots,y_n)$ a Walras equilibrium price $p = \pi(y_1,\ldots,y_n) \in \mathbb{R}^l_+$, and moreover, doing this in a continuous way. The existence of a price function of this type clearly presupposes that the set of Walras equilibria $W[\mathcal{E}(y_1,\ldots,y_n)]$ is non-empty for each feasible array $(y_1,\ldots,y_n)$ of production plans. This is, however, a minor problem, and non-emptiness is largely guaranteed by the standard assumptions on $\mathcal{E}$. The continuity assumption on π is more problematic: actually, it is extremely restrictive. We shall return to this point at the end of this section.

Given the price function π describing what comes out of the trading process in the market for given choices of production plans, the firms are able to evaluate the consequences in terms of changed profits of a change in their production plans. We assume that firm j, for $j = 1,\ldots,n$, chooses its production y_j so as to maximize profits, i.e. such that

$$\pi(y_j, y_{-j}) \cdot y_j \geq \pi(y_j', y_{-j}) \cdot y_j', \text{ all } y_j' \in Y_j(y_1,\ldots,y_n).$$

(2)

Now we define a *Cournot–Walras equilibrium* (relative to π) as an allocation $z^0 = (x^0, y^0)$, such that

(a) (x^0, y^0) is feasible,
(b) $(x_1^0,\ldots,x_m^0, \pi(y_1^0,\ldots,y_n^0))$ is a Walras equilibrium for $\mathcal{E}(y_1^0,\ldots,y_n^0)$,

(c) for each j, y_j^0 is chosen so as to satisfy (2).

With sufficiently strong assumptions on the economy, it can be shown that it has a Cournot–Walras equilibrium.

Theorem 8.2. *Let* $\mathcal{E} = ((X_i, P_i, \omega_i)_{i=1}^m, (Y_j)_{j=1}^n, (\theta_{ij})_{i=1\,j=1}^{m\ n})$ *be a private ownership economy satisfying the standard well-behavedness assumptions of Theorem 2.1. Suppose, furthermore, that there is a continuous price function $\pi : \widetilde{Y} \to \mathbb{R}_+^l$, where*

$$\widetilde{Y} = \left\{ (y_1, \ldots, y_n) \,\middle|\, y_j \in Y_j(y_1, \ldots, y_n), j = 1, \ldots, n \right\},$$

such that

(i) *$\pi(y_1, \ldots, y_n)$ is a Walras equilibrium price for the economy $\mathcal{E}(y_1, \ldots, y_n)$,*

(ii) *for each $(y_1, \ldots, y_n)$ and $j = 1, \ldots, n$, the set*

$$\{y_j' \in Y_j(y_1, \ldots, y_n) \mid \pi(y_j', y_{-j}) \cdot y_j' \geq \pi(y_1, \ldots, y_n) \cdot y_j\}$$

is convex.

Then there exists a Cournot–Walras equilibrium in $\mathcal{E}$ (relative to π).

Proof. First of all, we show that $\widetilde{Y}$ is convex: Let $(y_1, \ldots, y_n)$ and $(y_1', \ldots, y_n')$ be in $\widetilde{Y}$, and let $\lambda \in [0, 1]$. Then we have for all i that

$$\omega_i + \sum_{j=1}^n \theta_{ij} y_j \in X_i, \quad \omega_i + \sum_{j=1}^n \theta_{ij} y_j' \in X_i,$$

and consequently

$$\lambda\left(\omega_i + \sum_{j=1}^n \theta_{ij} y_j\right) + (1 - \lambda)\left(\omega_i + \sum_{j=1}^n \theta_{ij} y_j'\right) \in X_i.$$

But

$$\lambda\left(\omega_i + \sum_{j=1}^n \theta_{ij} y_j\right) + (1 - \lambda)\left(\omega_i + \sum_{j=1}^n \theta_{ij} y_j'\right) = \omega_i + \sum_{j=1}^n \theta_{ij}(\lambda y_j + (1 - \lambda) y_j'),$$

General Equilibrium and Imperfect Competition 243

and it follows that $(\lambda y_1 + (1 - \lambda)y_1', \ldots, \lambda y_n + (1 - \lambda)y_n')$ belongs to $\widetilde{Y}$. Thus, $\widetilde{Y}$ is convex.

Next, we show that Y is compact. Clearly, $\widetilde{Y}$ is closed, and boundedness of $\widetilde{Y}$ is proved as follows: Since $\sum_{j=1}^{n} Y_j$ satisfies Assumption. 0.3 there is a number $K > 0$ such that

$$\sum_{i=1}^{m} \omega_{ik} + \sum_{j=1}^{n} y_{jk} \leq K, \quad k = 1, \ldots, l,$$

for any feasible allocation $(x_1, \ldots, x_m, y_1, \ldots, y_n)$ (total output plus endowment is bounded from above). Since also each individual Y_j is well-behaved, there is $K_j > 0$ such that for all $y_j \in Y_j$, if $y_{jk} \geq -K$ for all k, then $y_{jk} \leq K_j$, all k (if no input is larger than the total available amount, then the firm cannot produce more than K_j). It follows that if $K^* = \max\{K, K_1, \ldots, K_n\}$, then

$$\widetilde{Y} \subseteq \{(y_1, \ldots, y_n) \in \mathbb{R}^{nl} | y_{jk} \leq K^*, \ j = 1, \ldots, n, \ k = 1, \ldots, l\},$$

so $\widetilde{Y}$ is bounded. Finally, $\widetilde{Y}$ is non-empty since $(0, \ldots, 0) \in \widetilde{Y}$.

Let $(y_1, \ldots, y_n) \in \widetilde{Y}$ be arbitrary. Suppose that there is some j and y_j' such that $y_j' \in Y_j(y_1, \ldots, y_n)$, meaning that $(y_1, \ldots, y_{j-1}, y_j', y_{j+1}, \ldots, y_n) \in \widetilde{Y}$, and

$$\left\{ y_j' \mid (y_1, \ldots, y_{j-1}, y_j', y_{j+1}, \ldots, y_n) \in \widetilde{Y}, \right.$$

$$\left. \pi(y_1, \ldots, y_{j-1}, y_j', y_{j+1}, \ldots, y_n) \cdot y_j' > \pi(y_1, \ldots, y_n) \cdot y_j \right\} \neq \emptyset.$$

By our assumptions, the latter set is convex, and since π is continuous, there is an open neighborhood U_y of $y = (y_1, \ldots, y_n)$ in $\widetilde{Y}$ such that the map $\varphi_y : U_y \to \widetilde{Y}$ taking $(y_1'', \ldots, y_n'')$ in U_y to

$$(y_1'', \ldots, y_{j-1}'', y_j', y_{j+1}'', \ldots, y_n'') \in \widetilde{Y}$$

(i.e. φ_y is the identical mapping on each coordinate except the jth which is taken to y_j') sends each production plan y_k'' either in itself or in a production plan which gives a higher profit.

244 *Theory of General Economic Equilibrium*

Now, if there was no Cournot–Walras equilibrium relative to π, then the family $(U_y)_{y\in\widetilde{Y}}$ with associated maps $(\varphi_y)_{y\in\widetilde{Y}}$ would be an open covering of $\widetilde{Y}$. Using partitions of unity and Brouwer's fixed-point theorem, we obtain a contradiction. □

On the face of it, we have now shown that a model of monopolistic competition may be constructed where firms act on correct conjectures of consumers' reactions. But this should be considered in the light of the assumptions made, and the model as such is open to criticism. On our way to the result, we have used assumptions which are stronger than what we would be willing to accept in general. In particular, we have assumed that one can select a Walras equilibrium price for exchange economies $\mathcal{E}(y_1,\ldots,y_n)$ in a continuous way, and this has not been substantiated so far. Even with continuity of the equilibrium price selection in hand, we still needed a convexity assumption on the sets of production plans yielding a higher profit than a given one, and even if the well-behaved demand and revenue functions drawn in microeconomics textbooks have this property, it is easy to construct examples where the property fails.

2.2   *The profit maximization problem*

In addition to the technical problems, there are certain conceptual ones as well. We have assumed that firms maximize profits relative to a given profit function π. Now π selects some equilibrium price for $\mathcal{E}(y_1,\ldots,y_n)$; therefore, it involves a (normalization) rule for deciding which of the infinitely many scalar multiples of a given equilibrium price should be used in firms' profit calculations. The point is that it does make a difference: profits depend on the normalization in a non-trivial way.

In the following example, due to Haller (1986), there are three goods, two consumers and one firm. Consumption sets are $\mathbb{R}_+^l$, and preferences are given by utility functions

$$u_1(x_{11},x_{12},x_{13}) = x_{11} + x_{12}, \quad u_2(x_{21},x_{22},x_{23}) = x_{21} + x_{23},$$

General Equilibrium and Imperfect Competition 245

and the initial endowments are $\omega_1 = \omega_2 = (2,0,0)$. The firm produces goods 2 and 3 using good 1 as input, with input requirements $C(y_2, y_3) = \frac{1}{4}y_2^2 + y_3^2$, and the choices of production are assumed restricted to the set

$$\widehat{Y} = \{(y_1, y_2, y_3) \mid -3.999 \le y_1 \le -C(y_2, y_3), 0.1$$
$$\le y_2 \le \sqrt{4 - y_3^2}, \sqrt{3.5} \le y_3 \le \sqrt{3.99}\}.$$

The consumers have equal shares of size $1/2$.

If the firm produces $y = (y_1, y_2, y_3)$, then each consumer enters the market with $\widetilde{\omega}_i = (2,0,0) + \frac{1}{2}(y_1, y_2, y_3) = \frac{1}{2}(4 + y_1, y_2, y_3)$, so that $\widetilde{\omega}_i \in \mathbb{R}^3_{++}$, $i = 1, 2$. Since $y \in Y$, we have that $0 \le 4 - y_3^2 \le \frac{1}{2}$ and $0 \le y_2 \le \frac{\sqrt{2}}{2}$. Again using the properties of Y, it can be checked that $\widetilde{\omega}_1 + \widetilde{\omega}_2 < \widetilde{\omega}_3$ and that the allocation (x_1, x_2, p) is a Walras equilibrium in the economy $\mathcal{E}(y)$ if and only if

$$x_1 = (2\widetilde{\omega}_1, 2\widetilde{\omega}_2, 0), x_2 = (0, 0, 2\widetilde{\omega}_3), p = \lambda\left(1, 1, \frac{\widetilde{\omega}_1 + \widetilde{\omega}_2}{\widetilde{\omega}_3}\right), \quad \lambda > 0.$$

$$(3)$$

Once we have the full description of all possible equilibria, we may consider the implication of different choices of production plan y.

Utility maximization. The utility of consumer 1 can be written using (3) as $u_1(y) = 4 + y_1 + y_2$, which attains its maximum over allocations satisfying (3) at $y_3 = \sqrt{3.5}$, $y_2 = \frac{\sqrt{2}}{2}$ and $y_1 = -C(y_2, y_3)$. For consumer 2, the utility can be written as $u_2(y) = 2\widetilde{\omega}_3 = y_3$, so that all y with $y_3 = \sqrt{3.99}$ maximize the utility for consumer 2. That allocations maximizing the utility of one consumer will be different depending on the consumer selected is of course not surprising.

Price normalization and profit maximum. As always, when considering Walras equilibria, the equilibrium price may be multiplied by any $\lambda > 0$ without changing the Walras equilibrium allocation. In this sense, the normalization rule for prices is of no importance for the equilibrium. But for profit maximization, the normalization matters:

(a) If commodity 1 is chosen as numeraire, so that $p_1 = 1$ for all admissible price systems, then for the equilibrium price found in (3), we get that

$$p \cdot y = \left(1, 1, \frac{\widetilde{\omega}_1 + \widetilde{\omega}_2}{\widetilde{\omega}_3}\right) \cdot y = 4 + 2(y_1 + y_2) = 2u_1(y) + 4,$$

and profit is maximized when $u_1(y)$ is maximal.

(b) Taking commodity 2 as numeraire will give the same result, but is prices are normalized in other ways than by choosing a numeraire, and profit maximization may give very different results. Thus, if prices must be chosen in such a way that $p_2 = p_1^2$, then it can be shown that the profit maximizing production plan is $y_1 = -3.999$, $y_2 = 0.1$, $y_3 = \sqrt{3.99}$ (cf. Haller, 1986, p. 724).

Going somewhat deeper into the matter, we note that the very objective of profit maximization is dubious. In the world of perfect competition which has been our main concern in the previous chapters, profit maximization is the best that the firm can do from the point of view of its shareholders. With imperfect competition, such shareholder unanimity cannot be expected to hold. Since the acts of the firm have an effect on prices, shareholders may have an interest in making the firms act so as to increase the value of their individual endowments. In that case, neither profit maximization nor indeed any other simple decision rule for the firm can be said to represent shareholders' preferences.

In view of this, it has been suggested that alternative objectives of the firm should be considered. If shareholders maximize profits, they extract monopoly gains from the consumers, but doing so, they may hurt themselves as consumers. To avoid this situation, Dierker

General Equilibrium and Imperfect Competition 247

and Grodal (1999) proposed to replace the objective of profit maximization by *real wealth maximization*.

Assume that consumers $i = 1, \ldots, m$ have well-defined *demand functions* ξ_i selecting for each price $p \in \mathbb{R}^l_{++}$ a specific bundle $\xi_i(p)$ which is maximal for the preferences of the consumer at the income obtained from selling the initial endowment and possible shares of the profits of the firms. Then $\xi_i(p)$ is positively homogeneous of degree 0, that is $\xi_i(\lambda p) = \xi_i(p)$ for each $\lambda > 0$. For a given normalization, for example the one obtained by choosing a particular good, say the lth, as numeraire, we let $w_i(p)$ be the income of consumer i at price p. If the subset I_j of $\{1, \ldots, m\}$ corresponds to the shareholders of firm j, then $\xi_{I_j}(p) = \sum_{i \in I_j} \xi_i(p)$ and $w_{I_j}(p) = \sum_{i \in I_j} w_i(p)$ denote the aggregate demand and income of firm j's shareholders.

Suppose now that as a result of the strategy choices τ^1 of firm j, which as in the model of quantity-setting oligopoly considered above were the production plans, but which in general might also be something else, an equilibrium with prices p^1 is established in the economy. Then the strategy τ^2 is rw-better (real wealth-better) than τ^1 with resulting prices p^2, written as $\tau^2 >_{rw} \tau^1$, if

$$p^1 \cdot \xi_{I_j}(p^1) < w_{I_j}(p^2),$$

so that taken together, the shareholders have sufficient income after using the strategy τ^2 to buy the consumption bundles obtained using τ^1, at least after some redistribution among shareholders. The firm maximizes real wealth if it chooses a strategy τ such that there is no other strategy τ' with $\tau' >_{rw} \tau^*$.

Clearly, the normalization of prices is irrelevant for a real–wealth maximizing firm, since multiplying p by $\lambda > 0$ means that both $p \cdot \xi_{I_j}(p)$ and $w_{I_j}(p)$ are multiplied by λ. Unfortunately, getting rid of one of the disturbing features of the models of imperfect competition does not mean that other problems disappear as well, so that even the use of alternative objectives does not yet seem to open up for general results.

248   *Theory of General Economic Equilibrium*

3. Exercises

(1) Let $\mathcal{E}$ be an economy with two commodities. There are three consumers with consumption set $\mathbb{R}^2_+$ and preferences described by utility functions

$$u_1(x_1, x_2) = x_1^2 x_2^3, \quad u_2(x_1, x_2) = x_1^3 x_2, \quad u_3(x_1, x_2) = x_1^3 x_2^4.$$

The initial endowments are

$$\omega_1 = (2, 4), \quad \omega_2 = (3, 7), \quad \omega_3 = (4, 5).$$

There is a single producer with a production set

$$Y = \{(y_1, y_2) \in \mathbb{R}_- \times \mathbb{R} \mid y_2 \le \sqrt{-y_1}\},$$

and the consumers have equal shares of $1/3$ in the profits of the firm.

The firm expects that the consumers' demand for its output is a decreasing function of its relative price, so that

$$y_2^* = \frac{p_1}{p_2}.$$

Find a monopolistically competitive equilibrium in $\mathcal{E}$.

(2) Consider an exchange economy $\mathcal{E} = (X_i, P_i, \omega_i)_{i=1}^m$, where for each i,

(i) $X_i = \mathbb{R}^l_+$,

(ii) P_i can be described by a Cobb–Douglas utility function,

(iii) $\omega_i \in \mathbb{R}^l_+$.

Check that there is a unique Walras equilibrium in $\mathcal{E}$.

We now assume that each consumer i may choose to withhold some of the endowment before engaging in market activities. Formulate a Nash equilibrium in the resulting game where each consumer may select endowments $\omega_i' \le \omega_i$. Does Nash equilibria exist? Are the allocations obtained in a Nash equilibrium Pareto optimal?

(3) Find a Cournot–Walras equilibrium in the economy described in Exercise 1.

General Equilibrium and Imperfect Competition 249

(4) Let $\mathcal{E} = ((X_i, P_i, \omega_i)_{i=1}^m, (Y_j)_{j=1}^n, (\theta_{ij})_{i=1\,j=1}^{m\ \ n})$ be a given economy and suppose that $\mathcal{E}$ is replicated to obtain economies $\mathcal{E}^k$ (where the consumers have shares only in the producers of their own "copy" of $\mathcal{E}$), for $k = 2, 3, \ldots$.

Suppose that there is a Cournot–Walras equilibrium in each of the replica $\mathcal{E}^k$. What happens to the equilibrium prices when k gets large?

(5) (From Bonanno (1990)). Suppose that there are two producers each choosing prices of a single output good. The demand function for the first producers have the form

$$D_1(p,p_2) = -p_1^3 + 12p_1^2 - 52p_1 + 93 + p_2,$$

$$D_2(p_1, p_2) = \begin{cases} 3 + 0.74p_1 - p_2 & 0 \le p_2 \le 1 + 0.68p_1 \\ 2.5 + 0.4p_1 - 0.5p_2 & p_2 \ge 1 + 0.68p_1. \end{cases}$$

Are the profit functions quasi-concave? Find the reaction curves (each producer's optimal response on the other producer's price) and check whether a Nash equilibrium exists.

(6) (Price-setting oligopolies, Nikaido (1975)). We consider a linear production economy (cf. Section 2.1.2) with $l - 1$ goods and labor, which is not produced but used as input for each of the other goods, with production matrix A and input coefficients $a_j l$ for labor, $j = 1, \ldots, l - 1$. The prices $(p_1, \ldots, p_{l-1}, 1) = (p, 1)$ are normalized by setting the price of labor equal to 1. There are two classes, workers and capitalists, and both choose their consumption as price takers with demand functions $F(p)$ and $G(p, \pi)$, respectively, where $\pi = (\pi_1, \ldots, \pi_{l-1})$ of profits. Finally, we let $L(p)$ denote the supply of labor at the prices $(p, 1)$.

Let $y = (y_1, \ldots, y_{l-1})$ be a vector of gross output. Show that the profits π satisfy

$$\pi_j = \left(p_j - \sum_{h=1}^{l-1} p_h a_{hj} - a_{lj} \right) y_j, \quad j = 1, \ldots, l - 1. \tag{4}$$

If supply and demand are balanced, we must have

$$F(p) + G(p, \pi) = (I - A)y \tag{5}$$

for the produced goods and

$$\sum_{h=1}^{l-1} y_h = L(p) \tag{6}$$

for the labor market.

Show that under standard conditions for the functions F, G and L, for each p there exists a gross output vector y such that (4)–(6) are satisfied. Explain that this gives rise to a mapping $p \mapsto y(p)$ which can be interpreted as an objective demand function.

Formulate a (Bertrand–)Nash equilibrium, where for each good, the price is set in such a way that profit is maximal given the prices of the other goods.

Chapter 9

Incentives and Mechanisms in General Equilibrium

1. Implementing Economic Equilibria

The design of institutions in society naturally depends on what society puts forward as goals. What is desirable for society cannot be decided from theory alone, and we assume here that society's goals have been specified and that they can be formulated in a precise way. So, we define an economic *performance correspondence* as a map P which to each economy $\mathcal{E}$ assigns a set $P(\mathcal{E})$ of socially desirable allocations in $\mathcal{E}$.

What we shall be looking for in the following is institutions leaving as much freedom of choice as possible to the agents while still achieving the goals for which they are designed. Consequently, we shall search for sets of rules such that agents, pursuing their own interest, will arrive at exactly the allocations specified by the given performance correspondence. We shall investigate to what extent this is possible.

Let $\mathfrak{E}$ be a set of exchange economies $\mathcal{E} = (X_i, P_i, \omega_i)_{i=1}^m$, where consumers satisfy Assumptions 0.1 and 0.2. We may think of $\mathfrak{E}$ as the relevant set of economic environments, for which the institutions to be designed should "work". A *mechanism for* $\mathfrak{E}$ is a *game form* $G = ((S_i)_{i=1}^m, A, f)$ defined by

(i) a set S_i of *strategies* for each individual $i \in \{1, \dots, m\}$,
(ii) a set of *outcomes*, which most conveniently are defined as a set of *net trades* $(z_1, \dots, z_m) \in \mathbb{R}^{lm}$ satisfying $\sum_{i=1}^m z_i = 0$,

252 *Theory of General Economic Equilibrium*

Box 1. Performance correspondences. Here are some examples of performance correspondences:

The Pareto correspondence PO assigns to $\mathcal{E} = (X_i, P_i, \omega_i)_{i=1}^m$ the set of Pareto-optimal allocations in $\mathcal{E}$. This performance correspondence was discussed in Chapter 3 and reappeared in several later chapters. It expresses the requirement that the final state realized in society should be efficient.

The performance correspondence IR assigns to every exchange economy $\mathcal{E} = (X_i, P_i, \omega_i)_{i=1}^m$ the set of feasible allocations $x = (x_1, \ldots, x_m)$ which are *individually rational*, no consumer being worse off at x than at the initial bundle. The ethics behind IR is a respect for the individual: the demand that no agent is forced to participate in any economic activity against his wishes.

The egalitarian performance correspondence EG assigns to an exchange economy $\mathcal{E} = (X_i, P_i, \omega_i)_{i=1}^m$ the egalitarian allocation $(\frac{1}{m} \sum_{i=1}^m \omega_i, \ldots, \frac{1}{m} \sum_{i=1}^m \omega_i)$.

The fairness correspondence FA has $FA(\mathcal{E}) = \{\text{fair allocations in } \mathcal{E}\}$.

The Walras correspondence W assigns to every economy $\mathcal{E}$ the set $W(\mathcal{E})$ of Walras allocations in $\mathcal{E}$, i.e. the allocations x such that (x, p) is a Walras equilibrium for some price vector p.

(iii) an outcome function $f : \prod_{i=1}^m S_i \to A$ taking strategy arrays $s = (s_1, \ldots, s_m)$ to outcomes $f(s) \in A$.

For every $\mathcal{E} \in \mathfrak{E}$, specifying the preferences of the individuals on outcomes transforms the game form G to a game $G[\mathcal{E}] = ((S_i, P_i^S)_{i=1}^m$, where P_i^S denotes the preference relation on strategy arrays induced by f, with

$$s \in P_i^S(s') \text{ if and only if } f(s) \in P_i(f(s')).$$

Choosing a particular solution concept S for games (which to each game $\Gamma = ((S_i, P_i))_{i=1}^m$ assigns a subset $S(\Gamma)$ of S), we say that the performance correspondence $\Pi : \mathfrak{E} \rightrightarrows A$ is implemented in S-solutions by the mechanism $G = ((S_i)_{i=1}^m, A, f)$ if for all $\mathcal{E} \in \mathfrak{E}$,

(i) $S(G[\mathcal{E}]) \neq \emptyset$,

(ii) $f(S(G[\mathcal{E}])) \subset \Pi(\mathcal{E})$.

The conditions for implementation are illustrated in the following commutative diagram, where the double arrows indicate correspondences: For each economy $\mathcal{E}$, the set of allocations selected by Π can be reached as solutions of the game $G[\mathcal{E}] \in \mathcal{G}$.

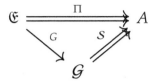

Instances of exchange economies, with precise specification of preferences and endowments, are transformed to games, whereby each individual tries to achieve the best possible final result given the rules of the game, and the performance correspondence is implemented if the solutions of the game result in allocations singled out by the performance correspondence.

2. Strategic Market Games

It should be noted that — in spite of the common use of language — the "competitive" or "market mechanism" assigning to an economy $\mathcal{E}$ its Walras equilibria is *not* a mechanism in our sense, since agents make choices depending on given prices (quoted by an auctioneer or a fictitious market agent). It is, however, possible to use the ideas behind trading in a market to construct a suitable mechanism, following the ideas of Schmeidler (1980) and Giraud (2003).

For each agent i, define the strategy set S_i as the set of all pairs (z, p), where $z \in \mathbb{R}^l$ is a net trade, and $p \in \Delta$ is a price, such that $p \cdot z \leq 0$. Thus, every agent chooses not only a net trade but also a price. The main function of prices in the mechanism will be to determine the possible trading partners, as only people choosing the same price are allowed to trade with each other. This is a consequence of the outcome function, to which we turn now.

Let $s = ((z_1, p_1), \ldots, (z_m, p_m))$ be an m-tuple of strategies. We define the outcome $f(s) = (f_1(s), \ldots, f_m(s))$ as follows: for $i = 1, \ldots, m$, let $M(p_i) = \{j | p_j = p_i\}$ denote the set of agents choosing the price p_i.

Then

$$f_i(s) = \left[z_i - \frac{\sum_{j \in M(p_i)} z_j}{|M(p_i)|} \right] + \omega_i,$$

where $|M(p_i)|$ denotes cardinality (number of elements) of the set $M(p_i)$.

Note that if $|M(p_i)| = 1$, then $f_i(s) = \omega_i$. If $|M(p_i)| > 1$, then the second term in the bracket acts as a correction factor on the sum $\sum_{j \in M(p_i)} z_i$ so as to obtain net trades summing to zero. Indeed, we have

$$\sum_{j \in M(p_i)} f_j(s) = \sum_{j \in M(p_i)} \left[z_i - \frac{\sum_{k \in M(p_i)} z_k}{|M(p_i)|} \right] + \sum_{j \in M(p_i)} \omega_i = \sum_{j \in M(p_i)} \omega_i.$$

Since the set $\{1, \ldots, m\}$ partitions into at most m mutually disjoint sets $M(p_i)$, it follows that $\sum_{i=1}^{m} f_i(s) = \sum_{i=1}^{m} \omega_i$ so that $f(s)$ belongs to the set A defined in (1) of Section 9.1.

Let $G = ((S_i)_{i=1}^{m}, A, f)$. We show that the mechanism G actually does the job of implementing PA $\cap$ IR. It does even more.

Theorem 9.1. *Let $m \geq l + 1$ and $\mathcal{E} \in \mathfrak{E}$.*

(a) *If s is a Nash equilibrium for $G[\mathcal{E}]$, then there is $p \in \Delta$ such that $(f(s), p)$ is a Walras equilibrium.*
(b) *For any Walras equilibrium (x, p), there is a strategy m-tuple s such that s is a Nash equilibrium in $G[\mathcal{E}]$ and $f(s) = x$.*

Proof. (a) Let $s = ((z_1, p_1), \ldots, (z_m, p_m))$ be a Nash equilibrium in $G[\mathcal{E}]$. Choose $i \in \{1, \ldots, m\}$ arbitrarily, and let i change her strategy to (z_i', p_j), where $j \neq i$ (so that the price part of the strategy of agent i coincides with that of individual j). Then she will obtain the bundle

$$\left[z_i - \frac{z_i'}{|M(p_j) \cup \{i\}|} - \frac{\sum_{k \in M(p_j) \setminus \{i\}} z_k}{|M(p_j) \cup \{i\}|} \right] + \omega_i$$

Incentives and Mechanisms in General Equilibrium 255

and therefore, by appropriate choice of z_i', she can get any bundle x_i' with $p_j \cdot x_i' \le p_j \cdot \omega_i$. Since s is a Nash equilibrium, we conclude that

$$x_i' \in P_i(f_i(s)) \text{ implies } p_j \cdot x_i' > p_j \cdot \omega_i. \tag{1}$$

If $|M(p_j)| > 1$ for some j, then it follows easily that $(f(s), p_j)$ is a Walras equilibrium: (1) holds for every $i \ne j$, and since $M(p_j)$ contains some $h \ne j$ with $p_h = p_j$, we have by the same reasoning as above that

$$x_j' \in P_j(f_j(s)) \text{ implies } p_j \cdot x_j' > p_j \cdot \omega_j.$$

Finally, $f(s) = (f_1(s), \ldots, f_m(s))$ must satisfy $p_j \cdot f_i(s) \ge p_j \cdot \omega_i, i = 1, \ldots, m$, since by Assumption 0.2, there is x_i' arbitrarily close to $f_i(s)$ such that $u_i(x_i') > u_i(f_i(s))$, and therefore, by (1), $p_j \cdot x_i' > p_j \cdot \omega_i$. From this and the fact that $f(s) \in A$ and therefore $\sum_{i=1}^m f_i(s) \le \sum_{i=1}^m \omega_i$, we get that $p_j \cdot f_i(s) = p_i \cdot \omega_i, i = 1, \ldots, m$.

If $|M(p_j)| = 1$ for each j, then since $m \ge l + 1$, by Caratheodory's theorem (cf., e.g. Rockafellar, 1970) one of the prices $p_1, \ldots, p_m$ is a convex combination of the others, say $p_m = \sum_{i=1}^{m-1} \lambda_i p_i$. Now $(f(s), p_m)$ is a Walras equilibrium: (1) holds with $j = m$ for each $i \ne m$, and

$$x_m' \in P_m(f_m(s)) \text{ implies } p_i \cdot x_m' > p_i \cdot \omega_m$$

for each $i \ne m$, whence, as $p_m = \sum_{i=1}^{m-1} \lambda_i p_i$, we must also have that

$$x_m' \in P_m(f_m(s)) \text{ implies } p_m \cdot x_m' > p_m \cdot \omega_m.$$

The remaining part of the argument follows exactly as before.

(b) Let (x, p) be a Walras equilibrium. Since $\mathcal{E}$ belongs to $\mathfrak{E}$, each consumer satisfies Assumption 0.2, whence

$$\sum_{i=1}^m x_i = \sum_{i=1}^m \omega_i.$$

Let $S = ((z_1, p), \ldots, (z_m, p))$, where $z_i = x_i - \omega_i, i = 1, \ldots, m$. Then $f_i(s) = x_i$, each i.

Suppose that for some agent, say agent 1, there is a strategy $s_1' = (z_1', p_1)$ such that

$$f_1(s_1', s_{-1}) \in P_1(f_1(s)) = P_1(x_1).$$

Then, we must have $p_1 = p$ since otherwise $f_1(s_1', s_2, \ldots, s_m) = \omega_1$. However, in this case it follows that $p \cdot f_1(s_1', s_2, \ldots, s_m) \leq p \cdot \omega_1$, and since x_1 is maximal for P_1 on $\{x_1' \mid p \cdot x_1' \leq p \cdot \omega_1\}$, we have a contradiction. $\qquad\square$

It should be mentioned that although the mechanism G considered here is balanced in the sense that $\sum_{i=1}^{m} f_i(s) = \sum_{i=1}^{m} \omega_i$ for all strategy choices $s = (s_1, \ldots, s_m)$, it is not necessarily individually feasible out of equilibrium: for some non-equilibrium strategy s, we have $f_i(s) \notin \mathbb{R}_+^l$. This feature of G is somewhat unsatisfactory: we should avoid institutions which for some acts of the agents produce very drastic effects, maybe — as in this case — threatening the very survival of the agents. It has, however, turned out to be difficult to construct mechanisms which perform the same task as G and are also individually feasible for all strategy choices.

3. Hurwicz's Theorem and the Walras Correspondence

In Section 2, we looked for an implementation of PO $\cap$ IR and we found one which also implements W. This is not a coincidence. We show that every implementation of PO $\cap$ IR is also an implementation of W.

For the statement of this result, we need specific types of economies: An economy $\mathcal{E} = (X_i, P_i, \omega_i)_{i=1}^{m}$ is *linear* if for each i, the preferences P_i can be described by a linear utility function, i.e. there is $q^i \in \mathbb{R}_+^l \setminus \{0\}$, such that

$$x_i' \in P_i(x_i) \text{ if and only if } q^i \cdot x_i' > q^i \cdot x_i,$$

all $x_i, x_i' \in \mathbb{R}_+^l$. A linear economy may be written as $\mathcal{E} = (\mathbb{R}_+^l, q_i, \omega_i)_{i=1}^{m}$. Now let $\mathfrak{E}'$ be a family of economies where consumer's preferences

Incentives and Mechanisms in General Equilibrium 257

satisfy Assumption 0.2 and which contains all linear economies such that $\omega_i \in \mathbb{R}^l_{++}, i = 1, \ldots, m$.

Theorem 9.2. *Let* $G = ((S_i)_{i=1}^m, f, A)$ *be a mechanism which implements* $PO \cap IR$ *on the family* $\mathfrak{E}'$. *Assume that for every* $s = (s_1, \ldots, s_m) \in \prod_{i=1}^m S_i$, *the set*

$$H_i(s) = \left\{ x_i \in \mathbb{R}^l \,\middle|\, x_i \leq f_i(s'_i, s_{-i}), \text{ some } s'_i \in S_i \right\}$$

is convex, $i = 1, \ldots, m$. *Then, for every* $\mathcal{E} \in \mathfrak{E}'$, *if the outcome* $f(s)$ *of a Nash equilibrium* $s \in N(G[\mathcal{E}])$ *is such that* $f_i(s) \in \mathbb{R}^l_{++}$, *each* i, *then it belongs to* $W(\mathcal{E})$.

Proof. Let $\mathcal{E} \in \mathfrak{E}'$, $\mathcal{E} = (\mathbb{R}^l_+, P_i, \omega_i)_{i=1}^m$, and $\bar{x} = f(s)$ for some $s \in N(G[\mathcal{E}])$, and suppose that $x_i \in \mathbb{R}^l_{++}$, each i.

For each i, we have $H_i(s) \cap P_i(\bar{x}_i) = \emptyset$, since s is a Nash equilibrium. Since both sets $H_i(s)$ and $P_i(\bar{x})$ are convex, and $H_i(s)$ contains a translate of the non-positive orthant, there is $q^i \in \Delta$, such that

$$q^i \cdot x'_i \leq q^i \cdot \bar{x}_i \quad \text{for } x'_i \in H_i(s)$$
$$q^i \cdot x''_i > q^i \cdot \bar{x}_i \quad \text{if } x''_i \in P_i(\bar{x}_i).$$

Now, consider the linear economy $\hat{\mathcal{E}} = (\mathbb{R}^l_+, q^i, \omega_i)_{i=1}^m$. It is easily seen that s is a Nash equilibrium for $G[\hat{\mathcal{E}}]$ also. Therefore, $\bar{x} = f(s) \in PO \cap IR(\hat{\mathcal{E}})$.

Since $\bar{x} \in PO(\hat{\mathcal{E}})$ and $\bar{x}_i \in \text{int } \mathbb{R}^l_+$, each i, by Theorem 3.3 there is $p \in \Delta$ such that $(\bar{x}, p)$ is a market equilibrium for some incomes $w_1, \ldots, w_m$. This means that $q^i = p$ for each i. Now, from individual rationality we get that $p \cdot \bar{x}_i = q^i \cdot \bar{x}_i \geq q^i \cdot \omega_i = p \cdot \omega_i, i = 1, \ldots, m$, and it follows that $(\bar{x}, p)$ is a Walras equilibrium in $\hat{\mathcal{E}}$, and consequently, in $\mathcal{E}$. $\qquad\square$

The theorem has a converse: If the mechanism G implements $PA \cap IR$ on $\mathfrak{E}'$ and, in addition, satisfies a weak continuity requirement, then every Walras allocation in $\mathcal{E}$ is the outcome of some Nash equilibrium in $G[\mathcal{E}]$ (cf. Hurwicz, 1979).

258   *Theory of General Economic Equilibrium*

These results confirm an impression obtained by now that Walras equilibria keep turning up at the most unexpected occasions. If we want allocations in society to be (1) efficient and respect property rights, and further (2) to be *incentive compatible* in the sense that they are implemented by equilibrium acts in certain institutions, then the allocations considered must be the Walras allocations. Essentially, there is no other choice.

4. Allocation Processes and Information

In the previous sections, we have considered implementation by game forms of allocations with particular properties, and the intuition behind the implementation approach has been that allocation in society depends on the characteristics of its agents, but the agents may occasionally have incentives to misinform about these characteristics. We shall be more explicit on this idea in what follows.

4.1 *Definition of an allocation process*

In the following, we consider economies with l commodities and m consumers, where each consumer has consumption set $\mathbb{R}^l_+$, and an irreflexive preference correspondence, which in our present context is more conveniently defined on the set $\mathbb{R}^l_+ - \{\omega_i\}$ of feasible net trades. The set of all conceivable economies is then

$$\mathfrak{E}^0 = \mathcal{P}^n \times \mathbb{R}^{ln},$$

where $\mathcal{P}$ is the set of irreflexive preferences on $\mathbb{R}^l_+$.

If $\mathcal{E} = (P_i, \omega_i)_{i=1}^m$ is an economy in $\mathfrak{E}^0$, then the set of *feasible net trades* in $\mathcal{E}$ is $F_n(\mathcal{E}) = \{z = (z_1, \ldots, z_m) \in \mathbb{R}^{lm} \mid z_i + \omega_i \in \mathbb{R}^l_+, i = 1, \ldots, m\}$. A feasible trade net z trade is *Pareto-optimal* in $\mathcal{E} = (P_i, \omega_i)_{i=1}^m$ if there is no feasible net trade $z' \in F_n(\mathcal{E})$, such that $z'_i \in P_i(z_i)$ for $i = 1, \ldots, m$.

We need some additional notions: For $\mathfrak{E} \subseteq \mathfrak{E}^0$, we let $L_i(\mathfrak{E})$ be the set of individual characteristics attained by consumer i in some economy $\mathcal{E}$ from $\mathfrak{E}$,

$$L_i(\mathfrak{E}) = \left\{ (P_i', \omega_i') \,\middle|\, \exists\, \mathcal{E} = (P_j, \omega_j)_{j=1}^m \in \mathfrak{E} : (P_i, \omega_i) = (P_i', \omega_i') \right\},$$

and we let $L(\mathfrak{E}) = \prod_{i=1}^m L_i(\mathfrak{E})$ be the product of these sets, so that an element of $L(\mathfrak{E})$ can be considered as an economy constructed from parts which all occur in some of the economies of $\mathfrak{E}$.

A subfamily $\mathfrak{E}$ of $\mathfrak{E}^0$ satisfying $\mathfrak{E} = L(\mathfrak{E})$ has the *uniqueness property* if for all $\mathcal{E} = (P_i, \omega_i)_{i=1}^m, \mathcal{E}' = (P_i', \omega_i')_{i=1}^m \in \mathfrak{E}$, if there is a net trade z which is Pareto-optimal in $\mathcal{E}$ and in all the economies

$$\mathcal{E}^{(i)} = \left((P_j, \omega_j)_{j \neq i}, (P_i', \omega_i') \right)$$

obtained by replacing the characteristics assigned to consumer i in $\mathcal{E}$ by those in $\mathcal{E}'$, $i = 1, \ldots, m$, then $\mathcal{E} = \mathcal{E}'$.

Now we turn to allocation processes, which we simplify somewhat compared to our previous discussions.

Definition 9.1. An allocation process for $\mathfrak{E}$ is a pair (v, h), where

(i) $v : \mathfrak{E} \rightrightarrows M$ is a correspondence assigning to each $\mathcal{E}$ a subspace $v(\mathcal{E})$ of M, an abstract space of messages,

(ii) $h : v(\mathcal{E}) \to N(\mathcal{E})$ is a map which to each of the messages in $v(\mathcal{E})$, assigns a net trade in $\mathcal{E}$.

We interpret $v(\mathcal{E})$ as a set of equilibrium messages, and h is called the outcome function of the mechanism. The allocation process (v, h) is

- *decisive* on $\mathfrak{E}' \subset \mathfrak{E}$ if $v(\mathcal{E}) \neq \emptyset$ for each $\mathcal{E} \in \mathfrak{E}'$,
- *non-wasteful* on $\mathfrak{E}'$ if for each $\mathcal{E} \in \mathfrak{E}'$ and $m \in v(\mathcal{E})$, $h(m)$ is Pareto optimal in $\mathcal{E}$,
- *privacy preserving* on $\mathfrak{E}'$ if there are $v_i : L_i(\mathfrak{E}') \rightrightarrows M$, $i = 1, \ldots, m$ such that

$$v(\mathcal{E}) = \cap_{i=1}^m v_i(P_i, \omega_i)$$

for each $\mathcal{E} = (P_i, \omega_i)_{i=1}^m \in \mathfrak{E}'$.

260   *Theory of General Economic Equilibrium*

An allocation process of particular interest is the *competitive resource allocation process:* To define the message space of an economy $\mathcal{E}$, we use the set $P = \mathbb{R}_{++}^{m-1}$ interpreted as the set of all positive prices of the l commodities where the last commodity has been chosen as a numeraire. Then the message space will be given as

$$M = P \times \mathbb{R}^{(m-1)(l-1)},$$

so that a message consists of a normalized price and a net trade proposal, specifying net trades of the first $m - 1$ consumers in each of the non-numeraire commodities. In order to get from a net trade proposal to a net trade (for all consumers and all commodities), we proceed as follows:

First of all, we introduce a specific way of balancing net trade proposals, assigning to each $(p, z) \in M$ the balanced array of net trades $h^c(p, z) \in \mathbb{R}^{lm}$ given by

$$h_i^c(p, z) = (z_i, -p \cdot z_i), \quad i = 1, \ldots, m - 1,$$

$$h_m^c(p, z) = \left(-\sum_{i=1}^{m-1} z_i, \sum_{i=1}^{m-1} p \cdot z_i \right), \tag{2}$$

where the net trade for consumer m is defined in such a way so as to achieve aggregate balance, $\sum_{i=1}^m h_i^c(p, z) = 0$. Now we can define

$$v_i^c(\mathcal{E}) = \Big\{ (p, z) \in M \Big| h_i^c(p, z) + \omega_i \in \mathbb{R}_+^l,$$

$$z' \in \mathbb{R}_+^{lm}, z_i' + \omega_i \in P_i(h_i^c(p, z) + \omega_i) \text{ implies } (p, 1) \cdot z' > 0 \Big\} \tag{3}$$

as the set of (p, z) which gives rise to balanced net trades that are optimal for consumer i. We then obtain the message space of $\mathcal{E}$ as

$$v^c(\mathcal{E}) = \cap_{i=1}^m v_i^c(\mathcal{E}),$$

We now want to compare the message spaces of general resource allocation processes to the competitive one. Suppose, therefore, that (v, h) is a resource allocation process which works on some family

Incentives and Mechanisms in General Equilibrium 261

$\mathfrak{E}$ of economies, which includes $\mathfrak{E}^*$. The behavior of (v, h) on the subfamily $\mathfrak{E}^*$ will reflect its informational properties.

Theorem 9.3. *Let $\mathfrak{E}$ be a family of economies containing $\mathfrak{E}^*$, and assume that (v, h) is decisive, non-wasteful and privacy preserving on $\mathcal{E}^*$. Then the message space of $v(\mathfrak{E}^*)$ is at least as large as $\mathbb{R}^{m(l-1)}$ in the sense that there exists a continuous surjective map from $v(\mathfrak{E}^*)$ to $\mathbb{R}_{++}^{m(l-1)}$.*

Proof. Let $\mathcal{E}^0 \in \mathfrak{E}^*$ be given by the parameter vector $\mathbf{c} = (c_i)_{i=1}^m$, and assume that the message $m^0 \in M$ is in $v(\mathcal{E})$ with $h(m^0) = x$ a Pareto-optimal allocation in $\mathcal{E}^0$. Then marginal rates of substitution between any pair of commodities must be equal for all agents, so that

$$
\frac{c_{ih}}{c_{ik}} \frac{x_{ik}}{x_{ih}} = \frac{c_{jh}}{c_{jk}} \frac{x_{jk}}{x_{jh}}, \quad i, j \in \{1, \ldots, m\}, \quad i \neq j, \ h, k \in \{1, \ldots, l-1\}, h \neq k,
$$

$$
c_{ih} \frac{x_{il}}{x_{ih}} = c_{jh} \frac{x_{jl}}{x_{jh}}, \quad i, j \in \{1, \ldots, m\}, \quad i \neq j, \quad h \in \{1, \ldots, l-1\}.
$$

$$(4)$$

It is seen from (4) that if the parameter of one individual, say i, is changed from c_i to $c_i' \neq c_i$, then the allocation x is no longer Pareto-optimal in the resulting economy (c_i', i), and if $m' \in v_i(c_i', i)$, then $m' \neq m^0$. If to each m in $v(c_i, i)$ we assign the set of coefficients for which $m \in \mu(c_i, i)$, then we have a surjective map from some subset of $v(\mathfrak{E}^*)$ to $\mathbb{R}_{++}^{l-1}$. Proceeding in the same way for all i, we obtain a surjective map from all of $\mathfrak{E}^*$ to $\mathbb{R}_{++}^{m(l-1)}$. $\qquad \square$

Using our previous discussion of the competitive allocation mechanism, we get the following.

Corollary. Let $\mathfrak{E}$ be a family of economies containing $\mathfrak{E}^*$, and assume that (v, h) is decisive, and non-wasteful, and privacy preserving on $\mathcal{E}^*$. Then the message space of (v, h) on $\mathfrak{E}^*$ is at least as large as that of the competitive allocation process in the sense that there is a continuous surjective map from $v(\mathfrak{E}^*)$ to $\mu(\mathfrak{E}^*)$.

262   *Theory of General Economic Equilibrium*

Box 2. Cobb–Douglas environments: As a prominent example of an environment, which will be used in the formulation of the main theorem, we consider a class $\mathfrak{E}^*$ of economies where aggregate endowment is ω^* and consumers have Cobb–Douglas utilities. We fix the initial endowment of each consumer to $\omega = (1, \ldots, 1)$, and we let $\mathcal{P}^*$ be the set of all preferences $P^c : \mathbb{R}_+^l \rightrightarrows \mathbb{R}_+^l$ for which there is $c \in \mathbb{R}_{++}^{l-1}$ such that

$$y \in P^c(x) \Leftrightarrow y_1^{c_1} \cdots y_{l-1}^{c_{l-1}} y_l > x_1^{c_1} \cdots x_{l-1}^{c_{l-1}} x_l.$$

Since the preferences can be identified by points $c = (c_1, \ldots, c_{l-1})$ in $\mathbb{R}^{l-1}$, the norm $\|x\|_\infty = \max\{|c_1|, \ldots, |c_{l-1}|\}$ in $\mathbb{R}^{l-1}$ induces a topology on $\mathcal{P}^*$ and hence on $\mathfrak{E}^*$.

Clearly, $\mathfrak{E}^* = L(\mathfrak{E}^*)$, since it is obtained as the product of characteristics of individual consumers, and it has the uniqueness property with respect to the relation $=$. Indeed, if for two economies $\mathcal{E} = (P_i, \omega_i)_{i=1}^n$, $\mathcal{E}' = (P_i', \omega_i')_{i=1}^n$ from $\mathfrak{E}^*$, the net trade z is Pareto-optimal in $\mathcal{E}$ and in $\mathcal{E}^{(i)}$ for all i, then by the equality of marginal rates of substitution, we must have that $\mathcal{E} = \mathcal{E}'$.

It may be checked (Osana, 1978) that the set of Cobb–Douglas economies $\mathfrak{E}^*$ has the uniqueness property in any $\mathfrak{E} \subseteq \mathfrak{E}^0$ which contains $\mathfrak{E}^*$.

The competitive allocation process (v^c, h^c) is clearly decisive, since each economy $\mathcal{E}$ in $\mathfrak{E}^*$ has a unique Walras equilibrium (cf. Box 3 in Chapter 1). It is non-wasteful on $\mathfrak{E}^*$ since Walras equilibria are Pareto-optimal, and it is privacy preserving by its very construction.

Since there is a unique Walras equilibrium for every Cobb–Douglas economy, and we know from Box 3 in Chapter 1 that there is a 1–1 correspondence, that is a bijective map, between economies $(c_i, \omega_i)_{i=1}^m$ in $\mathfrak{E}^*$ and Walras equilibria $((p_1, \ldots, p_{l-1}, 1), x)$ where $x_i = z_i + 1$, for all i and $\sum_{i=1}^l z_i = 0$. Using the topology on Cobb–Douglas economies introduced above, one easily sees that this map is a homeomorphism. Specifying the parameters $c_1, \ldots, c_m$ amounts to choosing a point in $\mathbb{R}_+^{m(l-1)}$, and indeed the competitive allocation process needs exactly this possibility of choice of message.

Proof. Compose the map from $\mathfrak{E}^*$ to $\mathbb{R}_+^{m(l-1)}$ with the homeomorphism between elements $\mathbf{c} = (c_i)_{i=1}^m$ of $\mathbb{R}_+^{m(l-1)}$ and messages $(p, z) \in \mu(\mathfrak{E}^*)$ found in Box 2. $\qquad\square$

The results obtained show that the competitive allocation process has a certain minimality property with regard to the use of information, since alternative methods for efficient allocation of resources need at least as much communication as the competitive process. This supports the intuitive approach to decentralizing the allocation using prices as a way of economizing with respect to the messages sent back and forth between individuals and the agency in charge of allocating the goods.

However, some care must be taken not to over-interpret the results. Indeed, the comparison of sets according to the existence of a continuous and surjective map from one larger to the smaller set is somewhat flawed. It makes perfect sense for comparison of finite sets but much less so when we are dealing with infinite sets. Indeed, Peano's space-filling curve is a continuous and surjective map from the unit interval to the unit square, showing that our method of comparing message spaces is unsatisfactory in general. In the literature (e.g. (Mount and Reiter, 1974); (Osana, 1978)), the notion of *informativeness* has been developed as a refinement of the comparisons performed here. We restrict our treatment to the preliminary notion of largeness for message spaces, which remains meaningful if the families considered are finite subsets of $\mathfrak{E}$ and $\mathfrak{E}^*$.

4.2 *Allocation processes with large message spaces*

We have seen that the Walrasian allocation process is informationally economical in the sense that it requires the smallest possible message space. That the message space of an allocation process can be very large is shown by the case considered below, which is due to Calsamiglia (1982).

The setting is very simple. There are only two commodities, an input commodity of which 1 unit is available, and a produced good. There are two producers, each with a production function of the form $g_i : \mathbb{R} \to \mathbb{R}$, $i = 1, 2$, transforming input commodity into output goods. There is a single consumer, only output goods matter in the preferences, which are assumed monotonic. Clearly, the allocation problem reduces to that of finding the proportion t of the endowment used by the first producer, leaving $1 - t$ to the other one, such that the total output is maximized. With notation

$h_1(t) = g_1(t)$ and $h_2(t) = g_2(1 - t)$, we can formulate the allocation problem as

$$\max_{t \in [0,1]} (h_1(t) + h_2(t)). \tag{5}$$

To introduce our allocation process, we must specify the environment in which it has to work, and we take this environment to be $\mathcal{H} = H_1 \times H_2$. Here, H_1 is the set of C^2 (twice differentiable with continuous partial derivatives) functions h_1 on $[0, 1]$ such that $h(0) = 0$ and $h'(t) \geq 0$, $h''(t) \geq 0$ for all $t \in [0, 1]$, so that production exhibits increasing returns to scale, and where H_2 is essentially the same set, apart from the notational convention, meaning that H_2 is the set of C^2 functions h_2 with $h_2(1) = 0$ and $h'(t) \leq 0$, $h''(t) \leq 0$ for all $t \in [0, 1]$.

The process to be considered consists of a message space $M = M_1 \times M_2$ of individual messages and a response output function f which in any given environment $h = (h_1, h_2)$ takes pairs $m = (m_1, m_2)$ of individual messages to revised messages (m'_1, m'_2). We assume that f is *privacy preserving* so that $f_j(m, h)$ has the form $f_j(m, h_j)$, that is it depends only on the jth component of h, $j = 1, 2$. A message m is an equilibrium message at h if

$$f(m, h) = m.$$

We shall assume that the process is *decisive* in the sense that an equilibrium message exists at all environments $h \in \mathcal{H}$. An outcome function π is defined on all equilibrium messages m, assigning an outcome $\pi(m)$ which is a partition t of the endowment between the two producers.

A particular subset of environments will be useful as we proceed, namely the set $\mathcal{H}^*$ of all $(h_1, h_2) \in \mathcal{H}$ for which $h_1 + h_2$ is a constant, independent of the value of t. We note that this set has a certain uniqueness property.

Lemma 9.1. *Let* $(h_1, h_2), (h'_1, h'_2) \in \mathcal{H}^*$ *and suppose that there is* $t \in [0, 1]$ *such that*

$$t \in PO(h_1, h_2) \cap PO(h'_1, h_2) \cap PO(h'_1, h'_2) \cap PO(h_1, h'_2).$$

Then $h_j = h'_j$, $j = 1, 2$.

Incentives and Mechanisms in General Equilibrium 265

Proof. Suppose to the contrary that $(h_1, h_2) \neq (h'_1, h'_2)$, and let k be the constant value of $h_1(t) + h_2(t)$ and let k' similarly be the constant value of $h'_1(t) + h'_2(t)$. We show that if t^* belongs to PO(h'_1, h_2), then t^* is not in PO(h_1, h'_2). Since the two environments are different, there must be some value t^0 of t for which $h_1(t^0) = h'_1(t^0)$. If $h_1(t^*) > h_1(t^0)$, we get that

$$h_1(t^*) + h'_2(t^*) > h_1(t^0) + (k' - h'_1(t^*)) = h'_1(t^*) + (k' - h'_1(t^*)) = k',$$

showing that the total output at the environment (h_1, h'_2) exceeds k'. If $t^* \in$ PO(h_1, h'_2), then output is at least k', so

$$k' > h_1(t^*) + h'_2(t^*) = k - h_2(t^*) + k' - h'_1(t^*),$$

and it follows that $h'_1(t^*) + h_2(t^*) < k$. It follows that t^* cannot be in PO(h'_1, h_2), since already for $t = 0$ we obtain that $h'_1(0) + h_2(0) = h'_1(0) + k - h_1(0) = k > h'_1(t') + h_2(t')$. $\quad\square$

We need another result about allocation processes in our specific context.

Lemma 9.2. *Suppose that* (M, f, π) *implements* PO$(\cdot)$ *on* $\mathcal{H}$, *let* M^* *be the set of equilibrium messages at environments in* $\mathcal{H}^*$, *and let the correspondence* $\lambda : M^* \rightrightarrows \mathcal{H}^*$ *be given by*

$$\lambda(m) = \{(h_1, h_2) \in \mathcal{H}^* \mid m \text{ is an equilibrium message of } (h_1, h_2)\}.$$

Then λ *is singles valued and may therefore be considered as a function from* M^* *to* $\mathcal{H}^*$.

Proof. By our assumption of decisiveness, $\lambda(m) \neq \emptyset$ for each $m \in M^*$. Suppose that some $m \in M^*$ is an equilibrium message of $(h_1, h_2), (h'_1, h'_2) \in \mathcal{H}^*$ with $(h_1, h_2) \neq (h'_1, h_2)$. Then the outcome $t = \pi(m)$ is the same for both (h_1, h_2) and (h'_1, h'_2), and since the process implements PO$(\cdot)$, we have that $t \in$ PO$(h_1, h_2) \cap$ PO(h'_1, h'_2). Moreover, by privacy preservation, we must have that $f_1(m, (h_1, h_2')) = f_1(m, (h_1, h_2)) = m$ and $f_2(m, (h_1, h_2')) = f_2(m, (h_1', h_2')) = m$, so that m is an equilibrium message at (h_1, h_2') and, by the same reasoning, at (h_1', h_2). We may therefore conclude that

$$t \in \mathrm{PO}(h_1, h_2) \cap \mathrm{PO}(h_1', h_2) \cap \mathrm{PO}(h_1', h_2') \cap \mathrm{PO}(h_1, h_2'),$$

and by Lemma 9.1, $(h_1, h_2) = (h_1', h_2')$. We conclude that $\lambda(m)$ can contain only one environment from $\mathcal{H}^*$. $\qquad\quad \beth$

We are now ready for the main result to be obtained in this model. Theorem 9.4 states that if allocation processes as those considered here should result in Pareto-optimal allocations, then the message space must be very large. In the formulation given below, this largeness is expressed by comparison to a particular space which intuitively has very many elements. To give a more precise formal description of this largeness of M, we would (as in (Calsamiglia, 1982)) need further notions from topology. We stick to the intuitive notion here to avoid additional formalism.

Theorem 9.4. *Suppose that (M, f, π) is an allocation process which is privacy preserving and decisive and which implements the Pareto correspondence over the set $\mathcal{H}$ of environments. Then the message space M is at least as large as H_1 in the sense that there is a surjective map from M to H_1.*

Proof. By Lemma 9.2, there is a surjective map λ from M^* to $\mathcal{H}^*$. Since $\mathcal{H}^*$ is isomorphic to H_1 (the surjective map taking each $(h_1, h_2) \in \mathcal{H}^*$ to $h_1 \in \mathcal{H}_1$ has an inverse, namely the map takes each $h_1 \in H_1$ to the pair $(h_1, h_1(1) - h_1)$), there is a continuous surjection from M to H_1. $\qquad\square$

5. The VCG Mechanism and Equilibrium Allocations

In this section, we consider economies where consumers' preferences P_i are described by quasi-linear utility functions. The lth commodity be called "money" and plays a special role, and for convenience, we consider preferences as defined on *net trades* rather

Incentives and Mechanisms in General Equilibrium 267

than on bundles. Let Z_i be the set of individually feasible net trades in the first $l-1$ commodities of the consumer $i \in \{1, \ldots, m\}$, then a pair $(z_i, t_i) \in Z_i \times \mathbb{R}$ is preferred to another such pair (z_i', t_i') if

$$u_i(z_i) + t_i > u_i(z_i') + t_i'.$$

The net trade 0, for which no trade in the first $l-1$ commodities are taking place, is assumed to have utility 0.

An array of net trades $z = (z_1, \ldots, z_m)$ in the first $l-1$ commodities is feasible if $\sum_{i=1}^{m} z_i = 0$ and *efficient* if

$$\sum_{i=1}^{m} u_i(z_i) \geq \sum_{i=1}^{m} u_i(z_i'), \quad \text{all } z' = (z_1', \ldots, z_m') \in Z_1 \times \cdots \times Z_m. \quad (6)$$

We note that in our present context, efficiency does not imply and is not implied by Pareto-optimality, which involves also the lth commodity. We use the terminology since it has become standard in the mechanism literature.

When considering feasibility and efficiency in this restricted sense, the net trades in the money component are not taken into consideration, so in the discussion of mechanisms for achieving efficiency, we implicitly assume that monetary transfers can be arranged in a way which does not disturb aggregate optimality, something which does not quite hold, as we shall see in what follows.

We proceed to consider mechanisms, and for this we need an environment of economies of the type introduced. We assume that the utility functions $u_i(\cdot, \theta_i)$ of consumer i belong to a family indexed by a parameter $\theta_i \in \Theta_i$, $i = 1, \ldots, m$, and the environment is characterized by the set $\prod_{i=1}^{m} \Theta_i$. We restrict attention to *direct* mechanism, where the individual messages are possible values of their types, so that consumer i chooses an element θ_i' of Θ_i, which may or may not be its true value.

The particular mechanisms to be considered have two parts, namely an *allocation component* $z(\cdot)$, which to each array $\theta = (\theta_1, \ldots, \theta_m)$ of messages assigns an efficient net trade allocation $z(\theta)$ satisfying (6) for the utility functions $u_i(\cdot, \theta_i)$, and a *transfer component* $t(\cdot) = (t_1(\cdot), \ldots, t_m(\cdot))$, where $t_i(\theta)$, the money allocated to individual

i given the message array θ, is determined by

$$t_i(\theta) = \sum_{j \neq i} u_j(z(\theta), \theta_j) + h_i(\theta_{-i}), \tag{7}$$

for $i = 1, \dots, m$ and $h_i : \prod_{j \neq i} \Theta_j \to \mathbb{R}$ is a function which does not depend on the type signaled by individual i. Such mechanisms are known as *Vickrey–Clarke–Groves* (VCG) mechanisms, and their attractiveness comes from the following general property (Groves and Loeb, 1975).

Lemma 9.3. *Let $(z(\cdot), t(\cdot))$ be a VCG mechanism. For any $\theta \in \Theta$, the true value θ_i are weakly dominating strategies for each individual i, and the outcome is efficient, so that it implements the efficiency correspondence in weakly dominating strategies.*

Proof. Let $\theta' = (\theta'_1, \dots, \theta'_m)$ be an arbitrary array of messages, and let θ_i be the true type of individual i. The final payoff of individual i at (θ_i, θ'_{-i}) is

$$u_i(z(\theta_i, \theta'_{-i}), \theta_i) + t_i(\theta'_{-i}) = \sum_{j=1}^{m} u_j(z(\theta_i, \theta'_{-i}), \theta_j) + h_i(\theta'_{-i})$$

$$\geq \sum_{j=1}^{m} u_j(z(\theta'), \theta_j) + h_i(\theta'_{-i})$$

$$= u_i(z(\theta'), \theta_i) + t_i(\theta'_{-i}),$$

where we have inserted from (7) and used the definition of $z(\cdot)$, showing that i is at least as well off using the true type θ_i as any other type no matter what was chosen by the other individuals. The efficiency follows now from (6). $\square$

The functions h_i are as yet unspecified. We shall consider a particular version of the VCG mechanism, with the specification

$$h_i(\theta_i) = -\max_{z \in Z^i} \sum_{j \neq i} u_j(z_j, \theta_j),$$

where $Z^i = \{z \in Z \mid \sum_{j \neq i} z_j = 0\}$, so that the transfer to individual i becomes $t_i(\theta) = \sum_{j \neq i} u_j(z(\theta), \theta_j) - \max_{z \in Z^i} \sum_{j \neq i} u_j(z_j, \theta_j)$, and rewriting the expression for final payoff, one gets that

$$u_i(z(\theta), \theta_i) + t_i(\theta) = \sum_{i=1}^{m} u_i(z(\theta), \theta_i) - \max_{z \in Z^i} \sum_{j \neq i} u_j(z_j, \theta_j). \tag{8}$$

The expression on the right-hand side of (8) can be seen as the *marginal contribution* of individual i to the overall efficiency score of the economy, so when using the pivotal VCG mechanism, every individual is assigned his/her marginal product.

Following Makowski and Ostroy (1987), we define a new class of mechanism, called a *marginal product mechanism*, by an outcome function $f = (f^0, f^1) : \Theta \to Z \times \mathbb{R}^m$, such that for each $\theta \in \Theta$ and each i,

$$u_i(f_i^0(\theta), \theta_i) + f_i^1(\theta) = \left[\sum_{i=1}^{m} u_i(z(\theta), \theta_i) - \max_{z \in Z^i} \sum_{j \neq i} u_j(z_j, \theta_j) \right] + g_i(\theta_{-i}),$$
$$\tag{9}$$

where (as before) $g_i : \prod_{j \neq i} \Theta_j \to \mathbb{R}$ is a function which does not depend on θ_i. The quantity in the brackets on the right-hand side can be considered as the marginal contribution of individual i to the aggregate utility; we denote it by $\mu_i(\theta)$. We shall be particularly interested in marginal product mechanisms. Furthermore, we restrict attention to *full appropriation mechanisms* which are marginal product mechanisms with g_i identically 0 for each i and for which $f(\theta)$ gives a Pareto-optimal allocation.

The full appropriation mechanism satisfies Lemma 9.3, and it has some additional properties.

Lemma 9.4. *A full appropriation mechanism f is individually rational in the sense that $u_i(f_i^0(\theta), \theta_i) + f_i^1(\theta) \geq 0$ for each $i \in \{1, \ldots, m\}$ and $\theta \in \Theta$. Moreover, the mechanism is budget balanced in the sense that $\sum_{i=1}^{m} f_i^1(\theta) = 0$ for each $\theta \in \Theta$.*

Proof. This follows from the definitions, since by (9), we get that

$$u_i(f_i^0(\theta), \theta_i) + f_i^1(\theta) = \max_{z \in Z} \sum_{i=1}^m u_i(z, \theta_i) - \max_{z \in Z^i} \sum_{j \neq i} u_j(z_j, \theta_j) \geq 0,$$

where we have consider the fact that the net trade 0 belongs to Z_i.

For the second statement, we note that $\sum_{i=1}^m f_i^1(\theta) \neq 0$ then $f(\theta)$ is not Pareto-optimal, a contradiction. $\qquad\square$

Now we turn to the third and final type of mechanism considered in this section, namely a *Walrasian mechanism*, which to each $\theta \in \Theta$ assigns a net trade allocation $W(\theta) = (W^0(\theta), W^1(\theta)) \in \mathbb{R}^{lm}$ with $\sum_{i=1}^m W_i(\theta) = 0$ and a price $p(\theta) \in \mathbb{R}_+^{l-1}$ such that $p(\theta) \cdot W_i^0(\theta) = W_i^1(\theta)$ and

$$u_i(z_i, \theta_i) + t_i > u_i(W_i^0(\theta), \theta_i) + W_i^1(\theta) \text{ implies } p(\theta) \cdot z_i + t_i$$

$$> p(\theta) \cdot W_i^0(\theta) + W_i^1(\theta) \tag{10}$$

each i, and for all $i \in \{1, \dots, m\}$, $\theta \in \Theta$ and $\theta_i' \in \Theta_i$,

$$u_i(W^0(\theta), \theta_i)) + W^1(\theta) \geq u_i(W_i^0(\theta_i', \theta_{-i})), \theta_i) + W_i^1(\theta_i', \theta_{-i}). \tag{11}$$

Thus, a Walrasian mechanism assigns to each θ a Walras equilibrium in the exchange economy defined by θ, and condition (11) states that no individual can manipulate the equilibrium prices.

The Walrasian mechanism is a special case of the previously considered mechanisms.

Theorem 9.5. *Assume that for each $\theta \in \Theta$ and each $i \in \{1, \dots, m\}$, there is $\theta_i^0 \in \Theta_i$, such that*

$$\mu_i(\theta_i^0, \theta_{-i}) = 0. \tag{12}$$

Then a Walrasian mechanism is a full appropriation mechanism.

Proof. Let W, $\theta \in \Theta$ be arbitrary, and let $p(\theta)$ be the price associated with $W(\theta)$. From the first welfare theorem, we know that $W(\theta)$ is

Incentives and Mechanisms in General Equilibrium 271

Pareto-optimal. It remains to check that it is a marginal product mechanism. From (12), we have that

$$\mu_i(\theta) = \sum_{j=1}^{m} u_j(W_j^0(\theta), \theta_j)$$

$$-\left[\sum_{j\neq i} u_j(W_j^0(\theta_i^0, \theta_{-i}), \theta_j) + u_i(W_i^0(\theta_i^0, \theta_{-i}), \theta_i^0) \right],$$

where we have used that $W^0(\hat{\theta})$ is efficient and $\sum_{i=1}^{m} W_i^1(\hat{\theta}) = 0$ for all $\hat{\theta} \in \Theta$. Now, for all $j \neq i$,

$$u_j(W^0(\theta), \theta_j) + W_i^1(\theta) = u_j(W^0((\theta_i^0, \theta_{-i}), \theta_j) + W_i^1(\theta_i^0, \theta_{-i}) \quad (13)$$

by property (11) of a Walrasian mechanism, and we conclude that

$$\mu_i(\theta) = u_i(W_i^0(\theta), \theta_i) + W_i^1(\theta) - \left[u_i(W_i^0(\theta_i^0, \theta_{-i}), \theta_i^0) + W_i^1(\theta_i^0, \theta_{-i}) \right].$$

We claim that $u_i(W_i^0(\theta_i^0, \theta_{-i}), \theta_i^0) + W_i^1(\theta_i^0, \theta_{-i}) = 0$. Since θ_i^0 is chosen so that

$$\sum_{j\neq i} u_j(W_i^0(\theta_i^0, \theta_{-i}), \theta_j) + u_i(W_i^0(\theta_i^0, \theta_{-i}), \theta_i^0) = \sum_{j\neq i} u_j(W^0(\theta_i^0, \theta_{-i}), \theta_j),$$

we get that $u_i(W_i^0(\theta_i^0, \theta_{-i}), \theta_i^0) = 0$. If $W_i^1(\theta_i^0, \theta_{-i}) > 0$, then $\sum_{j\neq i} W_j^1(\theta_i^0, \theta_{-i}) < 0$, and the coalition of all individuals except i can improve upon $W(\theta_i^0, \theta_{-i})$ by increasing their net trades in money by $W_i^1(\theta_i^0, \theta_{-i})$, a contradiction since $W(\theta_i^0, \theta_{-i})$ is a Walras allocation. We conclude that $W_i^1(\theta_i^0, \theta_{-i}) = 0$. $\square$

6. Exercises

(1) Give an example of a game form implementing the performance correspondence IR on the family $\mathfrak{E}$ of exchange economies $\mathcal{E}$ satisfying Assumptions. 0.1 and 0.2.

272   *Theory of General Economic Equilibrium*

(2) Let $\mathcal{E} = (X_i, P_i, \omega_i)_{i=1}^m$ be an economy with production. Suppose that both consumers and producers participate in a strategic market game, so that producers announce the desired net trade in the form of a production plan $z_j \in \mathbb{R}^l$, together with a suggested price p_j, such that

$$p_j \cdot z_j \leq \max\{p_j \cdot y_j \mid y_j \in Y_j\},$$

for $j = 1, \ldots, n$. Producer j then obtains the net trade

$$f_j(s) = \left[y_j - \frac{\sum_{k \in M(p_j)} z_k}{|M(p_j)|} \right]$$

given the strategy array s and to choose the strategy (z_j, p_j) in such a way that the result is a feasible and efficient production.

Investigate whether Nash equilibrium outcomes in the game $G[\mathcal{E}]$ coincide with Walras equilibrium allocations.

(3) Show that the assumption $\bar{x}_i \in \mathbb{R}_{++}^i$ can be omitted if all economies $\mathcal{E} = (X_i, P_i, \omega_i)_{i=1}^m$ in $\mathcal{E}'$ are such that $X_i = \mathbb{R}_+^l$, $i = 1, \ldots, m$.

(4) Let $G = ((S_i)_{i=1}^m, A, f)$ be the game form defining the strategic market game in Section 2. Check whether G gives rise to an allocation process in the sense of Definition 9.1, and if so, whether it is decisive, non-wasteful and privacy preserving.

(5) Let $\mathfrak{E}$ be the family of exchange economies $\mathcal{E} = (X_i, P_i, \omega_i)_{i=1}^m$ satisfying Assumptions 0.1 and 0.2. Show that the competitive resource allocation process defined by (2) and (3) is decisive, non-wasteful and privacy preserving on $\mathfrak{E}$.

(6) Consider the choice of allocations in an environment where consumers have quasi-linear preferences, and the types of the individuals are distributed according to the probability distribution F_i on Θ_i, $i = 1, \ldots, m$.

The AGV mechanism (d'Aspremont and Gerard-Varet, 1979) is given as $(z^A(\cdot), p^A(\cdot))$, where

- $z^A(\cdot)$ assigns to each $(\theta_1, \ldots, \theta_m)$ an (efficient) outcome maximizing $\sum_{i=1}^m u_i(z_i, \theta_i)$ over all feasible net trade arrays $(z_1, \ldots, z_m)$,

- the payment is defined as

$$p_i^A(\theta) = \frac{1}{m-1} \sum_{j \neq i} r_j(\theta_j) - r_i(\theta_i),$$

where for each i, $r_i(\theta_i) = \mathsf{E}_{\theta_{-i}}\left[\sum_{j \neq i}^m u_j(z_j(\theta_i, \theta_{-i}), \theta_j)\right]$ is the expected total utility of the other individuals.

Show that the AGV mechanism is budget balanced and that a $(\theta_1, \ldots, \theta_m)$ is a Bayesian equilibrium which gives rise to an efficient outcome.

Chapter 10

Market Failures

1. Introduction: The Classical Market Failures

Having dealt in Chapter 3 with the way in which allocations established by the market may be considered as socially optimal (at least in the sense of a Pareto-optimum), we turn here to some features, known as *market failures,* which may upset the fundamental theorems of welfare economics.

Market failures occur when some of the assumptions, explicitly or, as is more often the case, implicitly stated in the previous formulation of the theory, are no longer fulfiled. They may be classified in as follows:

I. Classical market failures, which come in the form of

(a) external effects,
(b) public goods,
(c) natural monopolies.

II. Non-classical market failures:

(a) Ponzi games,
(b) asymmetric information.

In what follows, we shall treat the three classical cases briefly (they will turn up again later) and postpone the other two to Chapters 12 and 13, respectively.

2. External Effects

Much of general equilibrium theory deals with the mutual dependencies between individuals and firms arising from their intercourse in the market, and such effects can be characterized as *intrinsic* or internal to the allocation method. We now turn to some important cases of interpersonal dependence which are present already in the characteristics of the economy, not arising from the process of allocation as such. These interdependencies will naturally have *external* effects on the way in which goods are allocated.

The standard example of a (positive) external effect in production is that of an orchard neighboring a beekeeper's hives. The negative external effects are more abundantly witnessed in the debate, counting examples such as pollution and carbon dioxide exhaust. The fact that external effects result in a breakdown of the fundamental theorems of welfare economics can be established using simple examples, see Box 1.

2.1 *General equilibrium with externalities*

Since the external effects are caused by the basic characteristics of the economy, their formal representation must be revised accordingly. In an economy with l commodities, m consumers and n producers, allocations are (as before) arrays $(x, y) = (x_1, \ldots, x_m, y_1, \ldots, y_n)$ of commodity bundles. Consumption and production sets as well as consumer preferences will now depend on the allocation. We let $A(\mathcal{E}) \subseteq \mathbb{R}^{lm} \times \mathbb{R}^{ln}$ be the set of possible allocations. The nature of this set will depend on the type of externalities in the model.

For *consumer i*, the consumption set X_i is now replaced by a *consumption correspondence* $\widehat{X}_i : A(\mathcal{E}) \rightrightarrows \mathbb{R}^l$, where $\widehat{X}(x, y)$ is the set of consumption bundles which are possible for i given that the allocation is (x, y). The preference correspondence P_i now assigns to each allocation (x, y) with $x_i \in \widehat{X}_i(x, y)$ a subset $P_i(x, y)$ of $\widehat{X}_i(x, y)$

Market Failures 277

Box 1. A simple example of an externality: We consider an economy with two goods, one consumer and one producer. The production set is

$$Y = \{y \in \mathbb{R}^2 \mid y_1 \leq 0, y_2 \leq g(y_1)\},$$

where g is a differentiable (production) function. The consumer has consumption set $X = \{(x_1, x_2) \in \mathbb{R}^2 \mid x_1 \geq -c, x_2 \geq 0\}$ with $c > 0$, and there is an external effect on consumption caused by the productive activity, here expressed as a dependence of the consumer's utility of the bundle (x_1, x_2) on the input level y_1 in production, giving a utility $u(x_1, x_2; y_1)$, which is assumed differentiable in all arguments, with $u'_3 > 0$. The initial endowment in the economy is $(0, 0)$.

An interior Pareto-optimal allocation maximizes u to the feasibility constraint, i.e. it solves

$$\max u(x_1, x_2; y_1) \text{ subject to:}$$
$$x_1 = y_1, \ x_2 = y_2 = g(y_1), \tag{1}$$

It may be verified that a solution must satisfy the equation

$$u'_1 + u'_2 g'_2 + u'_3 = 0. \tag{2}$$

None of the fundamental theorems of welfare economics hold in this economy. Indeed, the Pareto-optimal allocation cannot be obtained in a market equilibrium, since if $p = (p_1, p_2) \in \mathbb{R}^2_+$ is a market equilibrium price with $p_2 \neq 0$, then we would have $\frac{u'_1}{u'_2} = \frac{p_1}{p_2}$, and from (2) we obtain that

$$p_1 + p_2 g' + p_2 u'_3 = 0, \tag{3}$$

contradicting profit maximization by the firm, which would entail that $p_1 + p_2 g' = 0$, which contradicts (3).

Conversely, if $p = (p_1, p_2) \in \mathbb{R}^2_+$ are market equilibrium prices, then the individual optimality conditions imply that $\frac{u'_1}{u'_2} = \frac{p_1}{p_2} = -g'$, from which $u'_1 - u'_2 g' = 0$. If input of the commodity 1 is reduced by dy_1 and consumption is increased correspondingly, then the change in the consumer's utility level is

$$du = (u'_1 + u'_2 g'_2 + u'_3)dy_1,$$

(continued)

278   *Theory of General Economic Equilibrium*

Box 1. *(Continued)*

and inserting the equilibrium conditions, we get that

$$du = u_3' dy_1.$$

Since $dy_1 > 0$ (as $y_1 \leq 0$), $u_3' < 0$ and $g_3' < 0$, we get that $du > 0$, and we conclude that the equilibrium allocation is *not* Pareto-optimal.

---

interpreted as the bundles preferred by consumer i to x_i given the bundles of the other agents. For producer j, we have a *production correspondence* $\widehat{Y}_j : A(\mathcal{E}) \rightrightarrows \mathbb{R}^l$.

Now we may formally define an *economy with externalities* as an array $\mathcal{E} = ((\widehat{X}_i, P_i)_{i=1}^m, (\widehat{Y}_j)_{j=1}^n, \omega)$, where $\omega \in \mathbb{R}^l$ is the aggregate endowment of the l commodities. From here we may proceed to define feasibility, Pareto-optimality and market equilibria, using the approach of the previous sections while correcting for the interdependence. Thus, an allocation (x, y) is individually feasible if it satisfies individual feasibility

$$x_i \in \widehat{X}_i((x,y)_{-i}), \quad i = 1,\ldots,m, \quad y_j \in \widehat{Y}_j((x,y)_{-j}), j), \quad j = 1,\ldots,n.$$

For aggregate feasibility, we distinguish between two types of goods, since an important case of externalities is provided by *public goods* (to be treated separately in the text that follows), where all consumers get the available quantity simultaneously. The conditions for aggregate feasibility are then

$$\sum_{i=1}^m x_{ih} \leq \sum_{j=1}^n y_{jh} + \omega_h, \quad h = 1,\ldots,l_1,$$

$$x_{1k} = \cdots = x_{mk} = \sum_{j=1}^n y_{jk}, \quad k = l_1 + 1,\ldots,l. \tag{4}$$

Here, the first l_1 commodities are ordinary goods and services as we have been used to so far, the remaining $l - l_1$ commodities are public goods, consumed in the same amount (equal to the amount produced) by all consumers.

Market Failures 279

An allocation is Pareto-optimal if there is no other feasible allocation (x', y') such that

$$x'_i \in \text{cl}\, P_i(x, y) \text{ all } i,\ x'_k \in P_k(x, y),\ \text{some } k \in \{1, \ldots, m\},$$

and finally, an array (x, y, p), where (x, y) is an allocation and $p \in \mathbb{R}^l_+ \backslash \{0\}$ is a market equilibrium if

(i) (x, y) is feasible,
(ii) for each i, $P_i(x, y) \cup \{x'_i \in \widehat{X}_i(x, y) \mid p \cdot x'_i \leq p \cdot x_i\} = \emptyset$,
(iii) for each j, $p \cdot y'_j \leq p \cdot y_j$ for all $y'_j \in \widehat{Y}_j(x, y)$.

It is seen that once the concepts have been suitably adapted to take care of the additional forms of dependence, the basic concepts can be defined in a straightforward way.

This, however, does not mean that the analysis will proceed with similar simplicity. It was seen in the example that market equilibria are not necessarily Pareto-optimal and that Pareto-optimal allocations may not be obtainable in market equilibrium.

2.2 *Arrow commodities and Pigou taxes*

Since the main welfare theorems linking efficiency to market equilibria break down in the case of externalities, there is a need for some additional institutional arrangements which may supplement the market. Traditionally, two alternative approaches have been considered. The first and oldest is that of *taxes and subsidies*, first proposed by Pigou (1929), which, if properly devised, together with the standard markets may sustain a given efficient allocation and in this way reestablish the second welfare theorem. We postpone this approach a little and consider first another one, more in the spirit of what we have done so far, which is to introduce *artificial commodities* that ideally should capture all the external effects in the economy, and then look at market equilibria in such economies with all the new commodities. This approach was initially suggested in Meade (1952) and elaborated upon in Arrow (1969), and the artificial new commodities are usually referred to as *Arrow commodities.*

The idea of transforming an economy with externalities into another one without externalities by increasing the set of commodities is one which has already been touched upon in the general discussion of commodity spaces, and it will be used again when we turn to economies over time and economies with uncertainty. In our present case of externalities, we introduce a commodity which captures the external effect of some specific type of agent α (consumer or producer) on agent β, (consumer or producer), giving us new variables $z_{\alpha,\beta,k}$, where α, β run over the sets of consumers and producers, and k specifies the kind of external effect considered. The dimension of the resulting commodity space, taking into consideration both standard and new commodities, will have increased considerably, even if in many cases it will not be necessary to distinguish among different receivers of the Arrow commodity. However, transforming the data from their rather inconvenient general form to the new one, we obtain an economy which can be treated with standard tools of equilibrium analysis. The external effects have disappeared, or rather, they have taken another form, and one which we are used to dealing with.

However, this transformation of an economy with externalities to one without externalities comes at a cost. Since the assumptions of convexity (of preferred sets of consumers, of production sets) play such an important role in equilibrium theory, it is crucial that these assumptions can be upheld in one version or another in the artificial economies constructed by adding new commodities. This is, however, not the case, non-convexities arise which are *fundamental* in the sense of Starrett (1972) that they are related to the very nature of externalities: Consider a producer which is affected negatively by an external effect caused by another producer. The relevant Arrow commodity is in this case an output for the first producer (which can be thought of as rights to affect the production of the other commodities) and an input to the other producer. It is reasonable to assume that the producer may cease to produce the ordinary commodities while still selling the rights, while with a given amount of rights, production may be severely affected but may still be arbitrarily large given sufficient inputs. But taken together and assuming convexity, this would contradict the very meaning of an external effect. The situation is illustrated in Box 2.

The convexity problem is less serious when we turn to the idea of using commodity taxes and subsidies instead of markets for Arrow commodities, since in this case, there is a given level of externalities, which cannot be transgressed by the consumers or producers. To see this, let $\mathcal{E} = \left((\widehat{X}_i, P_i)_{i=1}^m, (\widehat{Y}_j)_{j=1}^n, \omega \right)$ be an economy with externalities, and assume that there are K Arrow commodities $z_1, \ldots, z_K$ which capture all the external effects in $\mathcal{E}$; each z_k is an effect from one or several agents to one or several others, we have here omitted reference to senders and receivers to facilitate notation. An allocation is then an array of consumption bundles (x_i, z_i) for the consumers $i = 1, \ldots, m$ and productions (y_j, z_j) for $j = 1, \ldots, n$. The allocation is feasible if it satisfies the standard feasibility conditions together with the condition

$$\sum_{i=1}^m z_{ik} = \sum_{j=1}^n z_{jk}$$

for each Arrow commodity k. It should be noted that consumption z_{ik} of an Arrow commodity may be negative, namely in the case that the external effect in consumption is negative, while negative y_{jk} indicates that the firm j is the originator of a negative external effect. For positive external effects, a receiving consumer gets a positive amount, whereas it is considered as an input for the receiving producer.

Suppose now that $((x_i^0, z_i^0)_{i=1}^m, (y_j^0, z_j^0))$ is a Pareto-optimal allocation in this economy. In order to apply Theorem 3.3, we need convexity assumptions on the sets of preferred bundles of each consumer as well as on the production sets, and this is exactly where

282   *Theory of General Economic Equilibrium*

Box 2. Fundamental non-convexities: In order to exhibit an economy with no market equilibria involving Arrow commodities, we slightly modify the economy in Box 1. There is now a third commodity x_3 which can be produced from commodity 1 by a new producer. All three commodities can be consumed, so that now $X = \{(x_1, x_2, x_3)) \in \mathbb{R}^3 \mid x_1 \geq -c, x_2, x_3 \geq 0\}$, and the external effect caused by producer 1 does not affect the consumer, only the second producer. We formalize the externality as an Arrow commodity z ('pollution rights'), the input of which to producer 1 equals y_1, and acting as an output for the second producer, who has a production set of the form $Y_2 = \{(y_1, 0, y_3, z) \mid z \geq 0, f(y_1, y_3, z) \leq 0\}$. We assume that Y_2 exhibits constant returns to scale for any given value of z.

Consider now a market equilibrium where p_1, p_2, p_3 are the prices of the ordinary commodities and q the price of the Arrow commodity. If producers maximize profits, then for any given value $\bar{z}$, we must have that

$$p_1 y_1 + p_3 y_3 = 0, \tag{5}$$

that is, the net profit of commodity production is 0, since > in (5) means that profit could be increased by suitable proportional increases in input and output.

It follows that positive profit in firm 2 can only arise from sale of z. But if $q > 0$, then firm 2 will prefer to choose $y_1 = y_3 = 0$ and supply arbitrarily large amounts of z, more than what can be met by the demand from firm 1. It follows that $q = 0$, and we conclude that in the economy considered, there is no market equilibrium with positive prices on Arrow commodities.

It may be noted that the production plan $(0, 0, 0, z)$ belongs to Y_2 for arbitrarily large z, since firm 2 can choose to close down production and concentrate on the sale of pollution rights. On the other hand, for each fixed value $\bar{z} \geq 0$, there are non-zero inputs and outputs y_1, y_3 with $(\lambda y_1, 0, \lambda y_3, \bar{z})$, all $\lambda \geq 0$ (constant returns to scale at fixed $\bar{z}$). If Y_2 was to be convex, then $(y_1, 0, y_3, 2z)$ should be in Y_2, since

$$(y_1, 0, y_3, z) = \frac{1}{2}(2y_1, 0, 2y_3, \bar{z}) + \frac{1}{2}(0, 0, 0, 3\bar{z}).$$

Thus, convexity would imply that commodity production could be upheld independent of the level of pollution, clearly not a realistic situation, so the non-convexity of Y_2 is difficult and fundamental to the model.

---

the fundamental non-convexities give us problems. However, for supporting an allocation using prices supplemented by taxes and subsidies, less strong assumptions are needed: Indeed, it suffices that for each consumer, the set

$$\widetilde{X}_i^0 = \left\{ (x_i, z_i) \in P_i(x_i^0, z_i^0) \,\middle|\, z_{ik} \le z_{ik}^0 \text{ if } z_{ik}^0 \ge 0, z_{ik} \ge z_{ik}^0 \right.$$

$$\left. \text{if } z_{ik}^0 \le 0, k = 1, \dots, K \right\}, \tag{6}$$

where $P_i(x_i^0, z_i^0)$ is the set of extended bundles preferred by i to (x_i^0, z_i^0), is convex, and for each producer j, the restricted production set

$$\widetilde{Y}_j^0 = \left\{ y_j \in Y_j \,\middle|\, z_{ik} \le z_{ik}^0 \text{ if } z_{jk}^0 \ge 0, z_{jk} \ge z_{jk}^0 \text{ if } z_{jk}^0 \le 0, k = 1, \dots, K \right\} \tag{7}$$

is convex.

Applying Theorem 3.3 now, we get that there are prices $p_1, \dots, p_l$ on ordinary commodities and $t_1, \dots, t_K$ on Arrow commodities such that the pair $((p, t), ((x^0, z_i^0)_{i=1}^m, (y_j^0, z_j^0)))$ is a market equilibrium in the economy where consumers have preferred sets $\widetilde{X}_i^0$ given by (6) and the production sets $\widetilde{Y}_j^0$ are those in (7). This means that for each consumer i, if $(x_i, z_i) \in \widetilde{X}_i^0$, then

$$\sum_{h=1}^{l} p_h x_{ih} + \sum_{k=1}^{K} t_k z_{ik} \ge \sum_{h=1}^{l} p_h x_{ih}^0 + \sum_{k=1}^{K} t_k z_{ik}^0,$$

and for each producer j,

$$\sum_{h=1}^{l} p_h y_{jh}^0 + \sum_{k=1}^{K} t_k z_{jk}^0 \ge \sum_{h=1}^{l} p_h y_{jh} + \sum_{k=1}^{K} t_k z_{jk}$$

for all $(y_j, z_j) \in \widetilde{Y}_j^0$. For the Arrow commodities, the consumption of which has been constrained in (6) and (7), the prices t_k may be interpreted as tax rates, so that $t_k x_{ik}^0$ is the tax paid by the consumer i in the case that x_{ik}^0 is negative (so that i gives rise to a negative external effect on other agents or receives a positive external effect), if x_{ik}^0 is positive, the amount $t_k x_{ik}^0$ is the subsidy paid to i. For the producers, the tax or subsidy is similarly measured as $t_k z_{jk}^0$. The application of the welfare theorem shows that the Pareto-optimal allocation can be decentralized in the restricted sense of using prices for ordinary commodities and taxes/subsidies for external effects.

284   *Theory of General Economic Equilibrium*

3.   Public Goods

A *public good* is a commodity for which the amounts consumed are identical for all consumers, and thus equal to the total amounts available. Technically, the distinguishing feature between such a public good and ordinary ones, in this context called *private goods*, is non-exclusivity in consumption: The same unit of the good can be consumed by many consumers simultaneously.

Examples from real-life are abundant: parks, fireworks on national holidays, and police protection, to mention only a few.

Since all consume the same amount of a public good, there is an obvious external effect in the economy, and it could be analyzed as such, but the particular structure of this externality makes it possible to use our previously developed tools with only minor changes.

In the following, we consider an economy with l private commodities and r public goods. The consumption sets X_i are now subsets of $\mathbb{R}^{l+r}$ and consumer i's preferences on X_i involve both private and public goods. Similarly, producers are characterized by production sets $Y_j \subset \mathbb{R}^{l+r}$. To simplify, we assume public goods to be produced (i.e. they are net outputs) by some producers, and finally, the vector ω of initial endowments is $\omega = (\omega_1, \ldots, \omega_l, 0, \ldots, 0) \in \mathbb{R}^{l+r}$, so public goods will have to be produced, if demanded.

The concept of feasibility is adjusted to the new situation as follows. An allocation $(x_1, \ldots, x_m, y_1, \ldots, y_n)$ is *feasible* if it satisfies individual feasibility,

$$x_i \in X_i, \quad i = 1, \ldots, m, \quad y_j \in Y_j, \quad j = 1, \ldots, n, \tag{8}$$

and aggregate feasibility

$$\sum_{i=1}^{m} x_{ih} = \sum_{j=1}^{n} y_{jh} + \omega_h, \ h = 1, \ldots, l,$$

$$x_{ih} = \sum_{j=1}^{n} y_{jh}, \quad i = 1, \ldots, m, \quad h = l+1, \ldots, l+r, \tag{9}$$

Box 3. The Coase theorem. In a much cited paper, Coase (1960) gave several examples where regardless of the way in which the cost inflicted by an external effect was distributed among the individuals, it was possible to obtain an efficient outcome. In subsequent discussions by several other authors, this was reformulated as a "theorem" stating that

(i) Under perfect competition, the cost to society and the private cost to individuals will be the same,

(ii) Composition of output is independent of the way in which the externality cost is distributed among the agents.

Part (ii), which is known as *Coase independence*, has been investigated in detail. In Hurwicz (1995), it was noted that Coase independence holds if consumers have preferences which can be described by quasi-linear utility functions. This is a sufficient but not a necessary condition for Coase independence: Assume, cf. Bergstrom (2017), that the external effect z inflicts a cost $c(z)$ to the society. The distribution of this cost among m consumers with utility functions $u_i(z, w_i)$, where w_i is the cost covered by the individual i (out of a total of ω of this good), can be illustrated by the *utility possibility set* $\{(u_1(z, w_1), \ldots, u_m(z, w_m)) \mid w_i \geq 0, i = 1, \ldots, m, \sum_{i=1}^{m} w_i = \omega - c(z)\}$. If utilities are quasi-linear, the north-eastern boundary of this set has a straight line segment for each given z, cf. Fig. 1.

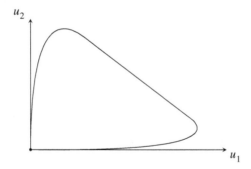

Fig. 1. The utility possibility set for continuous variation in the project z.

This may, however, also occur if utilities are *uniformly affine* in the sense that $u_i(z, w_i) = a(z)w_i + b_i(z)$, $i = 1, \ldots, m$, where $a(\cdot)$ and $b_i(\cdot)$ are real functions,

(continued)

Box 3. *(Continued)*
since in this case, the utility possibility set can be written as

$$\left\{ (u_1,\ldots,u_m) \geq (b_1(z),\ldots,b_m(z)) \;\middle|\; \sum_{i=1}^{m} u_i = a(z)(\omega - c(z)) + \sum_{i=1}^{m} b_i(z) \right\}.$$

While more general, the uniformly affine utility functions, where marginal rates of substitution do not depend on the private good, are still quite restrictive.

for private and public goods, respectively. With this definition of feasibility, Pareto-optimality is defined in the same way as for economies with only private goods. It is intuitively plausible and sustained by simple examples such as those in Box 4 that Pareto-optimal allocations cannot in general be realized in a market equilibrium.

The natural consequence is that we must consider institutional arrangements where different consumers face different prices, leading to the concept of a Lindahl equilibrium.

Definition 10.1. An array (x,y,p) consisting of an allocation $(x,y) = (x_1,\ldots,x_m,y_1,\ldots,y_n)$, a price $p \in \mathbb{R}_+^l$ on private commodities and a system of personal prices $q_1,\ldots,q_m \in \mathbb{R}_+^r$ is a *Lindahl market equilibrium* if

(i) the allocation (x,y) is feasible,
(ii) for each i, x_i is maximal for P_i on

$$\left\{ x_i' \in X_i \;\middle|\; \sum_{h=1}^{l} p_h x_h' + \sum_{k=1}^{r} q_{i,l+k} x_{i,l+k}' \leq \sum_{h=1}^{l} p_h x_h' + \sum_{k=1}^{r} q_{i,l+k} x_{i,l+k}' \right\},$$

(iii) for each j, y_j maximizes the profit $\sum_{h=1}^{l} p_h y_{jh} + \sum_{k=1}^{r} \left(\sum_{i=1}^{m} q_{i,l+k} \right) y_{j,l+k}$ on Y_j.

In the market underlying this equilibrium, private commodities have common prices, whereas public goods have agent-specific

Market Failures 287

Box 4. Marginal rates of substitution in Pareto-optima with public goods:
In the special case where consumers' preferences can be represented by differentiable utility functions, and production sets take the form

$$y_j = \left\{ y_j \mid f(y_j) \leq 0 \right\}$$

with f_j differentiable, $j = 1, \ldots, n$, a Pareto-optimal allocation with r public goods must solve the problem

$$\max \sum_{i=1}^{n} \lambda_i u_i(x_i), \quad such\ that$$

$$\sum_{i=1}^{m} x_{ih} = \sum_{j=1}^{n} y_{jh} + \omega_h, \quad h = 1, \ldots, l, \tag{10}$$

$$x_{1h} = \cdots = x_{mh} = \sum_{j=1}^{n} y_{jh}, \quad h = 1, \ldots, r.$$

First-order conditions for this constrained maximum imply that the marginal rate of substitution (MRS) $\dfrac{u'_{ih}}{u'_{ik}}$ between two private goods $h, k \in \{1, \ldots, l\}$ are identical and equal to the MRS in production $\dfrac{f'_{jh}}{f'_{jk}}$ for any firm with production involving the goods h and k. These conditions are the usual ones holding in economies with only private goods, but for any pair (h, k) with h a public good and k a private good, we get that

$$\frac{u'_{1h}}{u'_{1k}} + \cdots + \frac{u'_{mh}}{u'_{mk}} = \frac{f'_{jh}}{f'_{jk}},$$

so that the *sum* of the individual MRS in consumption should be equal to the MRS in production of the public good.

---

prices. The prices faced by a producer selling his/her production of a public good is the sum of the consumer-specific prices.

In this framework, we are able to reestablish our second fundamental theorem: a Pareto-optimal allocation can be realized in a Lindahl equilibrium.

Theorem 10.1. *Let $\mathcal{E} = ((X_i, P_i)_{i=1}^m, (Y_j)_{j=1}^n, \omega)$ be an economy with public goods, where all consumers satisfy Assumptions 0.1 and 0.2, and all*

288   *Theory of General Economic Equilibrium*

producers satisfy Assumption 0.3. If (x, y) is a Pareto-optimal allocation, where for all i, $x_i \in \text{int} X_i$, then there are prices p for private goods and personal prices $(q_i)_{i=1}^m$ for public goods, such that $(x, y, p, (q_i)_{i=1}^m)$, is a Lindahl market equilibrium.

The proof consists in reformulating the economy in a way which allows us to apply the second fundamental welfare theorem.

Proof of Theorem 10.1. We define suitable private commodities which will replace the public goods: for $k = 1, \dots, r$ and $i = 1, \dots, m$, there is a private good interpreted as the consumption by consumer i of the public good k. The consumption sets are then subsets $\widetilde{X}_i = X_i \times \mathbb{R}_+^{(m-1)r}$ of $\mathbb{R}_+^{l+mr}$, and the preference correspondence $\widetilde{P}_i : \widetilde{X}_i \rightrightarrows \widetilde{X}_i$ assigns to each bundle $\tilde{x}_i \in \widetilde{X}_i$ the set $\widetilde{P}_i(\tilde{x}_i) = P_i(x_i) \times \mathbb{R}_+^{(m-1)r}$. Then $\widetilde{P}_i$ has convex values.

For $j = 1, \dots, n$, let $\widetilde{Y}_j$ be the set consisting of all $(l + mr)$-vectors which coordinatewise are less than or equal to vectors of the form

$$(y_1, \dots, y_l, y_{l+1}, \dots, y_{l+r}, \dots, y_{l+1}, \dots, y_{l+r}) \tag{11}$$

with $(y_1, \dots, y_l, y_{l+1}, \dots, y_{l+r}) \in Y_j$ and where the last r coordinates of the production vector are repeated m times.

To each consumer's commodity bundle x_i^0 in the given Pareto-optimal allocation there now corresponds the new bundle

$$(x_1^0, \dots, x_l^0, 0, \dots, 0, x_{i,l+1}, \dots, x_{i,l+r}, 0, \dots, 0) \tag{12}$$

and likewise, for the production y_j^0, there is now a "new" production of the type specified in (11).

In this way, we obtain a new allocation which is Pareto-optimal, and Theorem 3.3 may then be applied to assure the existence of a price system (which is now a $(l + mr)$-vector)

$$(p_1, \dots, p_l, q_{1,l+1}, \dots, q_{m,l+1}, q_{2,l+1}, \dots, q_{m,l+r})$$

so that the allocation together with this price system and the specified incomes is an ordinary market equilibrium. $\square$

Market Failures 289

Box 5. A mechanism for allocating public goods: In order to obtain correct revelation of preferences for public goods, one may use a suitably designed mechanism. The one discussed here was introduced by Groves and Ledyard (1977).

We consider a simple economy with one private and one public good, where the latter is produced by a firm using the private commodity as input; there is constant returns to scale so that elements (y_1, y_2) satisfy

$$y_2 \leq -Ky_1 \quad \text{for some } K > 0,$$

where y_2 is the output of the public good and y_1 is the input of the private goods. Consumer i has consumption set $\mathbb{R}_+^2$ and preferences which are represented by differentiable utility functions u_i, $i = 1, \ldots, n$. The initial endowment ω_i consists only of the private good.

We now introduce a mechanism for deciding the production of the public good as well as the individual contributions to this production. Each consumer i sends a message $b_i \in \mathbb{R}_+$, and given the array $(b_1, \ldots, b_m)$ of messages, output of the public good is chosen as

$$x_2 = K \sum_{i=1}^{m} b_i$$

while consumer i's payment is

$$t_i = \frac{1}{m} \sum_{k=1}^{m} b_k + \left[\frac{m-1}{m} (b_i - \mu_i)^2 - \sigma_i^2 \right], \tag{13}$$

where $\mu_i = \frac{1}{m-1} \sum_{k \neq i} b_k$ and $\sigma_i^2 = \frac{1}{m-2} \sum_{k \neq i} (b_k - \mu_i)^2$, $i = 1, \ldots, m$. The second term on the right-hand side in (13) confirms that the total payments cover the cost of production, since

$$\sum_{i=1}^{m} (t_i - b_i) = \sum_{i=1}^{m} \left[\frac{m-1}{m} (b_i - \mu_i)^2 - \sigma_i^2 \right]. \tag{14}$$

We leave it to the reader to verify (14), which requires some manipulations of the sums involved. Note that if consumer i increases the message by a small amount, the payment will increase by

$$q_i := \frac{\partial t_i}{\partial b_i} = \frac{1}{m} + \frac{m-1}{m} 2(b_i - \mu_i),$$

whereby $\sum_{i=1}^{m} q_i = 1$.

(continued)

> **Box 5.** *(Continued)*
> Assume that $(b_1^0, \ldots, b_m^0)$ with the associated allocation $(x_1^0, \ldots, x_m^0)$ is a Nash equilibrium. We show that the associated allocation is Pareto-optimal.
> Since consumer i has chosen the message b_i^0 such that $u_i(\omega_i - t_i, K \sum_{i=1}^m b_i^0)$ is maximal, we have that
>
> $$\frac{\partial u_i}{\partial b_i} = -u'_{i1}q_i + u'_{i2}K = 0, \text{ from which } \frac{u'_{i2}}{u'_{i1}} = \frac{q_i}{K}$$
>
> so the fraction q_i/K corresponds to the marginal rate of substitution at the point x_i^0. Summing over i, we get that
>
> $$\sum_{i=1}^m \frac{u'_{i2}}{u'_{i1}} = \frac{1}{K},$$
>
> showing that the sum of the individual MRS in consumption equals the MRS in production, which is a necessary condition for Pareto-optimality, and assuming an interior maximum is a sufficient condition as well, cf. Box 4.

What we have demonstrated above is the following: if we want a result corresponding to the second welfare theorem to hold in economies with public goods, we can obtain this by introducing consumer-specific prices. The intuition from the classical welfare theorems can be carried over, since the Lindahl equilibrium connects the consumer's expenses with the level of satisfaction derived from the public goods.

Unfortunately, the concept of a Lindahl equilibrium has some weaknesses when it comes to establishing a counterpart of the pseudo-markets for public goods in the real world: The regulator in charge of setting the individual prices must know the different consumers' marginal rates of substitution. This is, however, private information, and the very connection between the individual tastes and the payments for the public good, which is anyway available for all, means that the individual has a strong incentive to misrepresent preferences.

4. Natural Monopolies

The last of the three "classical" instances of market failure, natural monopolies, occur in a natural way if there are increasing returns to scale in production or — what amounts to the same — there are large fixed costs in production. In such cases, the economy violates the very basic convexity assumptions for production sets, so that much of what we have done so far cannot be upheld any more. The usual competitive markets cannot establish equilibrium allocations, since profit maximization will induce choices of larger and larger levels of production, thus destabilizing any equilibrium of demand and supply.

Clearly, there is a need for another model of the market mechanism when production sets are not convex. This was realized long ago when setting up partial models of production and supply under increasing returns to scale. Here, the standard condition of equality of output price and marginal cost does not characterize a profit-maximizing production, so that depending on the case considered, and whether the viewpoint is that of a private firm or a benevolent social regulator, one may propose average cost pricing, marginal cost pricing (supplemented by some revenue transfers to cover a possible loss) or welfare optimizing prices. We shall not commit to any particulars of these pricing rules, but instead, using the approach of Cornet (1988) and Bonnisseau and Cornet (1991), we consider a general equilibrium where firms follow an unspecified pricing rule, formulated in such a way that the approaches treated in the literature emerge as special cases.

Let $\mathcal{E} = ((X_i, P_i)_{i=1}^m, (Y_j)_{j=1}^n, \omega)$ be an economy, where for each consumer i, the consumption set X_i is $\mathbb{R}_+^l$ and preferences P_i satisfy the usual conditions of irreflexivity ($x_i \notin P_i(x_i)$ for each $x_i \in X_i$), convexity ($P_i(x_i)$ is convex for each x_i), monotonicity ($P_i(x_i) + \mathbb{R}_+^l \subset P_i(x_i)$ for each x_i) and continuity (P_i has open graph), but where the production sets Y_j are *not* necessarily convex.

We assume that each production set $Y_j \subset \mathbb{R}^l$ satisfies the free disposal assumption

$$Y_j - \mathbb{R}_+^l \subset Y_j,$$

and that $0 \in Y_j$ (possibility of inaction). By δY_j we denote the boundary of Y_j, which has the weak efficiency property

$$\left[\delta Y_j + \mathbb{R}_{++}^l\right] \cap Y_j = \emptyset.$$

In order to model situations where prices are not profit maximizing, we introduce the notion of a price *compatible* with a given production $y_j \in \delta Y_j$. This is done using a price setting correspondence $\psi_j : \delta Y \rightrightarrows \Delta$, so that

$$p \in \psi_j(y_j)$$

indicates that the price p is compatible with the production y_j in firm j.

When producers do not maximize profits at the price p, the resulting net results $p \cdot y_j$ of the productive activity of firm j may be positive or negative, and the surpluses or deficits must be reflected in the budgets of the consumers. This is done here by assuming that there are *income transfer* functions $r_i : \Delta \times \prod_{j=1}^n \delta Y_j \to \mathbb{R}_+^l$ which to each price p and array $(y_1, \ldots, y_n)$ of weakly efficient productions assign an income $r_i(p, y_1, \ldots, y_n)$ to consumer i, satisfying the equality

$$\sum_{i=1}^m r_i(p, y_1, \ldots, y_n) = \sum_{j=1}^n p \cdot y_j + p \cdot \omega \tag{15}$$

for each $(p, y_1, \ldots, y_n) \in \Delta \times \prod_{j=1}^n \delta Y_j$. The equality (15) constitutes a counterpart of Walras' law for economies with increasing returns to scale, since for any price p, if x_i is maximal for P_i on the budget set $\{x_i' \in \mathbb{R}_+^l \mid p \cdot x_i' \le r_i(p, y_1, \ldots, y_n)\}$, and preferences are monotonic, then $p \cdot x_i = r_i(p, y_1, \ldots, y_n)$ for $i = 1, \ldots, m$, so that

$$p \cdot \left[\sum_{i=1}^m x_i - \sum_{j=1}^n y_j - \omega\right] = 0 \tag{16}$$

for all $p \in \Delta$, i.e. the value of the excess demand is always 0.

Market Failures 293

We may now introduce an *equilibrium with price-setting firms* as an array $((x, y), p) \in \prod_{i=1}^{m} X_i \times \prod_{j=1}^{n} \delta Y_j \times \Delta$ satisfying the following conditions:

(i) (x, y) is feasible, i.e. $x_i \in X_i, i = 1, \ldots, m, y_j \in Y_j, j = 1, \ldots, n$, and $\sum_{i=1}^{m} x_i \le \sum_{j=1}^{n} y_j + \omega$,

(ii) for each i, x_i is maximal for P_i on $\{x_i' \in \mathbb{R}_+^l \mid p \cdot x_i' \le r_i(p, y_1, \ldots, y_n)\}$,

(iii) for each j, $y_j \in \delta Y_j$ and $p \in \psi_j(p, y_1, \ldots, y_n)$.

The following result shows that equilibria with price-setting firms exist under what amounts to standard conditions (adapted to the situation at hand).

Theorem 10.2. *Let* $\mathcal{E} = ((X_i, P_i)_{i=1}^{m}, (Y_j)_{j=1}^{n}, \omega)$ *be an economy with price-setting correspondences* $\psi_j : \delta Y_j \to \Delta$, $j = 1, \ldots, n$, *and continuous income transfer functions* $r_i : \Delta \times \prod_{j=1}^{n} \delta Y_j \to \mathbb{R}_+^l$ *for* $i = 1, \ldots, m$. *Assume that*

(i) *for* $i = 1, \ldots, m, X_i = \mathbb{R}_+^l$, *and* $P_i : X_i \rightrightarrows X_i$ *satisfies Assumption 0.2*,

(ii) *for* $j = 1, \ldots, m, Y_j - \mathbb{R}_+^l \subset Y_j$,

(iii) *(a)* $\left[\sum_{j=1}^{n} Y_j - \{\omega\}\right] \cap \mathbb{R}_+^l \ne \emptyset$, *and (b)* $\mathsf{A}\left(\sum_{j=1}^{n} Y_j\right) \cap -\mathsf{A}\left(\sum_{j=1}^{n} Y_j\right) = \{0\}$,

(iv) $r_i(p, y_1, \ldots, y_n) > 0$, $i = 1, \ldots, m$, *for all* $(p, y_1, \ldots, y_n) \in \Delta \times \prod_{j=1}^{n} \delta Y_j$ *with* $p \in \psi_j(y_j), j = 1, \ldots, n$.

Then there is an equilibrium with price-setting firms in $\mathcal{E}$.

Proof. The first part of the existence proof is as usual a truncation of consumption and production sets. By (iii.a), the set of feasible allocations in $\mathcal{E}$ is non-empty, and by (iii.b) there is $K > 0$ such that $\|x_i\| < K, i = 1, \ldots, m$, and $\|y_j\| < K$ for $j = 1, \ldots, n$ whenever (x, y) is a feasible allocation. Define $\overline{X}_i$ by $\overline{X}_i = \{x_i \in \mathbb{R}_+^l \mid x_{ih} \le K, h = 1, \ldots, l\}$, $i = 1, \ldots, m$, let $a = (-K, \ldots, -K)$, and define for $j = 1, \ldots, n$ the truncated set of weakly efficient productions $B_j = \delta Y_j \cap \left[\{a\} + \mathbb{R}_+^l\right]$.

For $j = 1, \ldots, n$, we use a copy of Δ to select elements of B_j: For $q_j \in \Delta$, let $h_j(q_j) = \max\{\lambda > 0 \mid a + \lambda q_j \in Y_j\}$ be the unique element of $\overline{y}_j \in Y_j$ at which the half-line from a along q_j intersects the boundary of Y_j. Clearly, h_j is continuous and $h_j(q_j) \in B_j$ for each $q_j \in \Delta$. Define the correspondence $\eta_j : \Delta^2 \rightrightarrows \Delta$ by

$$\eta_j(p, q_j) = \begin{cases} \{q'_j \in \Delta \mid q'_j \\ \quad = \mu p + (1 - \mu)q''_j, q''_j \in \psi_j(h(q_j))\} & \text{if } p \notin \psi_j(h(q_j)), \\ \Delta & \text{otherwise.} \end{cases}$$

(17)

Then η_j is upper hemicontinuous with non-empty, closed and convex values, and $q_j \notin \eta_j(p, q_j)$ implies that $p \in \psi_j(h_j(q_j))$.

Write $q = (q_1, \ldots, q_n)$ and $h(q) = (h_1(q_i), \ldots, h_n(q_n))$. For $i = 1, \ldots, m$, define the budget correspondence $\gamma_i : \Delta^{n+1} \rightrightarrows \overline{X}_i$ by

$$\gamma_i(p, q) = \{x_i \in \overline{X}_i \mid p \cdot x_i < \gamma_i(p, h(q))\}.$$

The budget correspondence has convex, possibly empty values and an open graph. From this, we proceed to define the correspondence $\phi_i : \Delta^{n+1} \times \overline{X}_i \rightrightarrows \overline{X}_i$ by

$$\phi_i(x_i, p, q) = \begin{cases} \gamma_i(p, q) & x_i \notin \text{cl}\, \gamma_i(p, q), \\ \gamma_i(p, q) \cap P_i(x_i) & x_i \in \text{cl}\, \gamma_i(p, q). \end{cases}$$

Then ϕ_i has open graph and convex, possibly empty, values, and $x_i \notin \phi_i(x_i, p, q)$ for all (x_i, p, q).

To take care of feasibility, we define the correspondence $\phi_0 : \prod_{i=1}^{m} \overline{X}_i \times \Delta^n \rightrightarrows \Delta$ by

$$\phi_0(x, q) = \left\{ p \in \Delta \,\middle|\, p \cdot \left[\sum_{i=1}^{m} x_i - \sum_{j=1}^{n} h(q_j) - \omega \right] > 0 \right\},$$

where we have written x for $(x_1, \ldots, x_m)$. Again, this correspondence has convex, possibly empty, values and an open graph.

Using Theorem 1.7, we obtain the existence of $(x^0, p^0, q^0) \in \prod_{i=1}^{m} \overline{X}_i \times \Delta^{n+1}$ such that

$$\phi_i((x_i^0, p^0, q^0) = \emptyset, \quad i = 1, \ldots, m,$$
$$\phi_0(x^0, q^0) = \emptyset, \qquad\qquad (18)$$
$$q_j^0 \in \eta_j(p^0, q_j^0), \quad j = 1, \ldots, n.$$

To complete the proof we must check that (x^0, y^0, p^0) with $y^0 = h(q^0)$ is an equilibrium. Since $\phi_0(x^0, y^0) = \emptyset$, the vector $\sum_{i=1}^{m} x_i^0 - \sum_{j=1}^{n} y_j^0 - \omega$ must belong to $\mathbb{R}_-^l$, so that

$$\sum_{i=1}^{m} x_i^0 \leq \sum_{j=1}^{n} y_j^0 + \omega,$$

which means that (x^0, y^0) is a feasible allocation. For $j = 1, \ldots, n$, we get from $q_j^0 \in \eta_j(p^0, q_j^0)$ that $p^0 \in \psi_j(y_j^0)$. Using our assumption in (iv), we get that $\gamma_i(p^0, q^0) \neq \emptyset$ for each i, and since $\phi_i(x_i^0, p^0, q^0) = \emptyset$, we get that $p^0 \cdot x_i^0 = r_i(p^0, y^0)$ and that $x_i' \in P_i(x_i^0)$ implies that $p^0 \cdot x_i' > r_i(p^0, y^0)$. Thus, (x^0, y^0, p^0) is an equilibrium. $\square$

5. Exercises

(1) In the example of Box 1, show how to sustain a Pareto-optimal allocation either by introducing Arrow commodities or by Pigou taxes.

(2) In discussions of healthcare, it is often stated that a fundamental *Samaritan principle* is at work, according to which individuals suffering from diseases and life-threatening illness cause discomfort for the rest of the population.

   Formulate a model where this phenomenon occurs as an externality and investigate whether it can be handled by the introduction of suitable Arrow commodities to be traded in appropriate markets.

296   *Theory of General Economic Equilibrium*

(3) Consider an economy $\mathcal{E}$ with two consumers and one producer. There are two goods, of which the first one is private and the second a public good.

For the consumers $i = 1, 2$, we have that $X_i = \mathbb{R}_+^2$, and their preferences are described by utility functions

$$u_1(x_1, x_2) = x_1^{\frac{2}{3}} x_2^{\frac{1}{3}}, \quad u_2(x_1, x_2) = \min\{x_1, x_2\}.$$

The initial endowments are $\omega_1 = (6, 0)$, $\omega_2 = (8, 0)$. The firm has production set $Y = \{(y_1, y_2) \in \mathbb{R}^2 \mid y_2 \leq 2y_1, y_1 \leq 0\}$. Consumers 1 and 2 have equal shares in the firm.

Find a Lindahl equilibrium in $\mathcal{E}$.

(4) In the economy described in Box 5, consider the following mechanism for determining the provision of and payment for the public good (Walker, 1981).

Each individual sends a message $c_i \in \mathbb{R}$. Given the array of messages $c = (c_1, \ldots, c_m)$, the production of public goods is chosen as

$$x_2(c) = \sum_{i=1}^{m} c_i,$$

and the payment (in terms of the private good) is determined by the rule

$$\tau_i(c) = \left(\frac{1}{m} \frac{1}{K} + c_{i+2} - c_{i+1} \right) x_2(c),$$

where the subscripts $i+1$ and $i+2$ should be considered modulo m (so that, e.g. $m+1$ is equal to 1).

Show that a Nash equilibrium of the mechanism gives rise to a Lindahl equilbrium.

(5) The following dynamical procedure for deciding about public goods provision was proposed by Drèze and de la Vallee Poussin (1971): There are m individuals consuming bundles (x_i, y), where y is the provision with public goods. At each date t, an increase Δy of the provision is proposed, and each agent is asked to reveal the willingness to pay in terms of the private good, Δx_i. The cost

Market Failures 297

is determined from a production function

$$y = C(x)$$

giving the relation between the amount of private good used to produce public good and the resulting output. The marginal change in provision is carried out if $\sum_{i=1}^{m} \Delta x_i - \Delta C > 0$ and not otherwise. Each individual pays the announced amount, and the surplus generated will be distributed among the individuals according to fixed weights.

Show that announcing the truth is a maximin strategy, i.e. it is the best reply to the most unfavorable of the other individuals' strategy choices.

Suppose that this procedure is carried out at each date, thereby determining a trajectory $(x_1(t), \dots, x_m(t), y(t))$ starting at $y = 0$. What will be the final result?

(6) In the economy $\mathcal{E} = \left((\mathbb{R}^2_+, P_i, \omega_i)_{i=1}^{3}, Y, (\theta_i)_{i=1}^{3} \right)$, where for $i = 1, 2, 3$, the preferences P_i are representable by the utility function

$$u_i(x_1, x_2) = x_1 x_2^i,$$

and $\omega_1 = \omega_2 = (1, 2)$, $\omega_3 = (2, 2)$, and the single firm has production set

$$Y = \left\{ (y_1, y_2) \in \mathbb{R}_- \times \mathbb{R} \,\middle|\, y_2 \le (-y_1)^2 \right\}.$$

Each consumer has profit share $\theta_i = \dfrac{1}{3}$.

Find an equilibrium with price setting by the single firm.

Chapter 11

Time and General Equilibrium

1. Time and Economic Activity

In the preceding chapters, we have discussed the static general equilibrium model dealing with production and allocation of the produced commodities in a market. Time has not entered explicitly, except in our brief treatment of tâtonnement stability. This should not, however, be taken as an indication that time is irrelevant or that considerations of the future do not matter. Time is a notion of central importance in our everyday economic activities, so it should also play a role for our understanding of the economy. Consequently, we must extend our discussion of general equilibrium to take account of time, and indeed this is what we have considered in the remaining chapters.

We begin with the simplest possible way of letting the future influence the choices of the economic agents, extending the list of commodities to include also those which are specified by their *date of delivery*, an approach which was hinted at already in the Introductory chapter. This approach has the advantage of being simple, indeed it makes it possible to use all of what has been obtained so far, since what is involved is only the interpretation of the notion "commodity", not the formalism as such.

Unfortunately, it does not go very far, since what is covered are only cases where future transactions can be decided and contracted upon here and now, which is not the case in real-world markets, and therefore, we proceed to other ways of involving the future in the

300   *Theory of General Economic Equilibrium*

chapters to follow, where we shall also allow for uncertainty, which tends to be a standard companion of considerations regarding the future.

1.1 *Dated commodities and the price system*

Our first approach to the problems of treating economic activities over time uses only what we have already established, since it pertains only the interpretation of the model. A commodity should be characterized not only by its physical properties, its quality and the location on which it is available but also by its *date of delivery*. Thus, we assume that the l commodities traded in the market arise from an original set of K goods that can be delivered in any of the $T + 1$ possible dates $t = 0, 1, \ldots, T$, so that $l = K(T + 1)$.

If we want to keep track of the time structure of a commodity bundle x, over all the future dates $t = 0, 1, \ldots, T$, then we may write it as

$$x = (x_0, x_1, \ldots, x_T) = (x_{10}, \ldots, x_{K0}, x_{11}, \ldots, x_{qt}, \ldots, x_{KT}),$$

using the double index qt to indicate the commodity type as well as date. Accordingly, a consumption bundle specifies the consumption in each of the T dates, meaning that consumers plan their consumption for each of the periods up to the economic horizon T.

The consumer can set up such plans since trade can actually be carried out: There is a market given by a price system, which with the same conventions can be written as

$$p = (p_0, p_1, \ldots, p_T) = (p_{10}, \ldots, p_{K0}, p_{11}, \ldots, p_{qt}, \ldots, p_{KT}),$$

where p_{qt} is the price of the good of type q delivered at date t. We may think of the price p_{qt} as the amount of some numeraire commodity, say the Kth good delivered at date 0, which must be given up in order to obtain one unit of the good q delivered at date t. Therefore, p is called the system of *discounted* prices, which give us today's evaluation of future deliveries. Selling one unit of good q delivered at the date t_1 gives p_{qt_1} units of the numeraire commodity, and one may buy back some units of the good delivered at $t_2 > t_1$, at the cost of p_{qt_2} per unit acquired. This operation corresponds to a transfer of goods over time. Giving up one unit of good q at date t_1, one may obtain $\alpha_q(t_1, t_2)p_{qt_2} = p_{qt_1}$ of the good at t_2.

$$\alpha_q(t_1, t_2) = \frac{p_{qt_1}}{p_{qt_2}},$$

is known as the *accumulation factor* (of good q from date t_1 to t_2). In the case where $t_2 = t_1 + 1$, it can be used to define an interest rate

$$i_q(t_1) = \alpha_q(t_1, t_1 + 1) - 1,$$

called the *interest rate* (of commodity q at date t_1). Interest rates derived from the price system are specific for each good and each date, and there will in general not be equality between i_{q^1t} and i_{q^2t} when q^1 and q^2 are different goods.

If we think of the numeraire good K as some kind of money (we shall return to money and its role in general equilibrium theory in the chapters to follow), then it may be reasonable to measure prices at date t in terms of money *at the same date t*, so that we define the *non-discounted* price of good q delivered at time t as

$$\bar{p}_{qt} = \frac{p_{qt}}{p_{Kt}}.$$

Corresponding to the initial (discounted) price system, we have a system, of non-discounted prices

$$\bar{p} = (\bar{p}_{10}, \ldots, \bar{p}_{K0}, \bar{p}_{11}, \ldots, \bar{p}_{qt}, \ldots, \bar{p}_{KT}).$$

Turning to the production side of the economy, a production plan in the T period economy will be an array

$$y = (y_0, y_1, \ldots, y_T) = (y_{10}, \ldots, y_{K0}, y_{11}, \ldots, y_{qt}, \ldots, y_{KT}),$$

where y_{qt} is the net output of good q in period t. The production set Y_j of producer j is a collection of such production plans y. Consider a producer j with production set Y_j. If the objective of the producer is to maximize profits, then $y \in Y_j$ should be chosen in such a way that

$$p \cdot y = \sum_{q=1}^{K} \sum_{t=0}^{T} p_{qt} y_{qt} \tag{1}$$

is maximal. This means that the producer buys deliveries of input goods or sells promises to deliver output goods at some future dates. The profit expression (1) can be rewritten as

$$p \cdot y = \sum_{q=1}^{K} \sum_{t=0}^{T} p_{qt} y_{qt} = \sum_{q=1}^{K} \sum_{t=0}^{T} (p_{Kt} \bar{p}_{qt}) y_{qt} = \sum_{t=0}^{T} p_{Kt} (\bar{p}_t \cdot y_t)$$

$$= \sum_{t=0}^{T} [\alpha_K(0, t)]^{-1} \bar{p}_t \cdot y_t, \tag{2}$$

where we have used the notation $\bar{p}_t$ and y_t for the part of $\bar{p}$ and y referring to date t. This is the classical formula for discounted net value of an investment. If the interest rate of the numeraire good is time independent,

$$i_K(0) = i_K(1) = \cdots = i_K(T-1) = i,$$

so that $\alpha_K(1, t) = (1+i)^{t-1}$ for $t \leq T-1$, we get that

$$p \cdot y = \sum_{t=0}^{T} \frac{\bar{p}_t \cdot y_t}{(1+i)^t}. \tag{3}$$

1.2 *The role of money in general equilibrium*

As we have seen, the theory of prices in markets for future delivery gives us a theory of interest rates. In our theory, interest rates are determined in the same way as prices (since interest rates *are* prices), that is by the interplay of preferences (which describe — among other things — the willingness to postpone consumption), technology (including the technical possibilities of getting a larger output by increasing the time span between input and output, that is by using more "roundabout" methods of production) and scarcity of resources.

What is *not* so easy to obtain within the framework of our model is a monetary explanation of the interest rates. Indeed, money has played a very humble role in our theory so far. According to standard definitions, money has several functions, namely as

(1) a measure of value,
(2) a commonly accepted means of exchange,
(3) a store of value.

We have used money in the first role repeatedly, for example, whenever prices are measured relative to the price of a numeraire, which then becomes the standard of value. The second role of money will be studied in detail in Chapter 16 to follow. At this point, we comment briefly on the third role, which so far has not been used at all.

While stores of value make little sense in a timeless, static model, we cannot expect that the need for a store of value arises automatically once time is introduced. Indeed, if there are markets for future delivery of all goods, there will be no obvious reason for storing money. Even if some future markets do not exist, the remaining system of markets may still be large enough to enable traders to eliminate the need for a store of value: Suppose that there are markets for the numeraire good delivered at any of the dates $1, 2, \ldots, T$ (bought and sold against the numeraire good at date 1) with prices p_{Kt} and markets for the other goods at each date t traded against the numeraire good *at the same date*, with prices $\bar{p}_{qt}$. Then good h at date t can be traded in two steps, by first buying numeraire good for delivery at date t and then trading the numeraire good at this date against good h, so that the existence of these markets suffices for the traders to obtain same final net trades as in a perfect market.

Since forward markets for the numeraire good make a store of value superfluous, we consider a system of markets with no possibility of futures trading, with a numeraire good, which can be stored at no cost. We assume that all consumers have identical forecasts for the prices to prevail at future dates.

304   *Theory of General Economic Equilibrium*

An equilibrium in an economy of this type consists (as usually) of an allocation $x = (x_1^0, \ldots, x_m^0, y_1^0, \ldots, y_n^0)$ and a price system $p = (p_0, p_1, \ldots, p_T)$, where $p_0 = (p_{10}, \ldots, p_{K0})$ is the vector of current prices and $p_t = (p_{1t}, \ldots, p_{Kt})$ the common expectations of prices at date t, $t = 1, \ldots, T$, such that x is feasible and each consumer i has chosen a bundle maximal for P_i on the budget set

$$\gamma(p) = \left\{ x_i' \in X_i \,\middle|\, p_0 \cdot x_{i0}' \leq p_0 \cdot \omega_{i0}, \; p_t \cdot x_{iqt}' \leq p_t \cdot \omega_{it} \right.$$
$$\left. + p_{iKt} x_{iK,t-1}', t = 1, \ldots, T, \right\}.$$

The money stock carried over from the last period appears on the right-hand side of the budget equations for $t \geq 1$; this is the way in which the otherwise separated budget constraints of each period are interconnected.

This equilibrium is of a somewhat different nature than those previously encountered, since prices are no longer payments for future deliveries, but an expectation which happens to be the same for all. This is somewhat unsatisfactory, and indeed, we shall elaborate on this kind of equilibria in the following chapters. It suffices for our argument here, so we accept it as an intermediate concept. The point is that even in this economy, where the numeraire commodity has a well-defined storage property, nobody will want to keep a stock of money. The intuitive explanation is that since nothing happens after the Tth period, nobody will want to hold money at the end of this period. But if all consumers want to get rid of their money holdings, money cannot have a positive price, and then nobody will want to carry it over from period $T - 1$ to period T. Repeating this argument T times, we get that nobody will want to hold money at any period.

The argument is easily confirmed by our formal models.

Theorem 11.1. *Let $\mathcal{E} = (X_i, P_i, \omega_i)_{i=1}^m$ be an economy with storing facilities for commodity K, where all consumers i have $X_i = \mathbb{R}_+^l$, satisfy Assumption 0.1 and are such that*

$$[x_i, x_i' \in X_i, x_{ikt} = x_{ikt}', k \neq K, t = 1, \ldots, T]$$
$$\Rightarrow [x_i' \in \mathrm{cl}\, P_i(x_i), x_i \in \mathrm{cl}\, P_i(x_i')]. \tag{4}$$

Let $(x_1^0, \ldots, x_m^0, y_1^0, \ldots, y_n^0, p^0)$ be an equilibrium in $\mathcal{E}$. Then

$$p_{Kt}^0 x_{iKt}^0 = 0, \quad \text{all } i \text{ and } t.$$

The expression in (4) says that if two bundles differ only in the last (numeraire) coordinate of each period, then the consumer is indifferent. In other words, money has no intrinsic utility. Given this assumption, the proposition tells us that *nobody wants to hold money* in any period, and its price will be zero.

Proof of Theorem 11.1. We begin by considering the last date T: If $p_{KT} x_{iKT} > 0$, then $x_{iKT} > 0$ and by (4), consumer x_i would not be maximal for P_i on $\gamma(p)$. It follows that $p_{KT} x_{iKT} = 0$.

Next, suppose that we have shown that $p_{Kt} x_{iKt} = 0$ for all i and $t > \bar{t}$, where $\bar{t} < T$. If $p_{K\bar{t}} x_{iK\bar{t}} > 0$, then consumer i could obtain a preferred bundle by reducing $x_{iK\bar{t}}$ (since anyway $p_{K(\bar{t}+1)} x_{iK(\bar{t}+1)} = 0$) and increasing consumption of all other commodities. We conclude that $p_{K\bar{t}} x_{iK\bar{t}} = 0$. $\qquad\square$

Thus, the present model fails to give a meaningful theory of money as a store of value. A main reason is that in our model, the future ends at period T, and nobody is concerned with what comes after. This is artificial; in actual fact, the future has no such upper bound, even though single individuals do not live eternally. Consequently, we must consider a model of markets over time without such a limitation. This is the topic of Section 2.

2. Temporary Equilibrium

In our discussion of allocation over time in this chapter, we have treated the standard model of general equilibrium of markets for present and future delivery. This is appropriate when intertemporal allocation is our main concern, and it has the advantage that the theory developed so far can be exploited without changes of the formal structure. But it has shortcomings as well, indeed the very idea of deciding upon all future consumptions and productions today seems far from the intuitive conceptions of trading, even with due

306   *Theory of General Economic Equilibrium*

regard to the future. Most of the decisions regarding exchanges of goods and services are related to what happens in the current period.

A combination of the simplicity of the one-period model with the consistency of the infinite-horizon generations model is found in the *temporary equilibrium* model introduced by Hicks (1946) and Grandmont (1977). Here, individual optimization involves a consideration of the future, but market balance is supposed to obtain only for current-period goods and for money. Only the markets for commodities traded in the current period matter in this model. We assume that there are no markets for future delivery, so that the only way to obtain a particular commodity in the next period is to buy commodity l now and keep it till the next period when it can be used for buying the commodity.

Consumers are assumed to have consumption sets $X_i = X_i^0 \times X_i^1 \subset \mathbb{R}^{2l}$, with consumption bundles $x_i = (x_i^0, x_i^1)$ specifying the present consumption $x_i^0 \in X_i^0$ as well as a planned future consumption $x_i^1 \in X_i^1$, preference correspondences $P_i : X_i \rightrightarrows X_i$, and endowments $\omega_i = (\omega_i^0, \omega_i^1)$, $i = 1, \ldots, m$. Given the current prices p^0 and an expected price system p_i^1 for the next period, consumer i will choose a bundle (x_i^0, x_i^1) which is maximal for P_i given the constraints

$$p^0 \cdot x_i^0 \le p^0 \cdot \omega_i^0 \tag{5}$$

for the current period, and

$$p^1 \cdot x_i^1 \le p_i^1 \cdot \omega_i^1 + p_{il}^1 \max\{0, \omega_{il}^0 - x_{il}^0\} \tag{6}$$

for the second, where money left over from the first period can be used for buying consumption goods. A temporary equilibrium will be obtained when $\sum_{i=1}^m x_i^0 \le \sum_{i=1}^m \omega_i^0$. Since the temporary equilibrium depends on price expectations, we must specify how the expectations arise. We do this follow method, assuming that each individual i expects future prices to be determined according to a function $\psi_i : \Delta \to \Delta$, where, as usual, $\Delta = \left\{ p \in \mathbb{R}_+^l \,\middle|\, \sum_{h=1}^l p_h = 1 \right\}$.

Time and General Equilibrium 307

We are now ready for defining a temporary equilibrium.

Definition 11.1. Let $\mathcal{E} = (X_i, P_i, \omega_i)_{i=1}^m$ be a two-period economy, and for $i = 1, \ldots, m$, let ψ_i be the price expectation function of consumer i. A temporary equilibrium is a pair (x^0, p^0), where $x^0 = (x_1^0, \ldots, x_m^0)$ is an array of current period bundles and $p^0 \in \Delta$ is a current price system, such that

(i) x^0 is feasible, i.e. $x_i^0 \in \mathbb{R}_+^l$, $i = 1, \ldots, m$, and $\sum_{i=1}^m x_i^0 \leq \sum_{i=1}^m \omega_i^0$,

(ii) for each i, there is $x_i^1 \in \mathbb{R}^l$ such that (x_i^0, x_i^1) is maximal for P_i among all bundles $(\tilde{x}_i^0, \tilde{x}_i^1)$ satisfying the constraints (5) and (6) with $p_i^1 = \psi_i(p^0)$.

In what follows, we give an existence proof for temporary equilibria, following the approach laid down in Chapter 1. The assumptions on the characteristics of the economy are entirely standard, but we shall need an assumption on the price forecasts which guarantees the boundedness of planned future consumptions for each agent at any current price system. We assume that $X_i = \mathbb{R}_+^{2l}$ for simplicity.

Theorem 11.2. Let $\mathcal{E} = (X_i, P_i, \omega_i)_{i=1}^m$ be a two-period economy with price expectation functions $(\psi_i)_{i=1}^m$, where for $i = 1, \ldots, m$, $X_i = \mathbb{R}_+^{2l}$, P_i satisfies Assumption 0.2, and $\omega_i \in \mathbb{R}_{++}^{2l}$. Moreover, assume that for each i and each $p^0 \in \Delta$, $\psi_i(p^0) \in \Delta_\varepsilon = \{p \in \Delta \mid p_h \geq \varepsilon\}$ for some $\varepsilon > 0$.

Then there exists a temporary equilibrium in $\mathcal{E}$.

Proof. From the data of $\mathcal{E}$, we construct an exchange economy $\widehat{\mathcal{E}} = \left(\widehat{X}_i, \widehat{P}_i, \widehat{\omega}_i\right)_{i=1}^m$ with commodity space $\mathbb{R}^{l(1+m)}$. For each $i \in M$, the consumption set

$$\widehat{X}_i = \mathbb{R}_+^l \times \{0\} \times \cdots \times \{0\} \times \mathbb{R}_+^l \times \{0\} \times \cdots \times \{0\},$$

with elements $\hat{x}_i = (x_i^0, 0, \ldots, 0, x_i^1, 0, \ldots, 0)$ having non-zero coordinates only for $h \in \{1, \ldots, l\}$ and $h \in \{il + 1, \ldots, (i + 1)l\}$. The preference relation $\widehat{P}_i$ and the initial endowment $\widehat{\omega}_i$ of consumer i are obtained using the obvious identification of X_i and $\widehat{X}_i$.

An allocation in $\widehat{\mathcal{E}}$ is an array $\hat{x} = (\hat{x}_1, \ldots, \hat{x}_m)$ with $\hat{x}_i \in \widehat{X}_i$ for $i = 1, \ldots, m$. A price system for $\widehat{\mathcal{E}}$ is an $(m + 1)$-tuple of prices

$\hat{p} = (p^0, p^1, \ldots, p^m)$ with $p^0 \in \Delta$ and $p^i \in \Delta_\varepsilon$ for $k = 1, \ldots, m$. The allocation–price pair is feasible if

$$\sum_{i=1}^m x_i^0 \le \sum_{i=1}^m \omega_i^0, p^i = \psi_i(p^0), \quad i = 1, \ldots, m. \tag{7}$$

Proceeding as in the proof of Theorem 1.1, we define budget correspondences

$$\hat{\gamma}_i(\hat{x}, \hat{p}) = \left\{ \hat{x}_i \,\middle|\, p^0 \cdot \hat{x}_i^0 < p^0 \cdot \omega_i^0, p^i \cdot \hat{x}_i^1 < p^i \cdot \omega_i^1 + p_{il}(\omega_{il}^0 - \hat{x}_{il}^0) \right\}$$

and the derived correspondences

$$\varphi_i(\hat{x}, \hat{p}) = \begin{cases} \gamma(\hat{x}, \hat{p}) \cap \widehat{P}_i(\hat{x}_i) & \hat{x}_i \in \mathrm{cl}\, \gamma_i(\hat{x}, \hat{p}), \\ \gamma_i(\hat{x}, \hat{p}) & \text{otherwise.} \end{cases}$$

The correspondences φ_i have open graph and convex values, and $x_i^0 \notin \varphi_i(\hat{x}, \hat{p})$, all $(\hat{x}, \hat{p})$.

Let $K > 0$ be a real number with $K \ge \sum_{i=1}^m \omega_{ih}^0$ for $h = 1, \ldots, l$, and let X^K be the set of current bundles $x \in \mathbb{R}_+^l$ with $x_h \le K$ for all h. Then X^K is compact, and $x_i^0 \in X^K$, $i = 1, \ldots, m$, for any family $(x_i^0)_{i=1}^m$ satisfying (7). Similarly, there is M such that for each i and each $x_i^1 \in \mathbb{R}_+^l$, if $p^i \in \Delta_\varepsilon$ and $p^i \cdot x_i^1 \le p^i \cdot \omega_i^1 + p_{il}^1 \sum_{j=1}^m \omega_{jl}^0$, then $x_{ih}^1 \le M$ for all h. Let $X^M = \{x \in \mathbb{R}_+^l \mid x_h \le M, \text{ all } h\}$.

Let $\overline{X} = \prod_{i=1}^m X^K \times \prod_{i=1}^m X^M \times \Delta \times \Delta_\varepsilon^m$ with generic element

$$\overline{x} = (x_1^0, \ldots, x_m^0, x_1^1, \ldots, x_m^1, p^0, p_1, \ldots, p_m).$$

For each i, φ_i can be considered as a correspondence from $\overline{X}$ to X^K. Define the map $\varphi_0 : \overline{X} \rightrightarrows \Delta$ by

$$\varphi_0(\overline{x}) = \left\{ p_0' \in \Delta \,\middle|\, p_0' \cdot \sum_{i=1}^m (x_i^0 - \omega_i^0) > 0 \right\},$$

then φ_0 also has open graph, convex values and it satisfies $p^0 \notin \varphi_0(\overline{x})$ since $p^0 \cdot (x_i^0 - \omega_i^0) \le 0$ for $i = 1, \ldots, m$. Now, we may apply Theorem 1.7 to the collection of correspondences consisting of φ_i for

$i = 0, 1, \ldots, m$, the trivial correspondences sending each $\bar{x}$ to all of X^M for $i = 1, \ldots, m$, and the correspondences taking $\bar{x}$ to $\{\psi_i(p^0)\}$ for $i = 1, \ldots, m$, to give a current allocation $(x_1^0, \ldots, x_m^0)$ and a price p^0 satisfying the conditions for a temporary equilibrium. $\qquad \square$

The way in which we have formulated price expectations is very rudimentary, since it does not take into account the uncertainty which would presumably be connected with a forecast of prices. Also, our use of purely individual expectations means that we can justify almost any behavior by the particular way of forecasting future prices. To eliminate these shortcomings, we would need to elaborate on uncertainty and the method of treating uncertainty in general equilibrium models, which we do in later chapters (namely Chapters 13–15), and to consider possible ways of introducing some consistency in consumers' forecasts, something which will concern with in Chapter 12.

For the moment, we should however note that the influence of expectations with regard to the future on the current market conditions is a valuable contribution of the model, since this allows for a more refined vision of the workings of the market. Thus, undersupply or -demand does not necessarily eliminate themselves through changes in the market prices, indeed the expectations of future gains or losses can result in zero prices of goods which consumers want but which they do not buy since they prefer to buy the numeraire good (in a model with production, this phenomenon can be presented in a more striking version, cf. the work of Grandmont and Laroque (1977)). The view of the market mechanism as a well-functioning device for achieving balance of wants and possibilities of satisfying these wants should be taken with a pinch of salt.

The importance of the temporary equilibrium model lies in its pointing to the importance of the future for the market today. To elaborate on this, we need to improve the model in several respects, so that expectations become less arbitrary and the interdependence of market agents is restored. We shall leave the temporary equilibrium model at this point and turn to models where balance of supply and demand holds not only in the present period but also in those to follow.

3. Exercises

(1) Define the savings or loans of a consumer $i \in \{1, \ldots, m\}$ at date $t \in \{0, 1, \ldots, t\}$ in a Walras equilibrium (x, p), and check that all loans and savings over time balance for all consumers. Explain that the loans can be interpreted as bonds issued at date t and repaid at date $t + 1$ (possibly after new bond issues).

Assume that a fraction $\lambda_t < 1$ of the bonds issued at date t will not be repaid. The losses are fully reimbursed to the savers through a government intervention financed by tax which is proportional to the value of the endowment at $t + 1$. Define an equilibrium for this situation and check whether the equilibrium allocation is Pareto-optimal.

(2) For a production plan y_t in any of the periods $t = 1, \ldots, T$, (2) and (3) show that $p_t \cdot y_t$ expresses the net present value (NPV) of y_t. The *Macaulay* duration of the collection $y_1, \ldots, y_T$ of production plans is defined as

$$\text{MacD} = \frac{\sum_{t=1}^{T} t p_t \cdot y_t}{\sum_{t=1}^{T} p_t \cdot y_t}.$$

Give an interpretation of the Macaulay duration.

Suppose that the interest rate i of the numeraire good is time independent. Show that the Macaulay duration can be expressed as the elasticity of the present value with respect to $1 + i$.

(3) (Ljungqvist and Sargent, 2018). An economy consists of two infinitely lived consumers $i = 1, 2$. There is a single non-storable good. Consumer i consumes c_i^t at time t and ranks consumption streams by $\sum_{t=0}^{\infty} \beta^t u(c_i^t)$, where $0 < \beta < 1$ is a discount factor and u is increasing, strictly concave, and C^2, $i = 1, 2$.

Consumer 1 has the endowment $\omega_1 = (1, 0, 0, 1, 0, 0, 1, \ldots)$, and consumer 2 has the endowment $\omega_2 = (0, 1, 1, 0, 1, 1, \ldots)$.

Define and find a Walras equilibrium in this economy.

Suppose that one of the consumers markets an asset that promises to pay 0.1 units of the consumption good in each period. Find the price of this asset.

(4) Consider an economy with a current and a future period, and two goods, one of which (the second) is storable. There are two consumers, with consumption sets $\mathbb{R}^4_+$, and the preferences are representable by utility functions

$$u_1(x^0, x^1) = x^0_1 x^0_2 x^1_1 x^1_2, \quad u_2(x^0, x^1) = x^0_1 (x^0_2)^2 x^1_1 (x^1_2)^3.$$

The initial endowments are $\omega^0_1 = (4, 1)$, $\omega^1_1 = (0.4, 5)$ for consumer 1 and $\omega^0_2 = (2, 2)$, $\omega^1_2 = (2, 3)$ for consumer 2. Consumer 1 expects the price of the first good stated at $t = 0$ to be unchanged in the next period, whereas consumer 2 expects an inflation of 10% in all prices.

Find a temporary equilibrium.

(5) (Credit cycles, Kiyotaki and Moore, 1997) Consider an economy with a single non-storable good and a real asset, called land. Agents, which are infinitely lived, are either *entrepreneurs*, who have access to technology and own land, or *lenders*, who have an endowment of the good. The technology is such that one unit of the good combines with k units of land to produce y units of the good in the next period.

In each period, there is a market for land with price q_t per unit (measured in terms of the good at date t). Entrepreneurs can borrow the consumption good against full collateral to the amount of $q_{t+1}(1 + r)^{-1}$ per unit of land (where r is the intertemporal interest rate assumed here to be independent of t). To produce with this amount of the good, the entrepreneur uses $kq_{t+1}(1 + r)^{-1}$ of the land, and the rest, $(1 - k)q_{t+1}(1 + r)^{-1}$, is made available for the public against a rent h_t per unit of land, which is found in a market where the demand is a decreasing linear function of the rent.

Use the profit maximization condition for the entrepreneur to express q_t as a quadratic function of q_{t+1}. Show that this condition gives rise to a cyclical movement in land prices with period 2.

Chapter 12

Overlapping Generations Economies

1. The Overlapping Generations (OLG) Model

In Chapter 11, we saw that in a model of trade with a finite horizon, there is no need for money in the form of a specific good, with no intrinsic value to consumers, which should be used only for carrying value between one period and another. Doing away with the final date T for trade will take us to a model with an infinite number of commodities, since for each good there will be infinitely many dates at which it might be delivered. We have already seen that models with infinite-dimensional commodity spaces pose some new and often quite intricate problems of their own. However, there is a way of avoiding many (if not all) of these complications, basically by assuming that each individual will trade only in finitely many commodities, and each commodity is traded only by finitely many individuals.

Indeed, we consider an economy $\mathcal{E}$ where consumers live in only two consecutive periods. In each period t, there is a "newborn" generation (denoted I^t) consisting of m^t consumers, present also in the next period, but not in any of the following. This means that there are m^{t-1} old and m^t young consumers in the market at date $t, t \geq 0$. At $t = 0$ there are the m^0 new-born consumers and a set I^{-1} of old consumers, the origin of which need not concern us, at least for the moment.

For consumer $i \in I^t$, a consumption set $X_i \subset \mathbb{R}^{2l}$ is given, the elements of which are individually feasible consumption bundles

$x_i = (x_i^t, x_i^{t+1})$ specifying the consumption of the goods during the lifetime. The preferences of consumer $i \in I^t$ are given by a preference correspondence $P_i : X_i \rightrightarrows X_i$, and an initial endowment $\omega_i \in X_i$ is also given. The initial old consumers $i \in I^{-1}$ have consumption sets X_i which are subsets of $\mathbb{R}^l$, as we are not concerned about what happened before date 0.

So far, the model has been similar to those previously encountered, with the difference that consumers are only interested in some of the (dated) commodities, namely those delivered while the consumer is alive. An *allocation* in the overlapping generations model $\mathcal{E}$ is a sequence

$$x = ((x_i)_{i \in I^t})_{t=-1}^{\infty},$$

where $x_i \in X_i$ for each $i \in I^t, t = -1, 0, 1, \ldots$. The allocation x is *feasible* if it can be sustained by the individual endowments, i.e. if

$$\sum_{i \in I^{t-1}} x_i^t + \sum_{i \in I^t} x_i^t \leq \sum_{i \in I^{t-1}} \omega_i^t + \sum_{i \in I^t} \omega_i^t, \quad t = 0, 1, \ldots. \tag{1}$$

This means that the total consumption does not exceed the available resources in any period $t \geq 0$. Thus, goods cannot be transferred between periods.

A *Walras equilibrium* in the overlapping generations economy $\mathcal{E}$ is a pair (x, p), where x is an allocation and p is a price system. Here, a price system p is a sequence $(p^t)_{t=0}^{\infty}$ where $p^t \in \mathbb{R}_+^l$ specifies the price p_h^t of good h delivered at date t, for $h = 1, \ldots, l$. The equilibrium conditions are as follows:

(i) x is feasible, i.e. satisfies (1),
(ii) for $t = 0, 1, \ldots$ and each $i \in I^t, t \geq 0$, x_i is maximal for P_i on the set of all bundles $\hat{x}_i$ satisfying the budget constraint

$$p^t \cdot \hat{x}_i^t + p^{t+1} \cdot \hat{x}_i^{t+1} \leq p^t \cdot \omega_i^t + p^{t+1} \cdot \omega_i^{t+1}, \tag{2}$$

(iii) for $i \in I^{-1}$, x_i is maximal for P_i on the set of bundles $\hat{x}_i$ satisfying

$$p^0 \cdot \hat{x}_i \leq p^0 \omega_i. \tag{3}$$

Overlapping Generations Economies 315

It can be seen from the budget constraints that the Walras equilibrium presupposes perfect capital markets, or, otherwise put, there is a complete system of markets for future delivery. Therefore, each consumer faces a single overall budget constraint; the value of consumption through life must not exceed the value of the initial endowment, similarly summed through the whole lifetime.

While this assumption of a single budget constraint is natural as a continuation of the model in the previous chapter, it may not be too realistic. Indeed, the equilibrium is difficult to visualize as something established in a real-world market. If prices react in a way so as to achieve a balance of supply and demand, then the price mechanism must take into account the actions of consumers which are not yet born. We may, however, interpret the model and its equilibrium in a different way, not as a picture of an actual market coming to rest, but rather as a way to formalize trading between living persons taking into account that there will be future periods, future consumption and future generations. In this context, an equilibrium in the overlapping generation economy gives us a price system that balances supply and demand today even when due respect is payed to future generations, in the sense that the supply and demand to be expected from new-born in the next period is taken into consideration. Moreover, this expected supply and demand is *rational* in the sense of being the true supply and demand of the agents, at least if their characteristics are the same as the agents living now.

Turning to the question of existence of Walras equilibria, we first note that the overlapping generations model have infinitely many commodities *and* infinitely many consumers, making it a *large-square economy* in the sense of Section 3. Nevertheless, the existence question can be solved in largely the same way as for the standard general equilibrium model. We state the result and give an outline of its proof.

Theorem 12.1. *Let $\mathcal{E}$ be an overlapping generations economy satisfying the assumptions stated above, so that $\omega_i \in \text{int } X_i$ for each $i \in I^t$, $t = -1, 0, 1, \ldots$. Then there exists a Walras equilibrium in $\mathcal{E}$.*

Proof (outline). For each $T \geq 1$, we introduce a *truncated economy* $\mathcal{E}^T$ as follows: The consumers in $\mathcal{E}^T$ are the consumers in the generations I^t for $t \leq T - 1$ (i.e. all the consumers whose life span terminated before or at T); the consumption sets X_i are considered as subsets of $\mathbb{R}^{l(T+1)}$ (a copy of $\mathbb{R}^l$ for each date $t = 0, 1, \ldots, T$), which is the commodity space of the economy $\mathcal{E}^T$. The utility function u_i of any consumer in I^t, $t \leq T - 1$, is considered as defined on this X_i, and bundles are identified with elements of the new commodity space by adding zeros in the coordinates not otherwise specified.

Finally, we add a single firm to the economy; it has the production set

$$Y^T = \left\{ y \in \mathbb{R}^{l(T+1)} \,\middle|\, y \leq \left(0, \ldots, 0, \sum_{i \in I^T} \omega_i^T \right) \right\};$$

Thus, "producing" in the economy $\mathcal{E}^T$ means simply to use the resources of the young generation at time T (which is not represented among the agents in $\mathcal{E}^T$).

In this way, we have obtained a standard (at least from the technical point of view) economy which has a usual Walras equilibrium $(x^{[T]}, p^{[T]})$ (where we have omitted the production). Here, $x^{[T]} = ((x_i)_{i \in I^T})_{t=-1}^{T-1})$ is the allocation in the economy $\mathcal{E}^T$ specifying a bundle for each consumer, and $p^{[T]} \in \Delta^{l(T+1)} = \{ p \in \mathbb{R}^{l(T+1)} \mid \sum_{h=1}^{l} \sum_{t=0}^{T} p_h^t = 1 \}$ is the price.

Clearly, $p^{[T]}$ belongs to the set $\{ p \in \mathbb{R}^l \mid \sum_{h=1}^{l} p_h \leq 1 \}^{T+1}$, which is compact. Adding an infinite sequence of zeros, we may consider $p^{[T]}$ an element of

$$\Delta_\infty = \prod_{T \geq 0} \left\{ p \in \mathbb{R}^l \,\middle|\, \sum_{k=1}^{l} p_k \leq 1 \right\},$$

the product of infinitely many copies of $\{ p \in \mathbb{R}^l \mid \sum_{k=1}^{l} p_k \leq 1 \}$. Similarly, the allocation $x^{[T]}$ may be identified with an element of the infinite product $\mathbb{R}^\infty$ (with 0 in all places corresponding to $t > T$). Now, the advantage of this is that Δ_∞ is compact (as a product of compact sets). Also, while $\mathbb{R}^\infty$ is not compact, the allocation $x[T]$

Overlapping Generations Economies 317

belongs to a fixed compact subset, namely the set of all allocations x in the overlapping generations economy where no consumer at any period gets a bundle greater than the total initial resources available at that period.

It follows that the sequence $(x^{[T]}, p^{[T]})$ of equilibria in the T-truncated economies for $T = 1, 2, \ldots$ (considered as allocation–price pairs in the overlapping generations economy), has a subsequence converging to some point (x, p). We leave it to the reader to check that (x, p) is a Walras equilibrium of $\mathcal{E}$. □

2. Pareto Optimality of Equilibria Over Time

Having discussed the existence problem, we turn to questions of optimality. It turns out that the first fundamental welfare theorem may fail and equilibrium allocations are not necessarily Pareto optimal. In order to show this, it suffices to consider a simple version of the general model of the previous section, namely a model with only one good in each period; if there is a counterexample in such a model (to the first fundamental theorem of welfare theory), then this shows that the result cannot hold generally. Restricting ourselves to the one-good case has the advantage of keeping the model tractable while still admitting all the essential complications of the general case.

Thus, in this section we discuss an overlapping generations economy $\mathcal{E}$ with only one good in each period. We shall assume further that consumers' preferences are monotonic in the sense that more of the good in any single period is preferred; otherwise, we need no further simplifying assumptions; thus, we have said nothing about the number of consumers in each generation, which may be arbitrary.

Let $x = ((x_i)_{i \in I^t})_{t=-1}^{\infty}$ be an allocation together with the price system $p = (p_t)_{t=0}^{\infty}$. In accordance with Chapter 3, we say that (x, p) is a *market equilibrium* in $\mathcal{E}$ if it satisfies the conditions

(i) x is feasible, i.e. satisfies $\sum_{i \in I^{t-1}} x_i^t + \sum_{i \in I^t} x_i^t \leq \omega^t$ for $t = 0, 1, \ldots,$

(ii) for $t = 0, 1, \ldots$ and each $i \in I^t$, $t \geq 0$, x_i is maximal for P_i on the set of all bundles $\hat{x}_i$ satisfying the budget constraint

$$p^t \cdot \hat{x}_i^t + p^{t+1} \cdot \hat{x}_i^{t+1} \leq p^t \cdot x_i^t + p^{t+1} \cdot x_i^{t+1}, \tag{4}$$

(iii) for $i \in I^{-1}$, x_i is maximal for P_i on the set of bundles $\hat{x}_i$ satisfying

$$p^0 \hat{x}_i \leq p^0 x_i, \tag{5}$$

so that preferred bundles are more expensive assessed at the price system p. We assume that all prices are positive, $p_t > 0$, all t.

A *reduced model* is a collection $\mathcal{S} = (S_t)_{t=0}^{\infty}$ of subsets S_t of $\mathbb{R}^2$, such that for each t,

$$(0, 0) \in S_t, \quad S_t + \mathbb{R}^2 \subset S_t.$$

We can associate with each equilibrium (x, p) in $\mathcal{E}$ a reduced model by letting S_t be the set of (sums of) preferred net trades for the generation born at time t. The construction is given below.

For $t = 0, 1, 2, \ldots$, let

$$X_t^\circ = \sum_{i \in I^t} \mathrm{cl}\, P_i(x_i)$$

be the set of aggregate consumption plans which are as good for each consumer in generation t as that prescribed by the equilibrium (x, p). From X_t°, we obtain the set S_t of aggregate net trades as those goods that arise from x as

$$S_t = X_t^\circ - \{\omega_t\}.$$

It is easily verified that $\mathcal{S} = (S_t)_{t=0}^{\infty}$ is indeed a reduced model.

Suppose that the equilibrium (x, p) in $\mathcal{E}$ is not Pareto optimal. Then there must be another feasible allocation x' such that $x_i' \in P_i(x_i)$ for all $i \in I^t$ and all t with at least one strict inequality. It follows that

Overlapping Generations Economies

there must be a sequence $(z_t)_{t=0}^{\infty}$ with

$$z_t = \sum_{i \in I^t} (x_i' - x_i), \quad t = 0, 1, \dots,$$

from which it follows that $z_t \in S_t$, each t. Moreover, by the feasibility of the allocation x', we have that

$$z_{t-1}^t + z_t^t \le 0$$

for all t. By our monotonicity assumption we may choose x' such that equality holds for all t, so that

$$z_{t-1}^t = -z_t^t, \quad t = 0, 1, \dots.$$

Since a Pareto improvement must give the old generation at period 0 at least as much of the good as the original allocation, we have that $z_0^0 \le 0$, consequently, $z_0^1 \ge 0$ and, by a similar argument, $z_{t-1}^t \ge 0$, all t. It is seen that we can describe these preferred net trades by an *improving sequence* $(\varepsilon_t)_{t=0}^{\infty}$, where $\varepsilon_t \ge 0$ is given by

$$\varepsilon_t = -z_t^t, \quad \text{all } t,$$

so that $z_t = (-\varepsilon_t, \varepsilon_{t+1})$ for $t = 0, 1, \dots$.

Summing up, we have that any equilibrium (x, p) gives rise to a reduced model S, and a Pareto improvement x' of x corresponds to an improving sequence $(\varepsilon_t)_{t=0}^{\infty}$ in S. The converse of the latter statement is easily seen to hold as well: If there is an improving sequence in the reduced model S, then there is a Pareto improvement of x. Consequently, the question of whether or not equilibria are Pareto-optimal reduces to that of whether or not there are improving sequences in the reduced model S.

The advantage of the latter formulation is that it lends itself easily to a systematical investigation. We need one further piece of

320   *Theory of General Economic Equilibrium*

notation: Subsets S and T of $\mathbb{R}^2$ may be composed by the following rule

$$S \circ T = \{(z_1, z_2) \mid \exists x \in \mathbb{R}, (z_1, x) \in S, (-x, z_2) \in T\}.$$

The composition rule $\circ$ is associative: if $S, T, T' \neq \emptyset$ are subsets of $\mathbb{R}^2$, then

$$(S \circ T) \circ T' = S \circ (T \circ T'),$$

so that we may use the notation $S \circ T \circ T'$ for the composition.

We have the following general result (where pr_1 denotes projection on the first coordinate, $\mathrm{pr}_1(z_1, z_2) = z_1$.

Theorem 12.2. *Let* $\mathcal{S} = (S_t)_{t=0}^{\infty}$ *be a reduced model. If*

$$\inf_{t \geq 0} \sup_{T \geq t} \inf \mathrm{pr}_1(S_0 \circ \cdots \circ S_T) = 0, \tag{6}$$

then there is no improving sequence in $\mathcal{S}$. *Conversely, if each* S_t *is a convex set, and there is no improving sequence, then* (6) *holds.*

Proof. Suppose that there is an improving sequence $(\varepsilon_t)_{t=0}^{\infty}$. Let $r = \min\{t \mid \varepsilon_t > 0\}$. Then

$$-\varepsilon_r \in \mathrm{pr}_1(S_0 \circ \cdots \circ S_T)$$

for all $T \geq r$; consequently,

$$0 > -\varepsilon_r \geq \sup_{T \geq r} \inf pr_1(S_r \circ \cdots \circ S_T),$$

contradicting (6).

Conversely, suppose that all S_t are convex and that there are $r \geq 0$ and $a_r > 0$ such that

$$\sup_{T \geq r} \inf \mathrm{pr}_1(S_r \circ \cdots \circ S_T) < 0.$$

Choose $\bar{\varepsilon}_r$ with $0 < \bar{\varepsilon}_r < a_r$. It is easily checked that the set

$$A_t = \{y \mid (-\varepsilon_r, y) \in S_r \circ \cdots \circ S_T\}, \quad T \geq r$$

is non-empty for each $T \geq r$. Define a sequence $(\varepsilon_t)_{t=0}^{\infty}$ by $\varepsilon_t = 0$ for $t < r$, $\varepsilon_r = \bar{\varepsilon}_r$, and $\varepsilon_t = \min A_t$ for $T \geq r$.

Clearly, $\varepsilon_t \geq 0$ for all $t \in N$. Also, $\varepsilon_t > 0$ for $t \geq r$ since otherwise $(-\varepsilon_r, 0) \in S_r \circ \cdots \circ S_T$ for some $T \geq r$, a contradiction. We show that $(\varepsilon_t)_{t=0}^T$ is an improving sequence, i.e. that $(-\varepsilon_t, \varepsilon_{t+1}) \in S_t$ for all t.

For $t \leq r - 2$, we have that $(-\varepsilon_t, \varepsilon_{t+1}) = (0, 0) \in S_t$, and we have $(-\varepsilon_{r-1}, \varepsilon_r) = (0, \varepsilon_r) \in \mathbb{R}_+^2 \subset S_{r-1}$. Further, $(-\varepsilon_r, \varepsilon_{r+1}) \in S_r$ since $\varepsilon_{r+1} = \min A_r$.

Suppose that we have shown $(-\varepsilon_t, \varepsilon_{t+1}) \in S_t$ for $t \leq q$, where $q \geq r$. Then

$$(-\varepsilon_r, \varepsilon_{q+1}) \in S_r \circ \cdots \circ S_q,$$

$$(-\varepsilon_r, \varepsilon_{q+2}) \in S_r \circ \cdots \circ S_q \circ S_{q+1}$$

(both by the definition of $(\varepsilon_t)_{t=0}^\infty$), and since

$$S_r \circ \cdots \circ S_{q+1} = (S_r \circ \cdots \circ S_q) \circ S_{q+1},$$

there is $z \in \mathbb{R}$ such that

$$(\varepsilon_r, z) \in S_r \circ \cdots \circ S_q, (-z, \varepsilon_{q+2}) \in S_{q+1}.$$

By the definition of ε_{q+1}, $z \geq \varepsilon_{q+1} > 0$. Since S_{q+1} is convex, we have

$$\frac{\varepsilon_{q+1}}{z}(-z, \varepsilon_{q+2}) = \left(-\varepsilon_{q+1}, \frac{\varepsilon_{q+1}}{z}\varepsilon_{q+2}\right) \in S_{q+1},$$

and consequently,

$$\left(-\varepsilon_r, \frac{\varepsilon_{q+1}}{z}\varepsilon_{q+2}\right) \in S_r \circ \cdots \circ S_{q+1}.$$

From the definition of ε_{q+2}, we conclude that $\varepsilon_{q+1} = z$, whence

$$(-\varepsilon_{q+1}, \varepsilon_{q+2}) \in S_{q+1}.$$

By induction, it follows that $(\varepsilon_t)_{t=0}^\infty$ is an improving sequence, a contradiction. $\qquad\square$

The result of Theorem 12.2 gives a condition characterizing those equilibrium allocations which are Pareto-optimal. On the other hand, the condition in (6) is not very intuitive. In order to apply the result to situations with more intuitive content, we may consider special cases and investigate the result of applying the composition ∘ repeatedly to various classes of sets.

Halfspaces. If each S_t has the form of a halfspace, $S_t = \{(z^t, z^{t+1}) | p^t z^t + p^{t+1} z^{t+1} \geq 0\}$ for some sequence $(p^t)_{t=0}^{\infty}$, then we have

$$S_t \circ S_{t+1} = \{(z^t, z^{t+2}) | p^t z^t + p^{t+2} z^{t+2} \geq 0\},$$

so in this case the composition works in a very simple way. Unfortunately, halfspaces are not very likely to occur as sets of preferred net trades. However, a slight modification has some interest.

Halfspace with left boundary: Suppose that each of the sets S_t has the form

$$S(b_t) = \{(z^t, z^{t+1}) | z^t + z^{t+1} \geq 0, z^t \geq -b_t\},$$

where b_t is a non-negative number. An example of such a set is shown in Fig. 1. It is easily seen that

$$S(b_t) \circ S(b_{t+1}) = \{(z^t, z^{t+2}) | z^t + z^{t+2} \geq 0, z^t \geq -\min\{b_t, b_{t+1}\}\} = S(\min\{b_t, b_{t+1}\}),$$

so that the composition of sets "translates" to a minimization operation on parameters.

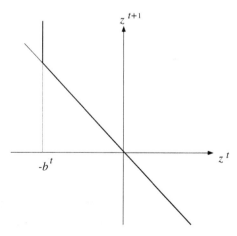

Fig. 1. A reduced model where the preferred net trades constitute a halfspace with left boundary.

Overlapping Generations Economies 323

If a reduced model S is composed of sets of this type, then each S_t is supported by the price $(1,1)$. This holds if the prices p^t of the equilibrium (x,p) are all 1, a case which in our present one-good world is not as special as it might seem: Indeed, a simple rescaling of the goods, so that 1 old unit of the good at period t becomes p^t new units, will achieve this purpose.

If the reduced model S consists of sets $S(b_t)$, then we get from Theorem 12.2 that non-existence of an improving sequence is equivalent to the condition

$$\liminf{}_{t \to \infty} b_t = 0.$$

Our analysis of this particular case may be put to general use: If the consumption set of each consumer of generation $t \geq 0$ is $\mathbb{R}^2_+$, then the preferred set at the bundle $x_i = (x_i^t, x_i^{t+1})$ is bounded to the left by $-x_i^t$; consequently, the aggregate preferred set of the consumers of generation t is bounded to the left by $-\sum_{i \in I^t} x_i^t$. Rescaling the units so that all prices are 1, we get that the resulting reduced model S constructed from the equilibrium satisfies the condition

$$S_t \subset S\left(\sum_{i \in I^t} p^t x_i^t\right).$$

Now we exploit that an improving sequence of S is also an improving sequence of the larger reduced model $(S(\sum_{i \in I^t} p^t x_i^t))_{t=0}^{\infty}$, and applying Theorem 12.2 we get.

Theorem 12.3. *Let $\mathcal{E}$ be a one-good overlapping generations model, where each consumer satisfies Assumptions 0.1 and 0.2 and preferences are monotonic. If (x,p) is an equilibrium in $\mathcal{E}$ with*

$$\liminf{}_{t \to \infty} \sum_{i \in I^t} p^t x_i^t = 0,$$

then the equilibrium is Pareto-optimal.

This result gives a sufficient condition for an equilibrium to be Pareto-optimal; the condition is not necessary, unless the preferred sets have the form $S(\sum_{i \in I^t} p^t x_i^t)$, which would only be the case if consumers had linear preferences. Other sufficient conditions may be derived in a similar way, considering particular reduced models where the composition $\circ$ works in an easy way. The approach can also be used to obtain necessary conditions, cf. Borglin and Keiding (1986).

Our treatment of the one-good model has shown that Pareto optimality is not any more an automatic consequence of equilibrium, so that one cannot, in general, rely on the market to produce the best possible allocation over time. There are also other surprising features of this one-good version of the overlapping generation model, which will be considered in Section 3.

3. Indeterminacy of Walras Equilibrium in the OLG Model

The simple one-good overlapping generations model, which served us in the discussion of Pareto-optimality, can be used to exhibit other particularities of the model. For this, we shall extend the set time line backward so as to include generations born at $t = -1$ as well (corresponding to the initial old generation of previous sections), $t = -2, -3, \ldots$. We assume that each generation contains exactly one consumer, and we use the birth date for indexing consumers, so that bundles of consumer t born at t are written as $x_t = (x_t^t, x_t^{t+1})$, $t \in \mathbb{Z}$. Similarly, the preferences and endowments of consumer t are written as P_t and $\omega_t = (\omega_t^t, \omega_t^{t+1})$, respectively.

An allocation is now a doubly infinite sequence $x = (x_t)_{t \in \mathbb{Z}}$ of such consumption bundles, which are feasible if $x_t = (x_t^t, x_t^{t+1}) \in X_t$, $t \in \mathbb{Z}$, whereby we choose the consumption set as $X_t = \mathbb{R}_+^2$ for all t, and

$$x_{t-1}^t + x_t^t = \omega_{t-1}^t + \omega_t^t, \quad t \in \mathbb{Z}.$$

A price system for this model is a doubly infinite sequence $p = (p^t)_{t \in \mathbb{Z}}$ for all t. Finally, a Walras equilibrium in our doubly infinite model is a pair (x, p), where x is an allocation and p is a price system, such that

(i) x is feasible, and
(ii) for each $t \in \mathbb{Z}$, x_t is maximal for P_t in the set

$$\{\hat{x}_t \mid (p^t, p^{t+1}) \cdot \hat{x}_t \leq p^t \omega_t^t + p^{t+1} \omega_t^{t+1}\}.$$

It can be checked rather easily that Walras equilibria exist for this model under the standard conditions, since the methods applied to the model with starting point at $t = 0$ can be adapted to treat the present situation as well. Here, we shall be interested in uniqueness properties of Walras equilibria, or rather, as we shall see, lack of uniqueness of Walras equilibria. Indeed, we show that one may choose an arbitrary relative price $\bar{r}_t = p^{t+1}/p^t$ at a fixed date t and construct a Walras equilibrium with exactly this value of the relative prices at date t. Consequently, there is a continuum of Walras equilibria in the model.

We shall need an assumption on the preferences P_t to carry out this construction.

Assumption 12.1. For each $t \in \mathbb{Z}$, P_t satisfies the following condition

(i) for each fixed value $\bar{x}_t^t > 0$, there is a strictly increasing map $\sigma(\bar{x}_t^t, \cdot) : \mathbb{R}_{++} \to \mathbb{R}_{++}$ taking x_t^{t+1} to a number r^{t+1}, such that $(1, r^{t+1})$ supports $P_t(\bar{x}_t^t, x_t^{t+1})$ at $(\bar{x}_t^t, x_t^{t+1})$, such that $r^{t+1} \to 0$ for $x_t^{t+1} \to 0$ and $r^{t+1} \to \infty$ for $x_t^{t+1} \to \infty$.
(ii) for each fixed value $\bar{x}_t^{t+1} > 0$, there is a strictly decreasing map $\sigma(\cdot, x_t^{t+1}) : \mathbb{R}_{++} \to \mathbb{R}_{++}$ taking x_t^t to a number r^{t+1}, such that $(1, r^{t+1})$ supports $P_t(x_t^t, \bar{x}_t^{t+1})$ at $(x_t^t, \bar{x}_t^{t+1})$, such that $r^{t+1} \to \infty$ for $x_t^{t+1} \to 0$ and $r^{t+1} \to 0$ for $x_t^{t+1} \to \infty$.

Assumption 12.1 is satisfied if P_t is represented by a utility function which is C^1 and for which the marginal rates of substitution behave in the way usually assumed in classical microeconomics, so that the consumers of our model are rather standard. The situation is illustrated in Fig. 2.

We can now show that there is a continuum of Walras equilibria in this model.

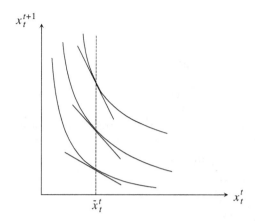

Fig. 2. The assumption of continuous marginal rates of substitution: For each fixed level $\bar{x}_t^t$ of consumption as young, the MRS at bundles with this consumption assume all values from 0 to ∞.

Theorem 12.4. *Let $\mathcal{E}$ be a one-good overlapping generations economy where each generation consists of one consumer satisfying Assumptions 0.1 and 0.2 and 12.1.*

Let $t \in \mathbb{Z}$ and $\bar{r}_{t+1} \in \mathbb{R}_{++}$ be arbitrary. Then there is a Walras equilibrium (x, p) in $\mathcal{E}$ such that $p^{t+1}/p^t = \bar{r}_{t+1}$.

Proof. Let $x_t = (x_t^t, x_t^{t+1}) \in \mathbb{R}_+^2$ be a bundle which is maximal for the preferences P_t of consumer t subject to the constraint

$$\hat{x}_t^t + \bar{r}_{t+1}\hat{x}_t^{t+1} \leq \omega_t^t + r_{t+1}\omega_t^{t+1}.$$

Define $p^t = 1$ and $p^{t+1} = r_{t+1}$. We use forward and backward induction from t to show that the price sequence (p^t, p^{t+1}) and the bundle x_t can be extended to double sequences $(p^\tau)_{\tau \in \mathbb{Z}}$ of prices and $(x_\tau)_{\tau \in \mathbb{Z}}$ of bundles such that for each τ,

(i) x_τ is maximal for P_τ at the budget constraint given by $(p^\tau, p^{\tau+1})$, and
(ii) $x_{\tau-1}^\tau + x_\tau^\tau \leq \omega_{\tau-1}^\tau + \omega_\tau^\tau$.

This will in its turn imply that (x, p) is a Walras equilibrium.

The induction hypothesis is clearly satisfied at date t. Suppose now that for some $\tau > t$, we have found prices $p^t, p^{t+1}, \ldots, p^\tau$ and

Overlapping Generations Economies 327

bundles $x_t, x_{t+1}, \ldots, x_{\tau-1}$ such that $x_{\tau'}$ is maximal for $P_{\tau'}$ under the budget constraint given by $(p^{\tau'}, p^{\tau+1})$ and

$$x_{\tau'-1}^{\tau'} + x_{\tau'}^{\tau'} \le \omega_{\tau'-1}^{\tau'} + \omega_{\tau'}^{\tau'}$$

for $\tau' = 1, \ldots, \tau-1$. Let $x_{\tau}^{\tau} = \omega_{\tau}^{\tau} - (x_{\tau-1}^{\tau} - \omega_{\tau-1}^{\tau})$ and apply Assumption 12.1(i) to find $x_{\tau}^{\tau+1}$ such that $\sigma(x_{\tau}^{\tau}, x_{\tau}^{\tau+1}) = r_{\tau+1}$ satisfies the budget equation

$$\hat{x}_{\tau}^{\tau} + r_{\tau+1}\hat{x}_{\tau}^{\tau+1} \le \omega_{\tau}^{\tau} + r_{\tau+1}\omega_{\tau}^{\tau+1}.$$

It follows that $x_{\tau+1}$ is maximal for $P_{\tau+1}$ at $(p^{\tau}, p^{\tau+1})$ with $p^{\tau+1} = r_{\tau+1}p^{\tau}$, and it satisfies the feasibility condition

$$x_{\tau-1}^{\tau} + x_{\tau}^{\tau} \le \omega_{\tau-1}^{\tau} + \omega_{\tau}^{\tau},$$

so that the induction hypothesis holds for τ.

Similarly, suppose that for some $v < t$, we have found prices $p^{v+1}, \ldots, p^t, \ldots$ and for $v' = v + 1, \ldots, t$, bundles $x_{v'}$ maximal for $P_{v'}$ under the budget constraint given by $(p^{v'}, p^{v+1})$ and

$$x_{v'-1}^{v'} + x_{v'}^{v'} \le \omega_{v'-1}^{v'} + \omega_{v'}^{v'}.$$

Let $x_v^{v+1} = \omega_v^{v+1} - (x_{v+1}^{v+1} - \omega_{v+1}^{v+1})$ and apply Assumption 12.1(i) to find x_v^v such that $\sigma(x_v^v, x_v^{v+1}) = r_{v+1}$ satisfies the budget equation

$$\hat{x}_v^v + r_{v+1}\hat{x}_v^{v+1} \le \omega_v^v + r_{v+1}\omega_v^{v+1}.$$

It follows that x_v is maximal for P_{v+1} at (p^v, p^{v+1}) with $p^v = p^{v+1}/r_{v+1}$, and it satisfies the feasibility condition

$$x_{v-1}^v + x_v^v \le \omega_{v-1}^v + \omega_v^v,$$

meaning that the induction hypothesis holds for v. □

The indeterminacy of Walras equilibrium once again points to the peculiarities in the model which are caused by the infinite horizon, this time stretching both forward and backward. Indeed, the openness of the model in the past was crucial for the approach, and given its consequences, it might be debated whether this is a useful way of modeling allocation over time. However, it cannot be discarded as being only a feature put into the model for considerations of mathematical perfectness: Indeed, in order to explain the formation of prices, we have to look into the past, and accepting the state of the economy as it is here and now does not provide us with much of an explanation, since the current state in its turn is a result of the past performance. This means that we must look backward as well, and for this purpose, we need the double infinity of the time set.

4. Cycles in the OLG Model

Another feature of the OLG model, this time one which for obvious reasons could not occur in the finite horizon models, is that of cyclical movements in the equilibrium prices and bundles. For this, we shall keep the simplified model of Section 3 and add the assumption of *stationarity,* meaning that the characteristics of the consumer in each generation are exactly the same.

In an OLG model of this type, there will be a *steady-state Walras equilibrium* where each generation t faces the same price in each period and obtains the same bundle maximal for P_t under the budget constraint given by the constant relative price 1. This equilibrium is not particularly interesting since the situation repeats itself without changes for every generation. What is much more interesting is that one can also find *cyclic* Walras equilibria, which points to possible applications in explaining business cycles, one of the standing preoccupations for economists.

A Walras equilibrium (x, p) is said to exhibit 2-cycles if there are $(\bar{x}, \bar{r}), (\hat{x}, \hat{r}) \in \mathbb{R}_+^2 \times \mathbb{R}_{++}$ such that

$$(x_t, r_t) = \begin{cases} (\bar{x}, \bar{r}) & t \text{ odd,} \\ (\hat{x}, \hat{r}) & t \text{ even,} \end{cases}$$

where, as before, $r_t = p^{t+1}/p^t$ denotes the relative prices of the good at the two dates relevant to the generation t. Thus, the relative prices shift from one to another in each period, and the bundle which is maximal for P_t under the budget constraint determined by these relative prices shift as well. The 2-cycle is *proper* if $\bar{r} \neq \hat{r}$.

Overlapping Generations Economies 329

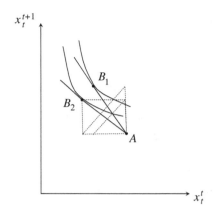

Fig. 3. A 2-cycle in the stationary OLG model. The initial endowment corresponds to A, and the relative prices at even dates give rise to the steeper budget line with resulting maximal bundle B_1. The net trade as young should be balanced by the net trade of the consumer at odd date as old, and the net trade as old by that of the young consumer at odd dates. This results in the bundle B_2 which is maximal under the budget constraint given by the less steep budget line.

The situation is illustrated in Fig. 3, where for simplicity the endowment is taken as the vector $(1,1)$. The net trade for even dates should be balanced by those of odd dates with coordinates interchanged. It is easily seen that if $\bar{r} < \hat{r}$ are the relative prices of a 2-cycle, which is not the steady state, then $\bar{r} < 1 < \hat{r}$.

The case exhibited in Fig. 3 does not require special forms of preferences and/or endowment; actually in what follow, it is shown that it is rather the typical situation, so that 2-cycles occur under rather weak assumptions. To show the existence of a Walras equilibrium with proper 2-cycles, we must make sure that we do not get the steady state. In the following theorem, we obtain this by assuming that relative prices which are close to 1 will give rise to an equilibrium only if they are equal to 1.

Theorem 12.5. *Let $\mathcal{E}$ be a one-good stationary overlapping generations model with only one consumer $(\mathbb{R}_+^2, P, \omega)$ in each generation. Assume that there is $\bar{\varepsilon} > 0$ such that*

(i) $|\bar{r} - 1| < \bar{\varepsilon}, |\hat{r} - 1| < \bar{\varepsilon}$,
(ii) $\bar{x}$ *is maximal for P under the constraint* $\bar{x}^t + \bar{r}\bar{x}^{t+1} \leq \omega^t + \bar{r}\omega^{t+1}$,

330 *Theory of General Economic Equilibrium*

(iii) $\hat{x}$ is maximal for P under the constraint $\hat{x}^t + \hat{r}\hat{x}^{t+1} \leq \omega^t + \hat{r}\omega^{t+1}$,

(iv) $(\bar{x}^t, \bar{x}^{t+1}) + (\hat{x}^{t+1}, \hat{x}^t) = (\omega^t + \omega^{t+1}, \omega^t + \omega^{t+1})$.

implies that $\bar{r} = \hat{r} = 1$.

Then there is a Walras equilibrium which exhibits a proper 2-cycle.

Proof. Let $\overline{X} = \{x \in \mathbb{R}^l_+ \mid x_h \leq K, h = 1, 2\}$, where $K > 2\max_h \omega_h$. Choose $0 < \varepsilon < \bar{\varepsilon}$ arbitrarily, let

$$\Delta_1 = \{(p_1, p_2) \in \mathbb{R}^2_+ \mid p_1 + p_2 = 1, p_1 - p_2 \geq \varepsilon\}$$

$$\Delta_2 = \{(p_1, p_2) \in \mathbb{R}^2_+ \mid p_1 + p_2 = 1, p_2 - p_1 \geq \varepsilon\},$$

define the correspondences $\gamma_i : \Delta_i \rightrightarrows \overline{X}$ by

$$\gamma_i(p_i) = \{x \in \overline{X} \mid p_i \cdot x < p_i \cdot \omega\}$$

and $\varphi_i : \Delta_i \times \overline{X} \rightrightarrows \overline{X}$ by

$$\varphi_i(p_i, x) = \begin{cases} \gamma(p_i) \cap P(x) & x \in \mathrm{cl}\, \gamma_i(p_i), \\ \gamma_i(p_i) & x \notin \mathrm{cl}\, \gamma_i(p_i), \end{cases}$$

for $i = 1, 2$. Finally, define for $i = 1, 2$ the correspondences $\varphi_0^i : \overline{X}^2 \rightrightarrows \Delta_i$ by

$$\varphi_i(x^1, x^2) = \{p_i \in \Delta_i \mid p_i \cdot (x_1^1 + x_2^2, x_2^1 + x_1^2) > (\omega_1 + \omega_2, \omega_1 + \omega_2).$$

Reasoning as in the proof of Theorem 1.1, one finds an array (x^1, x^2, p_1, p_2) such that $\varphi_i(p_i, x^i) = \emptyset$, $\varphi_0^i(x^1, x^2) = \emptyset$, for $i = 1, 2$. From $\varphi_i(p_i, x^i) = \emptyset$ we get that x^i is maximal for P_i on the budget set $\gamma_i(p_i)$, $i = 1, 2$, and from $\varphi_0^i(x^1, x^2)$ we have that $(x_1^1 + x_2^2, x_2^1 + x_1^2) \leq (\omega_1 + \omega_2, \omega_1 + \omega_2)$. It follows that (x^1, x^2, p_2, p_2) defines a Walras equilibrium in $\mathcal{E}$ exhibiting a 2-cycle, and by the assumptions (i)–(iv), we conclude that the 2-cycle is proper. $\square$

The presence of Walras equilibria exhibiting cyclical movements of relative prices is not restricted to cycles of period 2, and the above methods could be applied to show the existence of Walras equilibria with cycles of arbitrary length. We refer to, e.g. Geanakoplos (1989) and Tvede (2010) for a discussion of this topic.

Overlapping Generations Economies 331

5. Exercises

(1) Find a Walras equilibrium in the overlapping generations economy with two consumers in each generation, having consumptions set $\mathbb{R}^4_+$, preferences represented by utility functions

$$u^t_i(x^t_i, x^{t+1}) = (x^t_{i1})^i x^t_{i2} + 1.05^t (x^{t+1}_{i1})^i x^{t+1}_{i2},$$

and endowment $\omega^t_i = (3,3)$, $\omega^{t+1}_i = (1,1)$, for consumer i in generation t, $i = 1,2$, $t = 0,1,\dots$. The generation -1 has two consumers with consumption set $\mathbb{R}^2_+$, endowment $(1,1)$ and preferences given by utility functions $u^{-1}_i(x^0_{i1}, x^0_{i2}) = (x^0_{i1})^i x^0_{i2}$, $i = 1,2$.

(2) Consider a one-good *production model* $(f_t)^\infty_{t=0}$, where for each t, $f_t : \mathbb{R}_+ \to \mathbb{R}_+$ is concave, non-decreasing and satisfies $f_t(0) = 0$. A program $(x_t, y_t, c_t)^\infty_{t=0}$ is feasible if for all t, $y_t, x_t, c_t \in \mathbb{R}_+$, $y_t = x_t + c_t$, and $y_{t+1} = f_t(x_t)$. It is efficient if there is no other feasible program $(x'_t, y'_t, c'_t)^\infty_{t=0}$ with $c'_t \geq c_t$ for all $t \geq 1$ and $c'_t > c_t$ for some t.

A price system $(p_t)^\infty_{t=1}$ with $p_t > 0$, all t, supports the program $(x_t, y_t, c_t)^\infty_{t=1}$ if $p_{t+1} f_t(x_t) - p_t x_t \geq p_{t+1} x'_t - p_t x'_t$ for all x'_t, all t. Show that an efficient program is price-supported. Give an example of a production model where a price-supported program is not efficient.

(3) In the production model of Exercise 2, show that the family $S = (S_t)^\infty_{t=0}$ with

$$S_t = \{(x - x^0_t, y^0_{t+1} - y) \mid y \leq f_t(x)\}$$

for each t is a reduced model in the sense of Section 2, and that the program $(x_t, y_t, c_t)^\infty_{t=1}$ is efficient if there is no improving sequence for S.

Use this to characterize efficient programs in the production model.

(4) Consider a world with two countries, each described by a one-good overlapping generations model with a single consumer in

each generation. Show that there can be two different Walras equilibria, namely

(i) **Autarchy:** Trade is intergenerational between the young and the old consumer in each country.

(ii) **International trade:** All trade is between consumers of the same generation, belonging to different countries.

Is international trade always at least as good as autarchy?

(5) Consider an OLG economy with l commodities at each date, of which $l - 1$ are non-storable, whereas the lth commodity can be stored. The initial old generation has an endowment μ of the lth good, but none of the following generations is endowed with this commodity.

Show that the commodity can be used as money in the sense that intergenerational trades take the form of purchase or sale of money against the other commodities. What happens to the economy if the initial endowment μ is increased at date $t > 0$, for example by an extraordinary endowment given to the old consumers at date t?

(6) An OLG economy is stationary if the generations are identical (with the same number of consumers having the same characteristics). Suppose that in a one-good stationary OLG economy, there is a well-defined demand function $\xi : \mathbb{R}_{++} \to \mathbb{R}^2_{++}$, where $(\xi^0(r), \xi^1(r))$ is the aggregate demand for the good in the first and second life-year, respectively.

Show that if ξ is C^1 and $D\xi^1(1) > D\xi^0(1)$, then the economy has a 2-cycle.

Chapter 13

Uncertainty and General Equilibrium

1. Introduction

Uncertainty is a feature of everyday life and it has long ago been recognized as a crucial ingredient in models of economic behavior. A formal treatment of uncertainty using probability theory has been available for a long time, and it has been used, e.g. in economic models of insurance. A systematic extension to general equilibrium theory is however of more recent date.

We approach the treatment of uncertainty in general equilibrium from another angle, using the concept of a contingent commodity to be introduced in Section 2, thus interpreting the already established theory in a new way taking uncertainty into account. Convenient as this approach may be, it has costs in terms of realism, and therefore, we consider alternatives, either in the form of less demanding concepts of equilibrium, to be investigated in Section 3, or as absence of some of the markets necessary for the standard model to work. Moreover, we introduce the concept of *information* which has important implications for the workability of markets. Many of the problems introduced will resurface in the following chapters.

We conclude the chapter in Section 4 with an introduction to the theory of *sunspot equilibria* displaying the interplay of information, expectations and prices.

2. Contingent Commodities

One of the prerequisites of a formal treatment of uncertainty in economics is a discussion of the sources of this uncertainty.

We assume that *uncertainty is exogenous* to the model in the sense that what is uncertain is something in the environment or the agents' characteristics. The endowments may be subject to unpredictable fluctuations, the technology may be affected by uncertain factors and the preferences of the consumers may change in the future due to unpredictable shifts in taste and fashion. We formalize this by assuming that the environment, or "nature" as it is called in this context, may be in exactly one of the states in the set

$$S = \{s_1, \ldots, s_r\},$$

and that the uncertainty of the agents pertains to the exact state of nature at the moment when the economic decisions are taken. As it can be seen, we have assumed that there is only a finite number of states, and this may turn out to be too restrictive as we proceed, but for the moment, it will serve our purposes.

The state space S may have some further structure of which we shall take advantage later. First of all, uncertainty is usually (though not necessarily) connected with *time*. This is conveniently illustrated by a graph as shown in Fig. 1, which is called a *tree*. The present date ($t = 0$) is at the bottom (the root of the tree); the future periods are depicted as the layers above the root. At the start, nothing about the future is known with certainty, since "nature" can move along each of the edges out of the root, but as time passes, the knowledge of the actual path increases, until at the final date T it will be known in its entirely. The different states of nature $s_1, \ldots, s_r$ correspond to the different ways of moving through the tree starting at the root and ending at some vertex at time T (the set of these paths may be identified with the set of ("terminal") vertices at time T in this case; however, one might conceive of situations where there would be no final date T, so that the tree would continue indefinitely, and in this case, the identification of states of nature

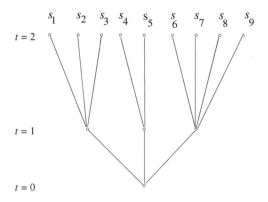

Fig. 1. An event tree. This way of illustrating uncertainty has turned out to be useful in the literature on uncertainty.

with paths is more appropriate). At some intermediate date t, the uncertainty is given by the set of paths starting at the actually realize vertex.

As we have seen in previous cases, the theory which we have already established may be interpreted to also cover the situation of uncertainty, provided that the concept of a commodity is understood in the right way. In our present case, we will let the specification of the state of nature s, in which the commodity will be delivered, enter into its description. A commodity specified in this way is called a *contingent commodity*; we shall write it in its dependency on the state of nature as

$$x(s) = (x(s_1), \ldots, x(s_r)),$$

where $x(s_h) \in \mathbb{R}^l$ is the bundle delivered if the state s_h obtains. This extension comes, as always, at a cost, but we postpone this discussion slightly in order to highlight the advantages of this approach into the theory of allocation under uncertainty. In particular, we can use the economic welfare theory to state that Pareto-optimal allocation under uncertainty can be achieved by the market mechanism applied to contingent commodities.

336   *Theory of General Economic Equilibrium*

Box 1. Expected utility: A simple way of modeling decisions under uncertainty is obtained using the *expected utility hypothesis*. Assume that there is given a set $S = \{s_1, \ldots, s_r\}$ of states of nature and a set X of commodity bundles.

A *risky prospect* is a pair $(x(s), \pi)$, where $x(s) = (x(s_1), \ldots, x(s_r))$ is a contingent commodity bundle with $x(s_k) \in X$ for $k = 1, \ldots, r$, and $\pi = (\pi(s_1), \ldots, \pi(s_r))$ a probability distribution on states. If X is the set of feasible contingent commodity bundles, we let Ξ denote the set of all risky prospects $(x(s), \pi)$ with $x(s) \in X$, π a probability distribution on S.

A preference correspondence $P : \Xi \rightrightarrows \Xi$ on risky prospects satisfies the *expected utility hypothesis* if there is a function $u : X \to \mathbb{R}$ such that

$$x(s), \pi) \in P(y(s), \pi') \iff \sum_{k=1}^{r} \pi(s_k) u(x(s_k)) \geq \sum_{k=1}^{r} \pi'(s_k) u(x(s_k)) \qquad (1)$$

for all $(x(s), \pi), (y(s), \pi') \in \Xi$. The function u is called a von Neumann–Morgenstern utility after von Neumann and Morgenstern (1944), who considered this kind of utility representation and characterized preference relations admitting the representation (1).

There is a very large literature on expected utility and its various extensions. We shall make use of von Neumann–Morgenstern utilities only in examples.

Admittedly, this version of the market mechanism with its abundance of contingent commodities is a far short from the markets of the real world. As we saw when dealing with allocation over time, not all the markets are necessary to achieve the same final allocations as the theoretical market. Suppose that for each agent in the economy, it is possible to trade a single good contingent on each of the states of nature; assume that good l has this property, so that there is a market of the type

$$\left\{ (z_l(s_1), \ldots, z_l(s_r)) \in \mathbb{R}^r \;\middle|\; \sum_{h=1}^{r} p_l(s_h) \cdot z_l(s_h) \leq 0 \right\}$$

for some price vector $(p_l(s_1), \ldots, p_l(s_r)) \in \mathbb{R}^r_+$. Furthermore if the agents can choose trades from each of the contingent markets given by a price vector $\bar{p}(s_h) \in \mathbb{R}^l_+$,

$$\mathcal{M}[s_h](\overline{p}(s_h)) = \{z = (z_1, \ldots, z_l) \in \mathbb{R}^l \mid p(s_h) \cdot z \leq 0\},$$

for $h = 1, \ldots, r$, then the total possibilities open to any agent are exactly the net trades from

$$\mathcal{M}(p(s)) = \left\{z(s) \in \mathbb{R}^{lr} \,\middle|\, \sum_{h=1}^{r} p(s_h) \cdot z(s_h) \leq 0\right\},$$

where $p(s) = (p(s_1), \ldots, p(s_r))$ with $p(s_h) \in \mathbb{R}^l$, $h = 1, \ldots, r$, is such that

$$p_k(s_h)\overline{p}_l(s_h) = \overline{p}_k(s_h)p_l(s_h) \text{ so that } p_k(s_h) = \frac{\overline{p}_k(s_h)}{\overline{p}_l(s_h)}p_l(s_h) \text{ if } \overline{p}_l(s_h) > 0$$

for each h. A system of markets where goods can be traded against each other given each of the states of nature, together with a single market where one (numeraire) commodity can be traded contingent on each of the states, is sufficient to make possible all the contracts which we need in the model.

3. Equilibrium of Plans, Prices and Price Expectations

As we mentioned in Section 2, the assumption that there are markets for all types of contingent commodities is not a realistic one. Most of these markets do not exist in practice. There are several reasons for that; one of the most convincing reasons is that contracts for delivery contingent on some event can be effected only when all the involved parties can observe if the event does actually occur.

In order to deal with such restrictions on the number of contingent markets, we introduce a refinement of the previous model, proposed by Radner (1972). The starting point is the event tree as shown in Fig. 1 and reproduced in Fig. 3. If we think of the uncertainty as being resolved successively with the passage of time, we get that the number of states (corresponding to the end nodes of the tree) which can be distinguished must increase with time. However, it might not be the case that each two of the nodes corresponding to a given date t may be distinguished, at least not to all of the agents. Some relevant event may have taken place but has not been

> **Box 2. Convex preferences and risk aversion:** Assumptions already known and accepted may have new implications. This concerns, for example, the convexity assumption: If convex preferences are representable by a utility function (now defined on bundles $x(s)$), then the consumer has a particular attitude toward risk (Fig. 2).
>
> Let $x^1(s)$ and $x^2(s)$ be two bundles that differ only in two coordinates corresponding to the same commodity contingent on two different events, $x^1_{h'}(s_k) = x^2_{h'}(s_k)$ for $h' \neq h$, $k \neq i, j$. Let us assume further that
>
> $$x^1_h(s_i) = x^2_h(s_j) = b > c = x^1_h(s_j) = x^2_h(s_i).$$
>
> Then convexity of preferences implies that
>
> $$u\left(\frac{1}{2}x^1(s) + \frac{1}{2}x^2(s)\right) > \min\{u(x^1(s)), u(x^2(s))\}.$$
>
>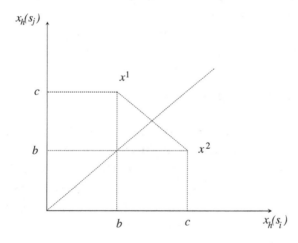
>
> Fig. 2. Convexity of preferences and risk aversion.
>
> Going back to the interpretation, we see that the "sure" (in the sense of being independent of whether state s_i or state s_j occurs) consumption of $(b + c)/2$ of the commodity is preferred to each of the originally uncertain prospects. We say in this situation that the consumer is *risk averse*.

observed yet. This means that we must specify exactly what can be distinguished at any date t. In Fig. 3, at the date $t = 1$ only two situations may be distinguished, one consisting of two vertices (which are indistinguishable), and another one consisting of a single vertex.

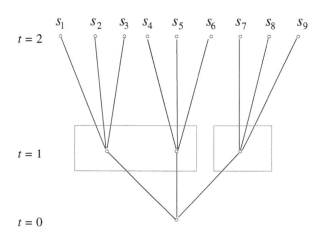

Fig. 3. The event tree with information sets.

The simplest way of doing this is by introducing for each date t, $t = 0, 1, \ldots, T$, a partition S_t of the set S of states into subsets, called the *events observable at* t. We assume that S_{t+1} is as fine as S_t for $t = 0, 1, \ldots, T - 1$, meaning that each set in S_t is a union of sets in S_{t+1}. Also, the initial partition S_0 is the trivial partition $S_0 = \{S\}$, and the final partition is

$$S_T = \{\{s_1\}, \ldots, \{s_r\}\}.$$

Given the family $(S_t)_{t=0}^T$ of partitions as a description of what can be observed at each date, we may describe the contracts which can be made at date t, for $t = 0, 1, \ldots, T$: Let $A \in S_t$. A trade contract at date t in event A specifies a delivery to the agent (positive or negative) of each of the l commodities at some date $u \geq t$ contingent on the event $B \in S_u$, $B \subset A$. The reason that B must be a subset of A is the obvious one: At the date of the contract, all traders are informed that the state of nature must belong to A, so uncertainty pertains only to subsets of A. We write a trade contract as $z^{tu}(A, B)$.

Trade contracts of the type described can be made for each t; thus, the agent has the possibility of trading on spot markets at each date (corresponding to contracts with $t = u$) or making forward trade contracts: If the event A has occurred at date t, and if furthermore event B occurs at date u, then a certain number of commodities is to be delivered. We assume that there is a price system $p^{tu}(A, B)$ for each type of trade contract. We use the notation

$$z = (z^{tu}(A, B))_{(t,u,A,B) \in \mathcal{A}},$$

where $\mathcal{A}$ is the set of (t, u, A, B), such that $t \leq u$, $A \in S_t$, $B \in S_u$, and $B \subset A$. Similarly, we write the price system as

$$p = (p^{tu}(A, B))_{(t,u,A,B) \in \mathcal{A}}.$$

The trade contracts are supposed to satisfy the system of budget constraints

$$\sum_{(u,B):(t,u,A,B) \in \mathcal{A}} p^{tu}(A, B) \cdot z^{tu}(A, B) \leq 0 \tag{2}$$

for each t and $A \in S_t$. This means that there is a separate budget constraint at each "contingent market" (t, A).

At this point, we may pause and consider the difference between our present setup and that of Section 2. The main difference is the possibility of postponing the engagement in contracts. In our present model, a delivery of some commodity at date u contingent on a certain (observable) event B may be obtained by signing a contract right away, i.e. at date 0 (at the price system $p^{0u}(S, B)$); this is the only way to get it in a model similar to that obtained in Section 2. But it may also be obtained through a contract of the type $z^{tu}(A, B)$ for some $t > 0$ and given an event $A \in S_t$. In this case, the contract itself is contingent (on the event A at date t). The greater flexibility of the present model is stressed by the fact that we might allow for different information in forward contracts at different dates and for different commodities.

The system $z_i = (z_i^{tu})_{(t,u,A,B) \in \mathcal{A}}$ of trade contracts concluded by a consumer i determines his/her final bundle. However, since we have introduced a sequence of markets, we must also keep record of the time path of trade and consumption. Let $\mathcal{A}_1$ be the set of pairs (t, A) with $t \in \{0, 1, \ldots, T\}$, $A \in \mathcal{S}_t$. A consumption plan of consumer i is a family

$$x_i = (x_i^t(A))_{(t,A) \in \mathcal{A}_1},$$

where $x_i^t(A) \in \mathbb{R}_+^l$ is the bundle available to consumer i at date t given that event A occurs. Let X_i be the set of feasible consumption plans, and let the initial endowments of consumer i be given as $\omega_i = (\omega_i^t(A))_{(t,A) \in \mathcal{A}_1}$. The system of contracts z_i is *feasible* for consumer i if it results in a feasible consumption, i.e. if the family of bundles given by

$$x_i^u(B) = \omega^u(B) + \sum_{(t,A):(t,u,A,B) \in \mathcal{A}} z_i^t(A) \tag{3}$$

for all $(u, B) \in \mathcal{A}_1$, belongs to X_i. The sum on the right-hand side in (3) is taken over all the date–event pairs which are previous to (u, B), since the delivery at date u in event B is the result of all trade contracts made at dates $t \le u$ given events A at t with $B \subset A$.

With the notions introduced so far, we may introduce the notion of equilibrium, known as a *Radner equilibrium* in our model. A *Radner equilibrium* (called an equilibrium of plans, prices and price expectations in Radner (1972)) is an array $(z_1^0, \ldots, z_m^0, p^0)$, where for each consumer i, z_i^0 is a system of trade contracts, and where p^0 is a price system, such that

(i) for each consumer i, the system of trade contracts z_i^0 is individually feasible, i.e. it satisfies (1)–(3) at the price system p^0, and it is individually optimal in the sense that there is no other system of trade contracts satisfying (1)–(3) such that the derived consumption plan yields a higher utility,

342   *Theory of General Economic Equilibrium*

(ii) the array of trade contracts is aggregate feasible in the sense that each market is in balance,

$$\sum_{i=1}^{m} (z_i^0)^{tu}(A, B) = 0$$

for all $(t, u, A, B) \in \mathcal{A}$.

These equilibrium conditions do not differ in any fundamental way from what we have seen in previous chapters. However, the interpretation of the price system is changed; for markets (t, u, A, B) with $t > 0$, corresponding to contracts which have not come into force at the date 0, we must visualize the trade contract $(z_i^0)^{tu}(A, B)$ as a trade *plan* and the price $(p^0)^{tu}(A, B)$ as a *price expectation*. Consequently, the equilibrium described above is one not only of actual trades (on spot and forward markets at date 0) but also an equilibrium of plans for future contracts *and* of price expectations.

This interpretation poses some questions of its own; how can we imagine that the economy will move toward an equilibrium in price expectations from some initial situation where agents may have non-equilibrium — and perhaps even differing — expectations? We shall leave this question open, instead we look at another problem arising in the present framework: An equilibrium may not exist under the standard conditions when not all possible markets are open.

3.1   *An economy for which there are no Radner equilibria*

In the following, we show by an example, due to Hart (1975), that Radner equilibria may fail to exist in otherwise well-behaved economies. We assume that there are two dates, $t = 0, 1$, and two states s_1, s_2 at date 1. There are two goods, with spot markets for each good in each period, and there are two forwards markets, here formulated as *securities*, whereby the first security yields one unit of good 1 at date 1 independent of the state, and the second security gives one unit of good 2 at date 1, also independent of state.

Uncertainty and General Equilibrium   343

There are two consumers who care only for consumption at $t = 1$. Consumption bundles of consumer i in the two states s_1, s_2 are written as $x_i(s_1) = (x_{i1}(s_1), x_{i2}(s_1)), x_i(s_2) = (x_{i1}(s_2), x_{i2}(s_2)) \in \mathbb{R}^2_+, i = 1, 2,$ and the preferences are assumed to be described by utility functions u_1, u_2 of the form

$$u_1(x_{11}(s_1), x_{12}(s_1), x_{11}(s_2), x_{12}(s_2))$$
$$= 2^{\frac{3}{2}} \sqrt{x_{11}(s_1)} + \sqrt{x_{12}(s_1)} + 2^{\frac{3}{2}} \sqrt{x_{11}(s_2)} + \sqrt{x_{12}(s_2)},$$
$$u_2(x_{21}(s_1), x_{22}(s_1), x_{21}(s_2), x_{22}(s_2))$$
$$= \sqrt{x_{21}(s_1)} + 2^{\frac{3}{2}} \sqrt{x_{22}(s_1)} + \sqrt{x_{21}(s_2)} + 2^{\frac{3}{2}} \sqrt{x_{22}(s_2)}.$$

Both consumers have non-zero endowment only at $t = 1$. Consumer 1 has the endowments

$$\omega_1(s_1) = \left(\frac{5}{2}, \frac{50}{21}\right) \quad \text{and} \quad \omega_1(s_2) = \left(\frac{13}{21}, \frac{1}{2}\right) \tag{2}$$

depending on whether state s_1 or s_2 occurs, and consumer 2 has endowments

$$\omega_2(s_1) = \left(\frac{1}{2}, \frac{13}{21}\right) \quad \text{and} \quad \omega_2(s_2) = \left(\frac{50}{21}, \frac{5}{2}\right). \tag{3}$$

Let $p(s_1), p(s_2)$ be the expected equilibrium prices at date 1 in the two states. Since the consumers are not interested in the goods at date 0, only the spot markets at date 1 and the two securities will be in use. The returns of the first security are $(p_1(s_1), p_1(s_2))$, and the second security has the return $(p_2(s_1), p_2(s_2))$. There are two cases to be considered.

(1) The state-dependent price vectors $p(s_1)$ and $p(s_2)$ are linearly independent. Then also the return vectors $(p_1(s_1), p_1(s_2))$ are linearly independent, and it is possible to buy all combinations of contingent goods at date 1 using the securities. If $q_h(s_j)$ is the cost of buying one unit of good h at date 1 given state s_j, then consumer i maximizes u_i

under the budget constraint

$$q(s_1) \cdot x_i(s_1) + q(s_2) \cdot x_i(s_2) \le q(s_1) \cdot \omega_i(s_1) + q(s_2) \cdot \omega_i(s_2), \qquad (4)$$

for $i = 1, 2$, where $\omega_i(s_j)$ are given in (2) and (3). In equilibrium, total consumption at $t = 1$ should be equal to the total endowment,

$$x_{1h}(s_j) + x_{2h}(s_j) = \omega_{1h}(s_j) + \omega_{2h}(s_j) = 3, \quad h = 1, 2, \quad j = 1, 2. \qquad (5)$$

There is an equilibrium in this forward market, that is a solution to (4)–(5), at

$$q_1(s_1) = q_2(s_1) = q_1(s_2) = q_2(s_2) = \frac{1}{4},$$

$$x_1(s_1) = x_1(s_2) = \left(\frac{8}{3}, \frac{1}{3}\right), \quad x_2(s_1) = x_2(s_2) = \left(\frac{1}{3}, \frac{8}{3}\right).$$

This is the only equilibrium, since the economy satisfies gross substitution (cf. Chapter 6, Exercise 3). However, since all the forwards prices are equal, we must have that

$$\frac{p_1(s_1)}{p_2(s_1)} = \frac{p_1(s_2)}{p_2(s_2)} = 1$$

contradicting linear independence of $p(s_1)$ and $p(s_2)$. We can therefore rule out this case as a candidate for a Radner equilibrium.

(2) We have that $p(s_1)$ and $p(s_2)$ are linearly dependent. Then also the returns of the two securities are linearly dependent, and this means that there is no way of transferring income from one state to another at date 1, and, consequently, no trading (in securities) takes place at date 0. We therefore look at what happens at date 1 when the states have been realized.

If state s_1 has occurred, consumer 1 maximizes utility $2^{\frac{3}{2}} \sqrt{x_{11}(s_1)} + \sqrt{x_{12}(s_1)}$ subject to the budget constraint

$$p(s_1) \cdot x_1(s_1) \le \frac{5}{2} p_1(s_1) + \frac{50}{21} p_2(s_1),$$

and consumer 2 maximizes $\sqrt{x_{21}(s_1)} + 2^{\frac{3}{2}} \sqrt{x_{22}(s_1)}$ under the constraint

$$p(s_1) \cdot x_2(s_1) \le \frac{1}{2} p_1(s_1) + \frac{13}{21} p_2(s_1),$$

and the equilibrium condition is $x_{1h}(s_1) + x_{2h}(s_1) = 3$ for $h = 1, 2$. Again, there is a unique equilibrium

$$p(s_1) = \left(\frac{2}{3}, \frac{1}{3}\right), \quad x_1(s_1) = \left(\frac{62}{21}, \frac{31}{21}\right), \quad x_2(s_1) = \left(\frac{1}{21}, \frac{32}{21}\right).$$

If instead state s_2 occurs, a similar computation will show that the unique equilibrium at this state is given by

$$p(s_2) = \left(\frac{1}{3}, \frac{2}{3}\right), \quad x_1(s_2) = \left(\frac{32}{21}, \frac{1}{21}\right), \quad x_2(s_2) = \left(\frac{31}{21}, \frac{62}{21}\right)$$

(the details are left to the reader). It is seen that the prices $p(s_1)$ and $p(s_2)$ are linearly independent, which contradicts our initial assumption of this case.

From the two cases we conclude that there can be no expected prices at date 1 which are consistent with a Radner equilibrium, and, consequently, a Radner equilibrium does not exist.

While the example shows that all the conditions of a Radner equilibrium can never be fulfiled, it does not give much hint to what goes wrong. However, it is clear that the problems come from the presence of forwards trading — consumers can contract for delivering goods in the future and use this for buying other goods for future delivery. If there is a limit to forwards trading, for example an upper bound K on the total delivery (positive or negative) of each commodity in the contracts, so that

$$|z_k^{tu}(A, B)| \leq K \quad \text{for all } k, \tag{6}$$

then Radner equilibria exist under usual conditions, cf. Radner (1972).

4. Sunspot Equilibria

Throughout this chapter, we have been dealing with uncertainty in an abstract way by assuming that economic activities are influenced by "nature" through its choice of state $s \in \{s_1, \ldots, s_r\}$. We did not explain in detail how, e.g. the agents' characteristics are influenced by nature's choice; what really mattered was the existence of markets for trades contingent upon these states. In other words, the crucial feature is that agents *believe* that the uncertainty under consideration does matter. This is in itself not a new idea, we have mentioned it repeatedly from the Introduction chapter onward. However, it acquires a new dimension in the present context: Phenomena which are *believed* to influence the economy will end up by actually influencing it, no matter how little reason there might be for those beliefs in the first place.

This is the so-called *sunspot phenomenon;* the name is derived from theory put forward by one of the pioneers of neoclassical economics, W. Stanley Jevons, who connected business cycles with sunspot activity (Jevons, 1878). While the physical part of the theory could not be substantiated, there is another connection, namely that based on the beliefs of the consumers, which merits a further investigation. We take a closer look at the sunspot phenomenon in a particular model of exchange, adapted from Cass and Shell (1983).

We assume that there are two ordinary consumption goods and two time periods. The set S of possible states of nature has only two elements, $S = \{s_1, s_2\}$. The economy works over two periods of time: In the first period, the uncertainty is not yet resolved; there is no trade in the consumption goods for immediate delivery, only in commodities delivered in the next period contingent on the state. In the second period, the state of nature is observed, and the contingent deliveries are made. In addition, there is a spot market for goods in this second period.

There are two types of consumers, namely those who are allowed to trade in the forward market in the first period, and those who are not; we may alternatively think of this as a distinction between generations born in the first or in the second period. We use the convention that only consumers $1, \ldots, m_1$ may trade in the market for contingent commodities. It is important for the conclusions that $m_1 < m$.

In defining the characteristics of the agents, we assume that *uncertainty is irrelevant:* We assume that consumers have preferences which can be represented by von Neumann–Morgenstern utility functions $u_i : \mathbb{R}_+^2 \to \mathbb{R}$ depending only on the consumption of the two ordinary goods and that the endowment of goods $\omega_i(s) = \omega_i \in \mathbb{R}_+^2$ does not depend on s. Finally, we introduce the beliefs of the agents as a probability distribution $\pi = (\pi_1, \pi_2)$ which is common for all agents. This means that for consumer i, the utility of a consumption bundle $(x_i(s_1), x_i(s_2)) = ((x_{i1}(s_1), x_{i2}(s_1)), (x_{i1}(s_2), x_{i2}(s_2)))$ can be written as

$$\pi_1 u_i(x_{i1}(s_1), x_{i2}(s_1)) + \pi_2 u_i(x_{i1}(s_2), x_{i2}(s_2)). \tag{7}$$

An equilibrium in this model must prescribe actions and prices such that the actions are individually optimizing at the prices and aggregate consistent. There are prices on each of the contingent commodities, and in addition, there are spot markets in the second period, given the realized state of nature. This makes a total of eight different prices; however, we assume that *second period spot prices are known in the first period,* so that the relative prices of the two goods given any of the states must be the same in the forward and spot market. Otherwise, it would be profitable to buy the good which is relatively cheap in the forward market and sell it in the spot market, at the same time selling the relatively expensive commodity in the forward market and buying it in the spot market. This is an example of no-arbitrage conditions, about which we shall have more to say in Chapter 14.

Now we may formulate the equilibrium conditions in our model: For given prices $(p(s_1), p(s_2)) = ((p_1(s_1), p_2(s_1)), p_1(s_2), p_2(s_2))$ in the forward market, a consumer with index $i \in \{1, \ldots, m_1\}$ will maximize (7) over all (state-dependent) bundles $(x_i(s_1), x_i(s_2))$ which satisfy the budget constraint

$$p(s_1) \cdot x_i(s_1) + p(s_2) \cdot x_i(s_2) \le p(s_1) \cdot \omega_i + p(s_2) \cdot \omega_i. \tag{8}$$

Clearly, we may assume the consumer to trade only in the forward market; since the prices in the spot markets are proportional to those of the forward market, he/she will have no additional advantage from trading in the spot market after the uncertainty has been resolved.

348   *Theory of General Economic Equilibrium*

For the consumers with index $i \in \{m_1 + 1, \ldots, m\}$, the state of nature is known when they enter the market. Consequently, such a consumer solves the problem

$$\max u_i(x_i(s)) \text{ over all } x_i(s) \text{ with } p(s) \cdot x_i(s) \leq p(s) \cdot \omega_i \qquad (9)$$

for $s = s_1, s_2$. We have used $p(s)$ (the prices in the forward market for deliveries contingent on state s) instead of the spot prices, but since they are proportional, the budget constraint will remain the same.

Now, an equilibrium consists of an array of bundles $(x_i(s_1), x_i(s_2))$ for each consumer and a price system $(p(s_1), p(s_2))$ such that each consumer solves his utility maximization problem (7)–(8) or (9), such that the demand does not exceed the supply in any period or state, meaning that

$$\sum_{i=1}^{m} x_{ih}(s) \leq \sum_{i=1}^{m} \omega_{ih}(s), \quad h = 1, 2, \quad s = s_1, s_2.$$

Essentially, the equilibrium introduced is of a type which we have seen before. There is a single unfamiliar feature, namely that spot prices in the second period are taken into account in the first period when the forward contracts are concluded. To justify this, we think of the equilibrium as one of *rational expectations:* Traders in the first period *know* the prices only in the markets which are actually open (the forward markets), and they form expectations about the spot prices in the next period. We then restrict our attention to such equilibria where the expectations will not be contradicted by the actual outcome of the model.

Since our main object is to show that uncertainty which does not affect any of the characteristics of the economy may neverthe-less influence allocation, we shall not worry too much about the interpretation. Instead, we want to formalize the notion that *uncer-tainty matters* in an equilibrium $((x_i(s_1), x_i(s_2))_{i=1}^{m}, (p(s_1), p(s_2)))$. This is the case if there is some consumer i such that the bundles dif-fer in the two states, $x_i(s_1) \neq x_i(s_2)$. Otherwise, uncertainty does not matter. It is useful to begin with a consideration of the latter case.

Theorem 13.1. *Assume that the utility functions u_i of consumers $i \in \{1, \ldots, m_1\}$ are strictly concave (so that these consumers are risk averse). Let*

$$(x, p) = ((x_i(s_1), x_i(s_2))_{i=1}^m, (p(s_1), p(s_2)))$$

be an equilibrium such that

$$\frac{1}{\pi_1} p(s_1) = \frac{1}{\pi_2} p(s_2).$$

Then uncertainty does not matter in the equilibrium.

Proof. It is clear that the equilibrium choices $(x_i(s_1), x_i(s_2))$ of consumers $i \in \{m_1 + 1, \ldots, m\}$ must satisfy $x_i(s_1) = x_i(s_2)$, since relative prices are the same at s_1 and s_2 and $x_i(s)$ is uniquely determined by the relative prices of $p(s)$, $s = s_1, s_2$. Suppose that there is $i \in \{1, \ldots, m_1\}$ with $x_i(s_1) \neq x_i(s_2)$. Let $x_i^\pi = \pi_1 x_i(s_1) + \pi_2 x_i(s_2)$. Then

$$u_i(x_i^\pi) > \pi_1 u_i(x_i(s_1)) + \pi_2 u_i(x_i(s_2))$$

by strict concavity of u_i. But

$$p(s_1) \cdot x_i^\pi + p(s_2) \cdot x_i^\pi = p(s_1) \cdot (\pi_1 + \pi_2) x_i(s_1) + p(s_2) \cdot (\pi_1 + \pi_2) x_i(s_2)$$
$$= p(s_1) \cdot x_i(s_1) + p(s_2) \cdot x_i(s_2),$$

where we have used the conditions on the equilibrium prices. This ensures that the sure bundle x_i^π is budget feasible and better for i than her equilibrium choice, a contradiction. $\square$

With the result of Theorem 13.1, we know that our search for an equilibrium where uncertainty matters — a proper sunspot equilibrium as it were — may be limited to equilibria where

$$\frac{1}{\pi_1} p(s_1) \neq \frac{1}{\pi_2} p(s_2).$$

But once we have this, it is rather easy to construct a model having sunspot equilibria: Choose two price vectors $p^1, p^2 \in \mathbf{R}_+^2$ which are not proportional (so that in particular, $\pi_2 p^1 \neq \pi_1 p^2$). Let the z^1 and z^2 be the excess demands of the consumers $i \in \{1, \ldots, m_1\}$, that is

$$z^1 = \sum_{i=1}^{m_1} (\xi_i[s_1](p^1, p^2) - \omega_i),$$

$$z^2 = \sum_{i=1}^{m_1} (\xi_i[s_2](p^1, p^2) - \omega_i).$$

If we can exhibit consumers $i \in \{m_1 + 1, \ldots, m\}$ such that their aggregate excess demand is $-z^1$ at p^1 and $-z^2$ at p^2, then we have an equilibrium at the prices (p^1, p^2). If, moreover, some of the consumers choose different bundles at the two price systems, we have an example of an equilibrium where uncertainty matters.

For this to happen, we clearly must have that $p^h \cdot (-z^h) = 0$ for $h = 1, 2$, since the consumers of index $i \in \{m_1 + 1, \ldots, m\}$ satisfy Walras' law at each state s. But this means that we must also have $p^h \cdot \sum_i [s_h](p^1, p^2) = 0$ for $h = 1, 2$. This does not necessarily hold with any given choice of p^1, p^2; on the other hand, it may be achieved by multiplying one of the prices p^1, p^2 by a sufficiently large number, keeping the other one constant. We leave the details to the reader.

The argument is illustrated in Fig. 4. Here, the two excess demand vectors $-z^1$ and $-z^2$ are supposed to have been obtained for a particular pair of prices (p^1, p^2) such that Walras' law is satisfied. In our example, there is no single consumer having the given excess demands; but if we allow for several consumers, they can be realized as excess demands.

The conclusion is that uncertainty which is extrinsic in the sense that none of the agents' characteristics depend on the state of nature, may matter for the process of allocation. This is an important insight, and the concept of sunspot equilibrium has found its way into applied economics (see, e.g. Exercise 5).

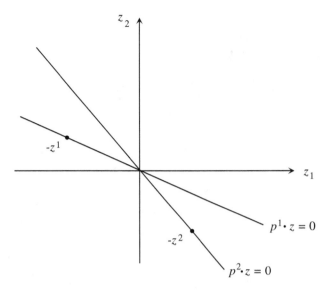

Fig. 4. Sunspot equilibria: excess demand at prices p^1 and p^2.

5. Exercises

(1) Consider an economy with one commodity, two consumers and two states of nature s_1, s_2. The consumers both assign the probability π_j to s_j, $j = 1, 2$, and the expected utility is given by

$$U_i(x_i(s_1), x_i(s_2)) = \pi_1 u_i(x_i(s_1)) + \pi_2 u_i(x_i(s_2)).$$

Suppose further that there is aggregate uncertainty given by $\omega_1(s_1) + \omega_2(s_1) = 2$, $\omega_1(s_2) + \omega_2(s_2) = 1$. Find the set of Pareto-optimal allocations. Compare the marginal rates of substitution with the ratio of probabilities.

(2) Let $\mathcal{E}$ be an economy with two periods $t = 0, 1$ and two goods in each period, but with two states at $t = 1$, where consumers maximize the expected value of $u_i(x_1^0, x_2^0) + u_i(x_1^1, x_2^1)$ with u_i given in Exercise 2 of Chapter 1. Endowments of consumer 1 is $(2, 3)$ in each period and state, whereas consumer 2 has endowment

$(1,1)$ at $t = 0$ and $(1,1)$ or $(4,4)$, each with probability $1/2$, at $t = 1$. Find a Walras equilibrium.

(3) (From Starr, 2011). Consider an economy in general equilibrium with a full set of contingent commodity markets. Explain how the economy deals with medical insurance. How does it work? Is medical insurance just another contingent commodity? Is there a moral hazard problem (over-spending when the insured event occurs because insurance will cover the bill)? Will every household be insured for every illness or injury?

(4) Consider a two-period exchange economy with one good and two consumers. There are two equally likely states s_1 and s_2. In s_1, consumer 1 has endowment 2 and consumer 2 has 2 units, and in s_2, consumer 2 has 2 units and consumer 1 only 1. Consumption takes place only in the second period, and both consumers maximize the expected value of $u(x) = \ln x$, where x is the consumption.

Find a Walras equilibrium in this economy.

Assume now that two financial assets A and B can be traded in the first period. Asset A pays out 1 in both states, and asset B pays 2 in state s_1, 3 in state s_2. If the assets have prices $(q_A, q_B) = (3, 5)$, then there is an opportunity of arbitrage, obtaining a surplus by suitably balancing sale and purchase of assets in the first period.

Find an equilibrium of consumption and asset holdings together with prices of assets and state-contingent goods such that the allocation is the same as that in the Walras equilibrium.

(5) Liquidity insurance and bank runs (Diamond and Dybvig, 1983). We consider a one-good economy over three periods, $t = 0, 1, 2$. There is a large number m of identical consumers with consumption paths $x = (x_0, x_1, x_2)$ and an endowment of one unit of the good at $t = 0$, and there is a technology with constant returns to scale transforming input at $t = 0$ of one unit into output $R > 1$ at $t = 2$. The consumption good can also be stored from each period to the next.

Uncertainty and General Equilibrium 353

For each consumer, the utility function is determined at $t = 1$ by an independent random event, so that

$$U(x_0, x_1, x_2) = \begin{cases} u(x_1), \ (\text{"impatience"}), & \text{with probability } \pi, \\ u(x_2) \ (\text{"patience"}) & \text{with probability } 1 - \pi, \end{cases}$$

where u is increasing, concave, and C^1 with $tu'(t) < 0$ for $t > 0$. The outcome of the random event is private knowledge for the consumer and cannot be observed by others.

Find a Pareto-optimal allocation and show that it can be sustained by a system of fractional banking, whereby consumers deliver the endowment to a centralized institution (a bank), after which they can claim an amount x_1^* in the case of impatience and x_2^* otherwise. Show that no consumer has an incentive to misinform about the state.

Reformulate the model so as to include beliefs on the part of patient consumers that the bank cannot fulfill its obligations at $t = 1$ and therefore is dissolved, so that their payment at $t = 2$ is lost. Show that bank runs (and subsequent bank failure) is a sunspot equilibrium.

Chapter 14

Financial Markets

1. One-Period Models of Financial Equilibrium

We saw in Chapter 13 that the analysis of allocation over time and the functioning of markets points to the important role played by the securities of various forms. Needless to say, this concern about financial transactions is not a feature of theoretical economics but is indeed shared by the general public. Financial markets have become one of the main public concerns, so that a closer look at these markets would make sense. In the following, we briefly survey the theory of financial markets, which may be considered as a theory of equilibrium with contingent markets for the numeraire good only, and where the spot markets for the other goods have been left out of concern.

1.1 Basic models of financial markets

The capital asset pricing model (CAPM) was introduced around 1960 by several authors independently. Due to its simplicity, it has found widespread application in practice, even though it is not easily verified empirically.

Consider a market where n securities are traded. The securities have returns $\tilde{r}_i$ with mean r_i and they are assumed to be correlated with covariance matrix $\Sigma = (\sigma_{ij})_{i=1}^{n}{}_{j=1}^{n}$. There is also a risk-free security with rate of interest r. A portfolio is a vector $x = (x_1, \ldots, x_n)$, where x_i is the amount invested in security i (here, x_i may be positive, negative or 0). If an investor with initial wealth w_0 chooses a portfolio $x = (x_1, \ldots, x_n)$, then wealth in the next period will be

$$\widetilde{w} = \left(w_0 - \sum_{i=1}^{n} x_i\right) r + \sum_{i=1}^{n} x_i \widetilde{r}_i$$

with mean

$$\mu(x) = \mathsf{E}\widetilde{w} = \left(w_0 - \sum_{i=1}^{n} x_i\right) r + \sum_{i=1}^{n} x_i r_i = w_0 r + \sum_{i=1}^{n} x_i (r_i - r)$$

and variance

$$\sigma^2(x) = \sum_{i=1}^{n} \sum_{j=1}^{n} \sigma_{ij} x_i x_j = x^t \Sigma x,$$

where x^t stands for the transpose of x (which is considered as a column vector).

We assume now that investor's preferences over portfolios depend only on mean and variance, so x is chosen so as to maximize a utility function $U(\mu, \sigma^2)$; this is quite restrictive as a general assumption, but it does hold if securities are normally distributed, or if preferences here are described by utility functions of a very particular type.

Assuming that x has been chosen optimally, it must satisfy the first-order conditions

$$U'_\mu \frac{\partial \mu(x)}{\partial x_i} + U'_{\sigma^2} \frac{\partial \sigma^2(x)}{\partial x_i} = U'_\mu (r_i - r) + 2U'_{\sigma^2} \sum_{j=1}^{n} \sigma_{ij} x_j = 0$$

for $i = 1, \ldots, n$, or, in matrix notation,

$$-\gamma \rho + \Sigma x = 0,$$

where $\gamma = -\dfrac{U'_\mu}{2U'_{\sigma^2}}$ is the marginal rate of substitution between mean and variance, and where $\rho = (r_1 - r, \ldots, r_n - r)$, and since Σ as a covariance matrix is symmetric and positive definite, it has an inverse Σ^{-1}, so that we can write the optimal portfolio as

$$x = \gamma \Sigma^{-1} \rho. \tag{1}$$

The solution depends on the particular investor only through the scalar γ, so that all investors in the market choose portfolios which are proportional to each other and, consequently, also proportional to the *market portfolio*, which is defined as the total amount of each of the securities held by all investors.

For an investor with $\gamma = 1$, say with utility function $U(\mu, \sigma^2) = \mu - \frac{1}{2}\sigma^2$, we get a simple expression for the market portfolio x^m (or, strictly speaking, a vector proportional to the market portfolio),

$$x^m = \Sigma^{-1}\rho. \tag{2}$$

We can use (2) to obtain an expression for the mean payoffs of the securities, expressed in terms of the payoff of the market portfolio, since

$$\rho = \Sigma x^m,$$

so that for each i, we have that

$$\rho_i = \sum_{j=1}^{n} \sigma_{ij} x_j^m,$$

where the right-hand side is the covariance of security i and the market portfolio, to be denoted σ_{im}. We then have the mean return of the market portfolio,

$$\rho_m = \sum_{i=1}^{n} \rho_i x_i^m = \sum_{i=1}^{n} \sum_{j=1}^{n} \sigma_{ij} x_i^m x_j^m = \sigma_m^2,$$

where we have introduced the notation $\sigma_m^2 = \sigma^2(x^m)$ for the variance of the market portfolio, so that we get the fundamental formula of CAPM,

$$\rho_i = r + (\rho_m - r)\beta_i, \tag{3}$$

with $\beta_i = \dfrac{\sigma_{im}}{\sigma_m^2}$ (known as the *betas* of the securities).

358   *Theory of General Economic Equilibrium*

The model has several important properties. One of these is *two-fund separation:* all portfolios can be conceived as a mixture of one particular (composite) security, namely the market portfolio, and the risk-free asset. In the interpretation, we have that there is only one source of risk, namely that of the market portfolio, all the individual risks connected with the n securities have been eliminated by the choice of a sufficiently diversified portfolio.

An alternative to the CAPM, the arbitrage pricing theory (APT) model, was proposed in Ross (1976). As before, there are n securities, but the returns take the form

$$\tilde{r}_i = r_i + \beta_{i1}\tilde{\delta}_1 + \cdots + \beta_{ik}\tilde{\delta}_k + \tilde{\varepsilon}_i,$$

where $\tilde{\delta}_h$ for $h = 1, \ldots, k$ are random payoff components, called *factors*, with mean 0, formalizing the non-diversifiable part of the security risk, and $\tilde{\varepsilon}_i$ with $\mathsf{E}\tilde{\varepsilon}_i = 0$, $\mathsf{E}\tilde{\varepsilon}_i\tilde{\varepsilon}_j = 0$ is the diversifiable risk associated with the ith security. The coefficients β_{ij} can be interpreted as covariances of security payoff and external risk, since $\beta_{ij} = \mathsf{E}\tilde{r}_i\tilde{\delta}_j$ for $i = 1, \ldots, n$, $j = 1, \ldots, k$. We refer to, e.g. Huberman (1982) for a further discussion of this model.

1.2   *Financial markets with bankruptcy*

Up to this point, we have considered markets for securities which could be bought and sold in a market. Buying and selling securities could be interpreted in the same way as buying and selling ordinary commodities, but when dealing with security markets, it is natural to think of the sale of a security not as the transfer of an acquired right to some payoff in the next period, but rather as taking a *short position* in this security, which amounts to selling a promise to deliver the same payoffs as those of the security in question. In other words, some of the trades in the market represent borrowing and lending rather than the ordinary transfer of commodities.

So far, this has been only a question of interpreting the model, but once loans are introduced, it seems natural to allow for the possibility that loans are not repaid. This would occur if an individual has made

Financial Markets 359

Box 1. An example of bankruptcy in equilibrium: The following example is due to Zame (1993). There are two periods and a countably infinite set $S = \{1, 2, \ldots\}$ of states in the second period. The economy has two individuals with identical preferences and their initial endowments are

$$\omega_1 = (1; 7, 1, 1, \ldots), \quad \omega_2 = (1; 1, 7, 1, 1, \ldots),$$

so that both have one unit in the first period and state-dependent endowments in the second period which, however, agree for all states except the first two. It is straightforward that the allocation

$$x_1 = x_2 = (1; 4, 4, 1, 1, \ldots)$$

constitutes a Pareto-optimum, and it can be sustained in an equilibrium where the individuals can transfer value between periods using the two (Arrow) securities which yield 3 units at date 1 in state 1 or state 2. Individual 1 should buy one such security with payoff in state 2 and sell a security with payoff state 1.

Things change, however, if only specific securities are available. Suppose that these securities are

$$A_1 = (1, 2, 0, 0, \ldots)$$
$$A_2 = (0, 1, 2, 0, \ldots)$$
$$\vdots$$
$$A_k = (0, \ldots, 0, 1, 2, 0, \ldots) \qquad \text{(with payoff 1 at state } k\text{)}.$$

If individual 1 wants to obtain $x_1 = (1; 4, 4, 1, 1, \ldots)$, she will need to sell 3 units of A_1 to obtain a net payoff of 3 in state 1, but this will entail an obligation to deliver 6 units in state 2, so that she must buy a total of 9 units of A_2, thus, having a surplus of 18 in state 3. This can be canceled by short selling 18 units of A_3, but the consequence will be an obligation to deliver 36 in state 4, etc. It is seen that there is no way of obtaining the allocation (x_1, x_2) using the securities $A_1, \ldots, A_k$ for any k.

If, however, the possibility of defaulting on the payments is taken into account, then the bundles x_1 and x_2 can be approximated arbitrarily close by security trades in $A_1, \ldots, A_k$, in the sense that default occurs only in states $k' > k$, and the probability of such states can be made arbitrarily small by choosing k large.

promises to pay, which cannot be fulfiled in some of the possible states in the next period. We assume that this results in *bankruptcy*, so that all assets of the individual are seized and used for paying the creditors. It may seem that bankruptcy would not occur in equilibrium, but this is not the case: it may be perfectly rational to take on a debt even knowing that it cannot be repaid in some specific state, and consequently, general equilibrium models of security markets should take this into account.

Models of financial markets where bankruptcy is possible have been considered by several authors, e.g. Zame (1993), Araujo and Páscoa (2002), Dubey *et al.* (2005). The model was proposed by Eichberger *et al.* (2014).

We consider a financial market with two dates, $t = 0$ and $t = 1$ and a finite set $S = \{1, \ldots, \bar{s}\}$ of uncertain states of the world. There are m individuals $i = 1, \ldots, m$ with endowments $(\omega_{i0}, \omega_{iS}) \in \mathbb{R}_{++} \times \mathbb{R}_{++}^S$, where $\omega_{iS} = (\omega_{is})_{s \in S}$. For each $s \in S$, let $\omega_s = \sum_{i=1}^m \omega_{is}$. We let the consumption set X of each individual be defined as

$$X = \{(x_0, x_S) \in \mathbb{R}_+ \times \mathbb{R}^S \mid x_s \geq \omega_s, s \in S\}.$$

Individuals choose consumption bundles $x_i = (x_{i0}, x_{i1}) \in X$, and their preferences are described by utility functions $u_i, i = 1, \ldots, m$ defined on X.

To connect the two periods, the consumers need to transfer the value using securities. There is a finite set of K securities, each specified by the payoff in each of the future states $y_k = (y_{k1}, \ldots, y_{k\bar{s}})$, $k = 1, \ldots, K$. Consumer i can buy or (short-)sell securities, thus obtaining a portfolio $\psi_i \in \mathbb{R}^K$, which will result in a net trade at $t = 1$ and state $s \in S$ of the size $z_{is} = \sum_{k=1}^K y_{sk} \psi_{ik}$, or, in matrix notation

$$z_i = Y \psi_i,$$

where Y is the matrix with elements y_{sk} for $s \in S, k = 1, \ldots, K$. It is assumed that trade in the securities does not lead to bankruptcy, so that

$$\omega_i + Y \psi_i \geq 0,$$

Financial Markets 361

all obligations to pay can be honored using the endowment in state s, for all $s \in S$. Bond prices at date 0 are denoted q_k, $k = 1, \ldots, K$.

To introduce the possibility of bankruptcy, we introduce one additonal security in the form of a bond with nominal repayment 1 in each state s at date 1. The bond can be used for contracting debts, in the sense that consumers can sell the bond at date 0 at some price p, thereby contracting an obligation to pay 1 to the bond owner at date 1. Conversely, buying a bond gives the owner the right to receive a bond payment at date 1, but due to the possibility of bankruptcy, the amount received may be less than one.

Bankruptcy will occur for consumer i at date s if the amount ζ_i^- of bonds sold, that is the payment due at $t = 1$, exceeds the amount available for the consumer to effectuate the payment, which is $\omega_{is} + Y_s \cdot \psi_i$ (here, Y_s is the sth row of Y). We assume that if bankruptcy occurs, the assets of the consumer will be seized and distributed among all bond holders in a proportional way, so that the return on one bond in state s will be

$$r_s = \frac{\sum_{i=1}^m \max\{\zeta_i^-, \omega_{is} + Y_s \cdot \psi_i\}}{\sum_{i=1}^m \zeta_i^-}.$$

Then r_s is the *recovery rate* of the bond in state i. When trading in the market, consumers take the recovery rate into consideration in the same way as the prices q of securities and the bond price p. To prevent excessive bankruptcy, we assume that the consumer has disutility of going bankrupt; formally, the consumption in the bankruptcy state s may be negative, and the utility function is defined on bundles with negative coordinates for some $s \in S$. The negative part of the consumption should be considered as a subjective burden of bankruptcy, which does not correspond to physical consumption; the latter is given by x_i^+, the non-negative part of x_i, with $x_i^+ = \max\{0, x_{i\sigma}\}$ for $\sigma \in \{0\} \cup S$.

We may now define an equilibrium in this exchange economy as an allocation–price pair $((x_1, \psi_1, \zeta_1), \ldots, (x_m, \psi_m, \zeta_m), (q, p))$ together with a recovery rate vector $r = (r_s)_{s \in S}$, which satisfies the standard

362 *Theory of General Economic Equilibrium*

conditions of aggregate feasibility and individual optimality,

(i) (a) $\sum_{i=1}^{m} x_i^+ = \sum_{i=1}^{m} \omega_i$

    (b) $\sum_{i=1}^{m} \psi_i = 0, \ \sum_{i=1}^{m} \zeta_i^- = \sum_{i=1}^{m} \zeta_i^+$,

(ii) for each i, x_i maximizes u_i on

$$\gamma_i(p, q, r) = \{(x_i, \psi_i, \zeta_i) \mid p\zeta_i + q \cdot \psi_i \le p\zeta_i^- + \omega_{i0},$$
$$x_{iS} = \omega_{iS} + \zeta_i^+ r - \zeta_i^- e + Y\psi_i, \omega_i + Y\psi_i \ge 0\} \tag{4}$$

(where e is the diagonal vector in $\mathbb{R}^S$).

In the equilibrium, bond and security markets balance at $t = 0$, and goods markets balance in each state s at $t = 1$. The equilibrium recovery rates, given by r, are taken as given by the consumers contemplating to buy the bond at $t = 0$. It is seen from the definition that we may assume that a given consumer is either a buyer or a seller in the bond market.

Following the tradition of previous chapters, we should now check that equilibria exist under standard assumptions on preferences. This will, however, be postponed to the next chapter, where we are more specifically dealing with problems of incomplete markets. At present, our main interest is in the income transfers across dates and individuals and the role of arbitrage. For this, we introduce probabilities of states $\rho = (\rho_s)_{s \in S}$ with $\rho_s > 0$ for all $s \in S$ and the associated inner product in $\mathbb{R}^S$,

$$\langle a, b \rangle = \sum_{s \in S} \rho_s a_s b_s, \quad a, b \in \mathbb{R}^S,$$

together with the inner product in $\mathbb{R} \times \mathbb{R}^S$,

$$\langle (\alpha, a), (\beta, b) \rangle = \alpha\beta + \langle a, b \rangle. \tag{5}$$

To define a no-arbitrage condition for the present model, we begin with the set of possible income transfers (between dates), which for bond and security prices p and q and recovery rate r is given by the set

$$T_{(p,q,r)} = \{\eta = (\eta_0, \eta_S) \in \mathbb{R} \times \mathbb{R}^S \mid \eta_0 = p(\zeta^+ - \zeta^-)$$
$$+ q \cdot \psi, \eta_S = \zeta_i^+ r - \zeta_i^- e + Y\psi\}, \tag{6}$$

where (ζ^+, ζ^-, ψ), the long and short positions in the bond and the portfolio weights, can take arbitrary values in $\mathbb{R}_+^2 \times \mathbb{R}^S$. It is seen from (6) that $T_{(p,q,r)}$ is a finitely generated convex cone.

Lemma 14.1. *Let (p, q, r) be given, and assume that there is no arbitrage in the sense that $T_{(p,q,r)} \cap [\mathbb{R}_{++} \times \mathbb{R}_{++}^S] = \emptyset$. Then there is $\pi \in \mathbb{R}_{++} \times \mathbb{R}_{++}^S$, such that $\langle \pi, \eta \rangle \leq 0$ for all $\eta \in T_{(p,q,r)}$.*

Proof. By separation of convex sets, there is a non-zero linear form $\widehat{\pi} \in \mathbb{R} \times \mathbb{R}^S$ such that

$$\widehat{\pi} \cdot \eta > 0 \quad \text{for } \eta \in \mathbb{R}_{++} \times \mathbb{R}_{++}^S,$$
$$\widehat{\pi} \cdot \eta \leq 0 \quad \text{for } \eta \in T_{(p,q,r)}.$$

Clearly, $\widehat{\pi}$ must be non-negative in all coordinates. Define π by $\pi_0 = \widehat{\pi}_0, \pi_s = \dfrac{\widehat{\pi}_s}{\rho_s}$ for $s \in S$. Then π satisfies the conclusions of the lemma. $\qquad\square$

Let $V = (e \vdots Y)$ be the matrix Y with the extra column e, and similarly, let $W = \left(r \vdots Y \right)$ be the matrix with the extra column r. Then the set of prices (p, q) for which there is no arbitrage must satisfy

$$\pi_S W \leq (p, q) \leq \pi_S V$$

for some $\pi_S = (\pi_s)_{s \in S} \in \mathbb{R}_{++}^S$.

If no bankruptcy occurs (in any state), then the state price π_S can be used to evaluate income streams generated by the available securities. If π_S^0 is the projection of π on the subspace E of $\mathbb{R}^S$ spanned by the securities, then π_S^0 is uniquely defined, and the value of an income transfer $\eta_S \in E$ measured in state prices can be written as

$$c(\eta_S) = \langle \pi_m^0, \eta_S \rangle = \langle \pi_S^0, \mathsf{E}\eta_S \rangle + \langle \pi_S^0, \eta_S - \mathsf{E}\eta_S \rangle,$$

where $\mathsf{E}\eta_S = \sum_{s \in S} \eta_s$ is the expected value of the transfer. Using (5), this can be transformed to

$$c(\eta) = \mathsf{E}\pi_S^0 \mathsf{E}\eta_S + \mathrm{cov}(\pi_S^0, \eta_S). \tag{7}$$

The expression can be further simplified if it is assumed that the vector e, which corresponds to a risk-free security, is in E, since the value of the risk-free asset is $c(e) = \mathsf{E}\pi_S^0 = \frac{1}{1+r^0}$, where r^0 is the risk-free rate of interest.

In the case where bankruptcy does happen in some state, let C be the set Π of state prices $\pi = (\pi_0, \pi_S)$ given by Lemma 14.1. Now we may have $\langle \pi_S, \eta_S \rangle < 0$ for some $\eta \in T_{(p,q,r)}$ (namely such as those obtained using the column r), so the value of the income stream η_S must be assessed by the largest of the inner products,

$$c(\eta_S) = \max_{\pi \in \Pi} \left\langle \frac{\pi_S}{\pi_0}, \eta_S \right\rangle = \max_{\pi \in \Pi} \left\{ \mathsf{E}\left[\frac{\pi_S}{\pi_0} \right] + \mathrm{cov}\left(\frac{\pi_S}{\pi_0}, \eta_S \right) \right\},$$

where we have used the same transformation as that leading to (7). It may be noted that the valuation of income streams using state prices is no longer linear, indeed $c(\cdot)$ is sublinear as a pointwise maximum of linear functions. Nevertheless, the basic structure remains the same and indeed bears close resemblance to the fundamental equations in the CAPM and APT models.

1.3 *Arbitrage and martingale measures*

In this section, we consider a general version of the models considered earlier, where securities are random variables defining the payoff at $t = 1$: There is a finite set of securities $S_0, S_1, \ldots, S_k$, each of which is a random variable defined on a probability space $(\Omega, \mathcal{F}, P)$ (whereby Ω is a set, $\mathcal{F}$ is a σ-algebra of subsets of Ω and

Financial Markets 365

P is a probability measure on $(\Omega, \mathcal{F})$. We assume that S_0 is a risk-free security with

$$S(\omega) = 1 + r_0$$

for all $\omega \in \Omega$ except possibly for a subset with probability 0, also written as P-a.e. (almost everywhere) or P-a.s. (almost surely), where r_0 is the risk-free rate of interest. We use the notation $S = (S_1, \ldots, S_k)$ for the array of risky securities and we write $\hat{S} = (S_0, S)$ for the full array. A portfolio is a vector $\hat{\xi} = (\xi_0, \xi_1, \ldots, \xi_k)$, also written as $\hat{\xi} = (\xi_0, \xi)$, in $\mathbb{R}^{k+1}$. We write $\hat{\xi} \cdot S = \sum_{j=0}^{k} \xi_j S_j$ for the random portfolio value.

Let $(p, q) \in \mathbb{R}_+^{k+1}$ be the vector of prices of the securities $j = 0, 1, \ldots, k$. We assume that the price vector is normalized so that the price of the risk-free asset S_0 is $p = 1$. A portfolio $\hat{\xi}$ is an *arbitrage opportunity* if

$$(p, q) \cdot \hat{\xi} \leq 0, \ \hat{\xi} \cdot S \geq 0 \ P\text{-a.e}, P[\hat{\xi} \cdot S] > 0.$$

Thus, an arbitrage opportunity is a mixture of short and long positions which is self-financing and which will always (except for an event with probability 0) give a non-negative payoff, and indeed a positive payoff with non-zero probability. If there are no arbitrage opportunities, then the market is *arbitrage-free*.

It may be noted that arbitrage opportunities can also be characterized without involving the risk-free asset. Indeed, suppose that $\hat{\xi} = (\xi_0, \xi)$ is an arbitrage opportunity. Then the vector $\xi \in \mathbb{R}^k$ is such that

$$\xi \cdot S \geq (1 + r_0)(p \cdot \xi) \ P\text{-a.s.}, \ P[\xi \cdot S > 0] > 0. \tag{8}$$

Conversely, given a vector ξ satisfying (8), then $(-q \cdot x_i, \xi)$ is an arbitrage opportunity.

We shall need an additional concept, namely that of a *risk-neutral* or a *martingale* probability measure on $(\Omega, \mathcal{F})$, which is a probability measure P^* with

$$p = \frac{S_0}{1 + r_0}, \quad q_j = \mathsf{E}^*\left[\frac{S_j}{1 + r_0}\right], \quad j = 1, \ldots, k, \tag{9}$$

where E^* denotes expectation w.r.t. P^*. Each of the securities $j = 1, \ldots, k$ are then fairly priced, in the sense that the price of the security corresponds to its expected payoff.

It seems intuitive that fair pricing and absence of arbitrage opportunities should go together, and indeed they do. The following is known as the fundamental theorem of financial economics (for the present one-period situation).[1]

Theorem 14.1. *A market is arbitrage-free if and only if the set*

$$\mathcal{P} = \{P^* \mid P^* \text{ is a risk-neutral measure}, P^* \sim P\}$$

of equivalent risk-neutral measures is non-empty, and in this case there is $P^* \in \mathcal{P}$ with a.s. bounded Radon–Nikodym derivative.

Proof. Suppose first that there is a risk-neutral probability measure P^* which is equivalent to P. If $\hat{\xi}$ is a portfolio with $\hat{\xi} \cdot \hat{S} \geq 0$ P-a.s. and $\mathsf{E}[\hat{\xi} \cdot \hat{S}] > 0$, then by equivalence we also have that $\hat{\xi} \cdot \hat{S} \geq 0$ P^*-a.s. and $\mathsf{E}^*[\hat{\xi} \cdot \hat{S}] > 0$, meaning that

$$(p, q) \cdot \hat{\xi} = p\xi_0 + \sum_{j=1}^{k} q_j \xi_j = \sum_{j=0}^{k} \mathsf{E}^*\left[\frac{S_j}{1 + r_0}\right]\xi_j = \mathsf{E}^*\left[\frac{(p, q) \cdot \hat{\xi}}{1 + r_0}\right] > 0,$$

and it is seen that ξ cannot be an arbitrage opportunity.

To prove that the no-arbitrage condition implies the existence of an equivalent risk-neutral measure, we consider for each of the

---

[1]In the sequel, we use some specific notions from measure theory: A probability measure P' on $(\Omega, \mathcal{F})$ is absolutely continuous w.r.t. P, written $P' \ll P$, if $P(E) = 0 \Rightarrow P'(E) = 0$ for all $E \in \mathcal{F}$. The two probability measures P' and P are equivalent if $P' \ll P$ and $P \ll P'$. If $P' \ll P$, then there is an $\mathcal{F}$-measurable function f such that $P^*(E) = \int_E f \, \mathrm{d}P$ for all $E \in \mathcal{F}$. The function f is called a Radon–Nikodym derivative of P^* w.r.t. P and is often written as $\frac{\mathrm{d}P^*}{\mathrm{d}P}$.

securities S_j, $j = 1, \ldots, k$, the discounted net gain

$$D_j = \frac{S_j}{1 + r_0} - q_j, \quad j = 1, \ldots, k,$$

and we let $D = (D_1, \ldots, D_k)$ be the random vector defined by the discounted net gains. Then absence of arbitrage means that

$$\xi \cdot D \geq 0 \ P\text{-a.s.} \Rightarrow \xi \cdot D = 0 \ P\text{-a.s.}$$

for any $\xi \in \mathbb{R}^k$. Since $D_j(\omega) \geq q_j$ for all $\omega \in \Omega$, we have that $\mathsf{E}[D_j]$ is well-defined (possibly $+\infty$) for every probability P^*, so that P^* is a risk-neutral measure if

$$\mathsf{E}^*[D_j] = 0, \quad j = 1, \ldots, k.$$

Assume first that $\mathsf{E}D_j < \infty$ for each j. Let Q be the set of probability measures $Q \approx P$ with bounded Radon–Nikodym derivative $\dfrac{dQ}{dP}$. It is easily seen that Q is convex. The same holds for the set

$$\mathcal{E} = \{(\mathsf{E}_Q D_1, \ldots, \mathsf{E}_Q D_k) \mid Q \in \mathcal{Q}\}.$$

Indeed, if $Q, Q' \in \mathcal{E}$ and $\lambda \in [0, 1]$, then $Q_\lambda = \lambda Q + (1 - \lambda)Q'$ has the bounded Radon–Nikodym derivative

$$\frac{dQ_\lambda}{dP} = \lambda \frac{dQ}{dP} + (1 - \lambda)\frac{dQ'}{dP},$$

and it follows that

$$\lambda \mathsf{E}_Q[D_j] + (1 - \lambda)\mathsf{E}_{Q'}[D_j] = \mathsf{E}_{Q_\lambda}[D_j], \quad j = 1, \ldots, k.$$

We claim that $0 \in \mathcal{E}$. Suppose not, then there is $\xi \in \mathbb{R}^k$ separating 0 and $\mathcal{E}$, so that $\xi \cdot y \geq 0$ for all $y \in \mathcal{E}$ with $\xi \cdot y_0 > 0$ for some $y_0 \in \mathcal{E}$, or equivalently,

$$\mathsf{E}_Q[\xi \cdot D] \geq 0, \quad \text{all } Q \in \mathcal{Q}, \tag{10}$$

$$\mathsf{E}_{Q_0}[\xi \cdot D] > 0, \quad \text{some } Q_0 \in \mathcal{Q}. \tag{11}$$

From (11), it follows that $P[\xi \cdot D > 0] > 0$; we shall show that (10) implies that $\xi \cdot D \geq 0$ P-a.a. Indeed, let A be the set of ω for which

$\xi \cdot D < 0$, let $\mathbb{1}_A$ be the indicator function of A and $\mathbb{1}_{A^c}$ that of its complement, and define for $n = 2, 3, \ldots$ the function

$$\varphi_n = \left(1 - \frac{1}{n}\right) \mathbb{1}_A + \frac{1}{n} \mathbb{1}_{A^c}.$$

Then the functions $\frac{\varphi_n}{\mathsf{E}[\varphi_n]}$ can be used to define probability measures Q_n with

$$Q_n(B) = \int_B \frac{\varphi_n}{\mathsf{E}[\varphi_n]} \, \mathrm{d}P, \quad \text{all } B \in \mathcal{F}.$$

It is seen that $Q_n \in Q$ for each n, so that

$$0 < \xi \cdot (\mathsf{E}_{Q_n}[D_1], \ldots, \mathsf{E}_{Q_n}[D_k]) = \frac{1}{\mathsf{E}[\varphi_n]} \mathsf{E}[\xi \cdot \varphi_n D].$$

Now the sequence $(\xi \cdot \varphi_n D)_{n=2}^{\infty}$ converges to $\xi \cdot \mathbb{1}_A D$, and since $\xi \cdot \varphi_n D \leq \xi \cdot D$ for $n = 2, 3, \ldots$, we get from the dominated convergence theorem that

$$\mathsf{E}[\xi \cdot \mathbb{1}_{\{\xi \cdot D < 0\}} D] \geq 0.$$

This shows that $\xi \cdot D \geq 0$ P-a.e., so that $0 \in \mathcal{E}$.

In the case where $\mathsf{E}D_j$ is infinite for some j, we replace P by an equivalent probability measure $\widetilde{P}$ with $\frac{\mathrm{d}\widetilde{P}}{\mathrm{d}P}$ bounded such that $\mathsf{E}_{\widetilde{P}}[D_j] \leq \infty$ for all j, for example such that

$$\widetilde{P}(B) = \int_B \frac{1}{1 + \max_j D_j} \left(\mathsf{E}\left[\frac{1}{1 + \max_j D_j}\right]\right)^{-1} \mathrm{d}P.$$

The details are left to the reader. $\qquad\square$

Let $\mathcal{V} = \{\hat{\xi} \cdot \hat{S} \mid \hat{\xi} \in \mathbb{R}^{k+1}\}$ be the set of all the payoffs that can be obtained by choosing suitable portfolios. An element of $\mathcal{V}$ is said to be an *attainable payoff*. It is seen that if the market is arbitrage-free, then portfolios with the same payoff will also have the same value at the prices (p, q): From $\hat{\xi} \cdot \hat{S} = \hat{\xi}' \cdot \hat{S}$ P-a.s. we get that $(\hat{\xi} - \hat{\xi}') \cdot \hat{S} = 0$

P^*-a.s. for all $P^* \in \mathcal{P}$, so that

$$(p,q) \cdot \hat{\xi} - (p,q) \cdot \hat{\xi}' = \mathsf{E}^* \left[\frac{(\hat{\xi} - \hat{\xi}') \cdot \hat{S}}{1 + r_0} \right] = 0$$

due to the absence of arbitrage.

This phenomenon is known as the *law of one price*, whereby the price of any $v \in \mathcal{V}$ is given by $\pi(v) = (p,q) \cdot \hat{\xi}$ for any $\hat{\xi}$ with $v = \hat{\xi} \cdot \hat{S}$. We have that $\pi(v) = \mathsf{E}^* \left[\frac{v}{1+r_0} \right]$ for all $v \in \mathcal{V}$. For each $v \in \mathcal{V}$ with $\pi(v) \neq 0$, the *return* of v is

$$r(v) = \frac{v - \pi(v)}{\pi(v)}.$$

For any attainable payoff $v = \hat{\xi} \cdot \hat{S}$ one has

$$r(v) = \frac{p\xi_0}{(p,q) \cdot \hat{\xi}} r_0 + \sum_{j=1}^{k} \frac{q_j \xi_j}{(p,q) \cdot \hat{\xi}} \pi(S_j).$$

Theorem 14.2. *Suppose that the market is arbitrage-free, and let $v \in \mathcal{V}$ be an attainable payoff with $\pi(v) \neq 0$.*

(a) *For any risk-neutral measure $P^* \in \mathcal{P}$, the expected return of v is the risk-free rate of interest.*

(b) *If $Q \approx P$ and all securities have finite expectation under Q, then the expected return of v satisfies*

$$\mathsf{E}_Q[r(v)] = r_0 - \mathrm{cov}_Q \left(\frac{\mathrm{d}P^*}{\mathrm{d}Q}, r(v) \right). \tag{12}$$

Proof. From $\mathsf{E}^*[v] = \pi(v)(1 + r_0)$, we get that

$$\mathsf{E}^*[r(v)] = \frac{\mathsf{E}^*[v] - \pi(v)}{\pi(v)} = r_0,$$

which is (a). To show (b), we let $h^* = \frac{\mathrm{d}P^*}{\mathrm{d}Q}$, so that $\mathsf{E}_Q[h^*] = 1$. It then follows that

$$\mathrm{cov}_Q = \mathsf{E}_Q[h^* r(v)] - \mathsf{E}_Q[h^*]\mathsf{E}_Q[r(v)] = \mathsf{E}_Q[r(v)] - \mathsf{E}_Q[r(v)],$$

from which we obtain (12). $\qquad\square$

2. Dynamic Models of Financial Markets

In Section, we have considered models of financial markets which were static or one-period models in the sense that securities ceased to exist after one period. In actual financial markets, securities have a longer life, so that their value in the next period depends not only on the uncertain events of this period but also on what will happen in the future. Obviously, models of financial markets must take this into account, which means that some extension of the formalism is called for.

We now consider the situation where the $k + 1$ assets $S_0, S_1, \ldots, S_k$ are available at the dates $t = 0, 1, \ldots, T$. The value of asset j at time t is formulated as a random variable on the probability space $(\Omega, \mathcal{F}, P)$, and each of the assets S_j^t are assumed to be measurable with respect to a sub-σ-algebra $\mathcal{F}_t$ of $\mathcal{F}$, interpreted as the set of events observable before or at date t. It is assumed that $\mathcal{F}_0 \subset \mathcal{F}_1 \subset \cdots \subset \mathcal{F}_T$, and the collection $(\mathcal{F}_t)_{t=0}^T$ is called a *filtration*. We assume (as is common) that $\mathcal{F}_0$ consists only of the two sets $\emptyset$ and Ω.

The collection $(\hat{S}_t)_{t=0}^T$ of vector-valued random variables constitutes a *stochastic process*. It is *adapted* to the filtration $(\mathcal{F}_t)_{t=0}^T$ if S_{jt} is $\mathcal{F}_t$-measurable for $j = 0, 1, \ldots, k, t = 0, 1, \ldots, T$, so that in particular, S_{j0} is a constant function for each j. It is *predictable* if $\hat{S}_t$ is measurable w.r.t. $\mathcal{F}_{t-1}$ for $t = 1, \ldots, T$.

Instead of a fixed portfolio of assets, we may now consider a sequence of portfolios. A *trading strategy* is a predictable stochastic process $\hat{\xi} = (\xi_0, \xi) = (\xi_{0t}, \xi_{1t}, \ldots, \xi_{kt})_{t=1}^T$. Here, ξ_{jt} denotes the quantity of shares in the jth asset which is held in the tth period, that is from date $t - 1$ to t. This means that $\xi_{jt} S_{j,t-1}$ is invested in the asset at date $t - 1$, and $\xi_{jt} S_{jt}$ is the value of this investment at date t. The total value of the portfolio at date $t - 1$ is then $\hat{\xi}_t \cdot \hat{S}_{t-1}$, and it has changed value to $\hat{x}_t \cdot \hat{S}_t$ at the next date. That $(\hat{\xi}_t)_{t=1}^T$ is predictable means that it must be decided upon based only on the observations available at the date at which it is implemented.

A trading strategy is *self-financing* if

$$\hat{\xi}_t \cdot \hat{S}_t = \hat{\xi}_{t+1} \cdot \hat{S}_t, \tag{13}$$

so that its portfolio value is preserved whenever it is rearranged. It follows from (13) that $\hat{\xi}_{t+1} \cdot \hat{S}_{t+1} - \hat{\xi}_t \cdot \hat{S}_t = \hat{\xi}_{t+1} \cdot (\hat{S}_{t+1} - \hat{S}_t)$, so that

$$\hat{\xi}_t \cdot \hat{S}_t = \hat{\xi}_1 \cdot \hat{S}_0 + \sum_{\tau=1}^{t} \hat{\xi}_\tau \cdot (\hat{S}_\tau - \hat{S}_{\tau-1}), \quad t = 1, \dots, T. \tag{14}$$

Assume that $S_{0t} > 0$ P-a.s. for all dates t. We may measure the value of any other asset relative to the 0th asset, thus using the latter as a numeraire. The relative value processes are written as $(Y_{jt})_{t=0}^{T}$, where Y_{0t} has the constant value 1 and

$$Y_{jt} = \frac{S_{jt}}{S_{j0}}$$

for $j = 1, \dots, k$. We may now define the value process $(v_t)_{t=0}^{T}$ associated with a trading strategy $\hat{\xi}$ by

$$v_0 = \xi_1 \cdot Y_0, \; v_t = \xi_t \cdot Y_t, \quad t = 1, \dots, T, \tag{15}$$

so that v_t expresses the value, measured relative to the numeraire asset, of the portfolio at the end of the period from t to $t + 1$. The expression (14) can then be reformulated as

$$v_t = v_0 + \sum_{\tau=1}^{t} \xi_t \cdot (Y_\tau - Y_{\tau-1}), \quad t = 1, \dots, T.$$

Our goal is to extend the fundamental result obtained in the one-period model to the present context of many periods. We, therefore, introduce arbitrage as follows: A self-financing trading strategy ξ with associated value process v is an *arbitrage opportunity* if $v_0 \leq 0$, $v_T \geq 0$ P-almost surely, with $P[v_T > 0] > 0$, and the market is arbitrage-free if there are no arbitrage opportunities. An arbitrage opportunity is a trading strategy which without initial investment gives a non-negative payoff at the end of the period, and positive payoff with non-zero probability. But if the market is arbitrage-free, such gains without investment cannot be obtained at any of the intermediate dates.

Lemma 14.2. *The market has an arbitrage opportunity if and only if there is t with $1 \leq t \leq T$ and an $\mathcal{F}_t$-measurable vector function $\eta = (\eta_1, \dots, \eta_k)$*

372   *Theory of General Economic Equilibrium*

such that

$$\eta \cdot (Y_t - Y_{t-1}) \geq 0 \ P\text{-a.e.}, \ P[\eta \cdot (Y_t - Y_{t-1}) > 0] > 0. \tag{16}$$

Proof. If $\hat{\xi}$ is an arbitrage opportunity with value process v, let t be the first date at which $v_t \geq 0$ P-a.e. and $P[v_t > 0] > 0$. Since the inequalities hold at date T, we have that $0 \leq t \leq T$. If $v_{t-1} = 0$ P-a.e., then $\xi_t \cdot (Y_t - Y_{t-1}) = v_t - v_{t-1} = v_t$ P-a.e., and if we put $\eta = \xi_t$, then (16) is satisfied. If not, we must have that $P[v_{t-1} < 0] > 0$, but then we may define η as $\mathbb{1}_{\{v_{t-1} < 0\}}$, and we get that

$$\eta \cdot (Y_t - Y_{t-1}) = (v_t - v_{t-1})\mathbb{1}_{\{v_{t-1} < 0\}} > -v_{t-1}\mathbb{1}_{\{v_{t-1} < 0\}},$$

and $-v_{t-1}\mathbb{1}_{\{v_{t-1} < 0\}}$ is non-negative and positive with non-zero probability, so again we obtain (16).

For the opposite implication, use η to define a trading strategy ξ by $\xi_\tau = \eta$ for $\tau = t$ and 0 otherwise, with $\xi_0 = (x_{0,t})_{t=1}^T$ determined so that $\hat{\xi}$ becomes self-financing (so that $\xi_{0,\tau} = 0$ for $\tau \leq t$ and $\xi_{0,\tau} = \eta \cdot Y_t$ otherwise). It is easily checked that the associated value process satisfies (15) so that ξ is an arbitrage opportunity. $\qquad\square$

A stochastic process $G = (G_t)_{t=0}^T$ on $(\Omega, \mathcal{F}, P)$, which is adapted w.r.t. the filtration $(\mathcal{F}_t)_{t=0}^T$, is a martingale if $\mathsf{E}_Q[|G_t|] < \infty$, $t = 0, 1, \ldots, T$, and

$$G_s = \mathsf{E}_Q[G_t \mid \mathcal{F}_s], \quad 0 \leq s < t \leq T.$$

This condition is a straightforward generalization of the condition used in the one-period case and it formalizes the notion of a fair gamble, since for each date s and horizon t, the expected future gain $M_t - M_s$ is zero.

In our present multi-period context, a probability measure Q on $(\Omega, \mathcal{F})$ is said to be a *martingale measure* if the process $(Y_t)_{t=0}^T$ is a Q-martingale, that is if

$$\mathsf{E}_Q[|Y_{jt}|] < \infty, \ Y_{js} = \mathsf{E}_Q[Y_{jt} \mid \mathcal{F}_s], \ 0 \leq s < t \leq T, \quad j = 1, \ldots, k.$$

Financial Markets 373

As expected there is a close connection between the absence of arbitrage opportunities and the martingale property.

Theorem 14.3. *The multi-period market is arbitrage-free if and only if the set $\mathcal{P}$ of all martingale measures equivalent to P is non-empty, and then there exists a martingale measure $P^* \in \mathcal{P}$ with bounded Radon–Nikodym derivative $\frac{\mathrm{d}P^*}{\mathrm{d}P}$.*

Proof. Suppose now that the market is arbitrage-free. For each date t with $1 \le t \le T$, let

$$U_t = \{\eta \cdot (Y_t - Y_{t-1}) \mid \eta_j \text{ is } \mathcal{F}_{t-1}\text{-measurable}, j = 1, \ldots, k\}.$$

By Lemma 14.2, we have that if a non-negative $\mathcal{F}_t$-measurable function is in U_t, then it must be 0 P-a.e. We can now apply Theorem 14.1 to the market at date t. For the final date T, we obtain a probability measure P'_T with bounded Radon–Nikodym derivative $\frac{\mathrm{d}P'_T}{\mathrm{d}P}$, such that

$$\mathsf{E}_{P'_T}[Y_T - Y_{T+1}] = 0.$$

Suppose now that for some $t \le T - 1$, we have found a probability measure P'_{t+1}, with bounded Radon–Nikodym derivative $\frac{\mathrm{d}P'_{t+1}}{\mathrm{d}P}$, such that

$$\mathsf{E}_{P'_{t+1}}[Y_\tau - Y_{\tau-1}] = 0, \quad t+1 \le \tau \le T. \tag{17}$$

Since $P'_{t+1} \approx P$, we have again that if a non-negative $\mathcal{F}_t$-measurable function is in U_t, then it must be 0 P'_{t+1}-a.e., and Theorem 14.1 yields a probability measure P'_t with bounded Radon–Nikodym derivative $Z_t = \frac{\mathrm{d}P'_t}{\mathrm{d}P}$, such that

$$\mathsf{E}_{P'_t}[Y_t - Y_{t-1}] = 0.$$

It is straightforward that P'_t is equivalent to P, and for $t+1 \le \tau \le T$, we have that

$$\mathsf{E}_{P'_t}[Y_\tau - Y_{\tau-1} \mid \mathcal{F}_{\tau-1}] = \frac{\mathsf{E}_{P'_{t+1}}[(Y_\tau - Y_{\tau-1})Z_t \mid \mathcal{F}_{\tau-1}]}{\mathsf{E}_{P'_{t+1}}[Z_t \mid \mathcal{F}_{\tau-1}]}$$

$$= \mathsf{E}_{P'_{t+1}}[Y_\tau - Y_{\tau-1} \mid \mathcal{F}_{\tau-1}] = 0,$$

374   *Theory of General Economic Equilibrium*

so that (17) holds also for P'_t. Proceeding with the recursion, we finally get the equivalent martingale measure $P^* = P'_1$.   □

3.   Exercises

(1) Consider an economy with just two assets. The details of these are given below.

	Number of shares	Price	Expected return	Standard deviation
A	100	1.5	15	15
B	150	2	12	9

The correlation coefficient between the returns on the two assets is $\frac{1}{3}$ and there is also a risk-free asset. The assumptions of the CAPM are taken to be satisfied.

Find the expected rate of return on the market portfolio as well as its standard deviation.

Find the beta of stock A and the risk-free rate of return.

(2) A financial institution has an initial capital K and uses it to acquire a portfolio x of n securities which have a joint normal distribution with mean vector ρ and covariance matrix Σ. The risk-free rate of interest is assumed to be 0, and the manager selects the portfolio so that it maximizes a utility function depending only on the mean and variance of the portfolio. The institution defaults if the value of its portfolio becomes negative.

It is decided that the financial institution must inform the public about is *weighted capital ratio* $\frac{K}{\alpha \cdot x}$ where α is a vector of non-negative weights (with at least one of them strictly positive). Show that the probability of default is a decreasing function of the capital ratio.

(3) Consider a situation where Ω has three elements $\omega_1, \omega_2, \omega_3$, and where the probability measure P is such that $P[\{\omega_i\}]$ for $i = 1, 2, 3$. There are three assets with prices $(p, q) = (2, 7, 1)$ and outcomes

Financial Markets 375

(From Föllmer and Schied (2016))

$$\hat{S}(\omega_1) = \begin{pmatrix} 3 \\ 9 \\ 1 \end{pmatrix}, \quad \hat{S}(\omega_2) = \begin{pmatrix} 1 \\ 5 \\ 1 \end{pmatrix}, \quad \hat{S}(\omega_1) = \begin{pmatrix} 5 \\ 13 \\ 1 \end{pmatrix}.$$

Show that the market is arbitrage-free and find all risk-neutral measures.

(4) A discount certificate on a portfolio $V = \hat{\xi} \cdot \hat{S}$ pays the amount $C = \min\{V, K\}$, where $K > 0$ is known as a *cap*. Show that buying a discount certificate amounts to purchasing $\hat{\xi}$ and selling a call option $\max\{0, V - K\}$. Explain that the discount certificate is cheaper than the portfolio itself.

(5) Let $G = (G_t)_{t=0}^{T}$ be stochastic process on $(\Omega, \mathcal{F}, P)$, which is adapted w.r.t. the filtration $(\mathcal{F}_t)_{t=0}^{T}$ and satisfies $\mathsf{E}_Q[|G_t|] < \infty$, $t = 0, 1, \ldots, T$. Show that the following are equivalent:

(i) G is a martingale,
(ii) $G_t = \mathsf{E}_P[G_{t+1} \mid \mathcal{F}_t]$ for $0 \le t \le T - 1$,
(iii) There is a function $F \in L^1(\Omega, \mathcal{F}_T, P)$, such that $G_t = \mathsf{E}_P[F \mid \mathcal{F}_t]$ for $t = 0, \ldots, T$.

Chapter 15

Economies with Incomplete Markets

1. Equilibria in Economies with Financial Assets

In this chapter, we return to the models of general equilibrium with incomplete markets briefly introduced in Chapter 13. We have noted that in the absence of markets for all possible contingent commodities, the consumers may be restricted in the transfer of income from one uncertain state to another. This in itself is, however, no obstacle for the workings of markets if the securities have a particular form, with yields in terms of accounting value at the particular state (or "money" value, if money is considered as unit of account). In what follows, we discuss a model of this type, proposed by Werner (1985).

We consider an economy which runs over two periods of time, indexed by $t = 0, 1$. In the second period, an uncertain state $s \in S = \{s_1, \ldots, s_r\}$ is realized, and assuming that there are n goods available at $t = 0$ and again at $t = 1$ in each state of nature, then there are $l = (r + 1)n$ commodities in our model. Consumers $i = 1, \ldots, m$ have consumption sets $X_i = \mathbb{R}_+^l$ and preference correspondences P_i defined on X_i, and they have endowments ω_i with ω_i^0, the endowment at $t = 0$ belonging to $\mathbb{R}_{++}^n$, and, similarly, $\omega_i^1(s) \in \mathbb{R}_{++}^n$ for all $s \in S$ (endowment at $t = 1$ is positive in all states of nature).

The market structure in our economy has the following structure: There are spot markets for all goods at $t = 0$ and at $t = 1$ in all $s \in S$. In addition, there are k securities or financial assets, whereby a security is a map $v : S \to \mathbb{R}_+$, giving the owner an income $v(s)$ at date 1 if state s occurs. We assume that there are k securities in the economy and that no consumer has endowments of securities.

A consumption plan for the consumer i is an array $x_i = (x_i^0, x_i^{s_1}, \ldots, x_i^{s_r})$ specifying the consumption of goods at date 0 and at date 1 in each state $s \in S$. To realize a consumption plan, the consumer can buy goods in the spot market and obtain a portfolio $\theta_i \in \mathbb{R}^k$ of securities at date 0, and then trade in the spot market at date 1 given the state which occurs. If the spot prices are $p = (p^0, (p^s)_{s \in S})$ and the k securities $v_1, \ldots, v_k$ have prices $q_1, \ldots, q_k$, then consumption plan x_i and the securities portfolio θ_i for consumer i must satisfy the budget constraints

$$p^0 \cdot x_i^0 + q \cdot \theta_i \leq p^0 \cdot \omega_i^0,$$
$$p^s \cdot x_i^s \leq p^s \cdot \omega_i^s + \theta_i \cdot v(s), s \in S, \tag{1}$$

where $v(s) = (v_1(s), \ldots, v_k(s))$.

A *Radner equilibrium* in the economy with financial assets is an array (x, θ, p, q), where $x = (x_1, \ldots, x_m)$ is an m-tuple of consumption plans and $\theta = (\theta_1, \ldots, \theta_m)$ is an m-tuple of portfolios, and where p and q are commodity and securities prices, such that

(i) for each consumer i, x_i is maximal for P_i among consumption plans satisfying the budget constraint (1) for some portfolio θ_i,
(ii) the plans are consistent at each date and state of nature,

$$\sum_{i=1}^{m} x_i^0 \leq \sum_{i=1}^{m} \omega_i^0, \sum_{i=1}^{m} x_i^s \leq \sum_{i=1}^{m} \omega_i^s, s \in S, \sum_{i=1}^{m} \theta_i = 0.$$

It should be noted that the transfers of value between states which can be effectuated using the securities are independent of the prices or price expectations. This simplifies the model so that existence of a Radner equilibrium can be proved using standard methods.

Theorem 15.1. *In the economy described above, assume that preferences satisfy Assumption 0.2 and strict monotonicity, then there exists a Radner equilbrium.*

Proof (outline). Let H be the linear subspace of $\mathbb{R}^k$ spanned by the vectors $v(s) = (v_1(s), \ldots, v_k(s))$ for $s \in S$. We may take $H_+ \cap \mathbb{R}_+^k$ as the space of security prices. Indeed, if $q \notin \text{int}\, H_+$, then there is a portfolio $\theta \in \mathbb{R}^k$ such that $\theta \cdot q \leq 0$ and $\theta \cdot v(s) \geq 0$ for all s with at least one strict inequality, meaning that there is a possibility of arbitrage at these prices. We may assume without loss of generality that consumers choose portfolios from H.

Let $\Delta^0 = \{(p^0, q) \in \mathbb{R}_+^n \times H_+ \mid \sum_{h=1}^n p_h^0 + \sum_{j=1}^k q_j = 1\}$ be the set of feasible prices of goods and securities at $t = 0$, and let $\Delta = \{p \in \mathbb{R}_+^n \mid \sum_{h=1}^n p_h = 1\}$ be the set of goods prices in each state at $t = 1$. We write elements $((p^0, q), (p^s)_{s \in S})$ as (p, q) with $p = (p^0, (p^s)_{s \in S})$.

For each consumer i, define the budget correspondence $\gamma_i : \Delta^0 \times \Delta^S \rightrightarrows \mathbb{R}^l \times H$ by

$$\gamma_i(p, q) = \Big\{(x, \theta_i) \big| p^0 \cdot x_i^0 + q \cdot \theta_i < p^0 \cdot \omega_i^0, p^s \cdot x_i^s < p^s \cdot \omega_i^s$$

$$+ \theta_i \cdot v(s), s \in S \Big\}. \tag{2}$$

The correspondence γ_i has open graph and convex values. Proceeding as in the previous existence proofs, we define the correspondence $\varphi_i : \Delta^0 \times \Delta^S \times X_i \rightrightarrows X_i \times H$ by

$$\varphi_i(p, q, x_i) = \begin{cases} \gamma_i(p, q) \cap [P_i(x_i) \times H] & (p, q, x_i, \theta_i) \in \text{cl}\,\gamma_i(p, q), \\ \gamma_i(p, q) & \text{otherwise.} \end{cases}$$

Choose $K > \max_h \sum_{i=1}^m \omega_{ih}$, define the truncated consumption sets $\overline{X}_i = \{x_i \in X_i \mid x_{ih} \leq K, h = 1, \ldots, l\}$, and let $\overline{X} = \overline{X}_1 \times \cdots \times \overline{X}_m$. To obtain a truncated set of portfolio choices, we restrict attention to the subset Δ_ν^0 of Δ^0, such that $q_j \geq 1/\nu$ for $j = 1, \ldots, k$, where ν is a large natural number, and define the set $\Theta_i^\nu = \{\theta \in H \mid \theta_j \cdot q_j \leq p^0 \cdot \omega_i^0, \text{ all } (p, q) \in \Delta_\nu^0\}$. Then Θ_i^ν is convex and compact. Let $\Theta^\nu = \Theta_1^\nu \times \cdots \times \Theta_m^\nu$.

We may now define correspondences $\varphi_0^\nu : \overline{X} \times \Theta^\nu \times \Delta_\nu^0 \times \Delta^S \rightrightarrows \Delta_\nu^0$ by

$$\varphi_0^\nu(x, \theta, p, q) = \Big\{(\hat{p}^0, \hat{q}) \in \Delta_\nu^0 \big| \hat{p}^0 \cdot (x_\circ^0 - \omega_\circ^0) + \hat{q} \cdot \theta_\circ$$

$$> p^0 \cdot (x_\circ^0 - \omega_\circ^0) + q \cdot \theta_\circ \Big\},$$

where $x_\circ^0 = \sum_{i=1}^m x_i^0$, $\omega_\circ^0 = \sum_{i=1}^m \omega_i^0$ and $\theta_\circ = \sum_{i=1}^m \theta_i$. Finally, for each $s \in S$, define the correspondence $\varphi_s^\nu : \overline{X} \times \overline{\Theta}^\nu \times \Delta_\nu^0 \times \Delta^S \rightrightarrows \Delta$ by

$$\varphi_s(x, \theta, p^s) = \left\{ \hat{p}^s \in \Delta \,\middle|\, \hat{p}^0 \cdot (x_\circ^s - \omega_\circ^s) > p^0 \cdot (x_\circ^s - \omega^s \circ) \right\},$$

where $x_\circ^s = \sum_{i=1}^m x_i^s$, $\omega_\circ^s = \sum_{i=1}^m \omega_i^s$.

Using Theorem 1.7, we get the existence of an array $(x^\nu, \theta^\nu, p^\nu, q^\nu) \in \overline{X} \times \overline{\Theta}^\nu \times \Delta_\nu^0 \times \Delta^S$, such that $\varphi_i(p^\nu, q^\nu, x_i^\nu) = \emptyset$ for $i = 1, \ldots, m$, $\phi_0^\nu(x^\nu, \theta^\nu, p^\nu, q^\nu) = \emptyset$ and $\phi_s^\nu(x^\nu, \theta^\nu, p^\nu, q^\nu) = \emptyset$ for each $s \in S$. Reasoning as in previous existence proofs, it can be shown that (x_i^ν, θ_i^ν) are maximal for P_i on the budget set defined by (p^ν, q^ν), that aggregate net trade is zero at $t = 1$ for any $s \in S$ and that aggregate net trade and aggregate portfolio are close to zero at $t = 0$. To obtain a Radner equilibrium, one must consider a convergent sequence of points $(x^\nu, \theta^\nu, p^\nu, q^\nu)$ and check that the sequence of portfolios can be assumed to be bounded. This final part is left as an exercise. $\qquad\square$

Once existence of equilibria with incomplete markets has been established, one may proceed to consider the welfare properties. It is straightforward that Pareto-optimal allocations may be difficult to establish using the market, since the latter does not allow for all possible reallocations of commodities, and therefore, what can be established is optimality within the framework of admissible reallocations. It turns out that even so the correspondence between constrained optimality and equilibrium allocations is far from perfect, cf. Werner (1985).

2. Pseudoequilibria in Economies with Real Assets

In this section, we consider economies where the securities take the form of *real assets*, which at each state specify a payoff in terms of the goods available in this state. This seemingly minor change has wide-ranging consequences, something which we might perhaps expect given the cases of non-existence of Radner equilibria. When assets are real, their value will depend on the goods prices in each of the states, so that the financial value of a security is determined by the prices. Depending on the prices, the securities may allow different extents of value transfer between states, technically the dimension of the value space spanned by the securities may be smaller than the number of securities. We, therefore, reformulate the notion of equilbrium, introducing *subspaces* of security payoffs as one of the parameters to be determined in the equilibrium. The resulting *pseudoequilibrium* will be discussed subsequently.

Traditionally, the existence problem for pseudoequilibria has been approached using differentiability assumptions, cf. Geanakoplos (1990). We shall use a method more in line with what has been done previously, involving, however, a specific parametrization of subspaces, known as Plücker coordinates, to be explained in detail as we proceed.

2.1 *Pseudoequilibria in economies with incomplete markets and real assets*

We retain the model of Section 2, except for the asset structure, where the payoff of the securities is defined in terms of the goods in the relevant state of nature. As before, there are two periods of time, $t = 0$ and $t = 1$, and a set $S = \{s_1, \ldots, s_r\}$ of uncertain states of the world at $t = 1$. There are n basic goods that can be traded at each date and state of the world, so that $l = (n + 1)r$.

An *asset structure* is given by k assets $a_1, \ldots, a_k \in \mathbb{R}^{rn}$, which define the transfers of value between states that are possible in the model. For $p = \left(p^0, (p^s)_{s \in S}\right)$ a price and $z = (z_0, z_1, \ldots, z_S) \in \mathbb{R}^l$ a vector of net trades in each state, the box product of p and z is defined by

$$p \,\square\, z = (p^s \cdot z^s)_{s \in S}.$$

The first component p^0 is irrelevant for the box product, but we include it in the notation for simplicity.

382 *Theory of General Economic Equilibrium*

In the sequel, it is simpler to work with *net trades* rather than with consumption plans, so that consumers $i = 1, \ldots, m$ choose net trades $z_i = \left(z_i^0, (z_i^s)_{s \in S} \right)$, which are individually feasible if $z_i \in X_i - \{\omega_i\}$. With a slight abuse of notation, we write $P_i(z_i)$ for the set of net trades (rather than bundles) preferred to z_i, whereby P_i inherits all the standard properties from the preferences defined on bundles.

We shall need to select k-dimensional subspaces L from $\mathbb{R}^r$. The set of all such subspaces is denoted $\mathbb{L}_k^r$. If the vectors $p \mathbin{\square} a_j$ for $j = 1, \ldots, k$ are linearly independent, they span an element of $\mathbb{L}_k^r$, but since this depends on the prices to be established in each state, we cannot be sure of this, and indeed we know from Chapter 13, Section 3.1 that this may not happen even in an equilibrium. As a consequence, we use the following less demanding notion of a pseudoequilibrium.

Definition 15.1. An array (z, p, L) with $z = (z_1, \ldots, z_m) \in Z = \prod_{i=1}^m Z_i$, $p \in \Delta = \left\{ p \in \mathbb{R}_+^l \mid \sum_{h=1}^l p_h = 1 \right\}$, and $L \in \mathbb{L}_k^r$, is a *pseudoequilibrium* if

(i) $\sum_{i=1}^m z_i = 0$,
(ii) for all i, $p \cdot z_i = 0$, $p \mathbin{\square} z_i \in L$, and $P_i(z_i) \cap \{z_i' \in Z_i \mid p \cdot z_i' = 0, p \mathbin{\square} z_i' \in L\} = \emptyset$,
(iii) $p \mathbin{\square} a_j \in L$, $j = 1, \ldots, k$.

Since k-dimensional subspaces of r-dimensional Euclidean space play a particular role, we shall make use of some concepts of multilinear algebra, briefly reviewed in what follows.

Let $v_1, \ldots, v_k \in \mathbb{R}^d$ be a basis of a k-dimensional subspace L of $\mathbb{R}^d$. Letting V be the $k \times d$ matrix having the basis vectors $v_1, \ldots, v_k$ as rows,

$$V = \begin{pmatrix} v_{11} & v_{12} & \cdots & v_{1d} \\ \vdots & \vdots & & \vdots \\ v_{k1} & v_{k2} & \cdots & v_{kd} \end{pmatrix},$$

we may consider the vector of determinants of $k \times k$ matrices consisting of k (not necessarily distinct) columns from V with array of indices $(i_1, \ldots, i_k)$. This gives us an array of d^k numbers, not all zero (since $v_1, \ldots, v_k$ is a basis). If $(u_1, \ldots, u_k)$ is another basis for L, there is a regular $k \times k$ matrix B such that $u_i = Bv_i$ for $i = 1, \ldots, k$, and the array of determinants of $k \times k$ submatrices of

$$
\begin{pmatrix}
u_{11} & u_{12} & \cdots & u_{1d} \\
\vdots & \vdots & & \vdots \\
u_{k1} & u_{k2} & \cdots & u_{kd}
\end{pmatrix},
$$

differs from the old array only in multiplication by the non-zero scalar $\det B$. Since the mapping taking a basis of a given subspace to d^k-tuples of numbers is a unique modulo multiplication by a real number, it defines a unique point in real projective space of dimension $d^k - 1$, called the Plücker (or Grassmann) coordinates of the subspace (see Griffiths and Harris (1978), pp. 209–211).

Let $y \in \mathbb{R}^d$ be arbitrary. Then y belongs to the subspace $\mathrm{span}(\{v_1, \ldots, v_k\})$ spanned by the vectors $v_1, \ldots, v_k$ if and only if all the determinants of the $(k + 1) \times (k + 1)$ submatrices of

$$
V =
\begin{pmatrix}
v_{11} & v_{12} & \cdots & v_{1d} \\
\vdots & \vdots & & \vdots \\
v_{k1} & v_{k2} & \cdots & v_{kd} \\
y_1 & y_2 & \cdots & y_d
\end{pmatrix}
$$

are zero. Since all these determinants can be expressed as linear combinations of the determinants of $k \times k$ submatrices of V with coefficients from w, we conclude that the Plücker coordinates of k-dimensional subspaces are uniquely determined, a fact which will be used later.

Define the mapping $W : (\mathbb{R}^d)^k \to \mathbb{R}^{d^k}$ by

$$
W_{(i_1, \ldots, i_k)}(v_1, \ldots, v_k) = \det
\begin{pmatrix}
v_{1 i_1} & \cdots & v_{1 i_k} \\
\vdots & & \vdots \\
v_{k i_1} & \cdots & v_{k i_k}
\end{pmatrix}
$$

and let Im $W = W((\mathbb{R}^d)^k)$. We shall use the following properties of W, formulated as a lemma. A proof is provided in Section 2.4 at the end of this Chapter.

Lemma 15.1. *For each* $(i_1, \ldots, i_k)$, *the map* $u^{(i_1, \ldots, i_k)} : \mathbb{R}^{d^k} \to (\mathbb{R}^d)^k$ *defined by*

$$u_j^{(i_1, \ldots, i_k)}(z) = (u_{j1}, \ldots, u_{jd}), u_{jh}$$

$$= \begin{cases} z_{i_1, \ldots, i_{j-1}, h, i_{j+1}, \ldots, i_k} & \text{if } h \notin \{i_1, \ldots, i_{j-1}, i_{j+1}, \ldots, i_k\}, \\ 0 & \text{otherwise,} \end{cases} \tag{3}$$

for $j = 1, \ldots, k$, *is linear and satisfies*

(i) $z_{i_1, \ldots, i_k} \neq 0 \Rightarrow u_1^{(i_1, \ldots, i_k)}(z), \ldots, u_k^{(i_1, \ldots, i_k)}(z)$ *span a k-dimensional subspace of* $\mathbb{R}^d$,

(ii) $W(u^{(i_1, \ldots, i_k)}(z)) = z$ *if* $z \in \text{Im } W$ *and* $z_{i_1, \ldots, i_k} = 1$.

The collection $u^{(i_1, \ldots, i_k)}$ will be referred to as the $(i_1, \ldots, i_k)$-basis induced by z. We note for later use that the basis depends on the array $(i_1, \ldots, i_k)$, and even the subspace spanned by this basis may depend on the choice of $(i_1, \ldots, i_k)$ if z is not in $\text{Im} W$.

2.2 Existence of pseudoequilibria I

In what follows, we prove the existence of a pseudoequilibrium under standard conditions of the underlying economy and fairly weak assumptions on the asset structure. For this purpose, we shall need the notion of a *price-dependent asset*, which is a continuous map $a(\cdot) : \mathbb{R}_+ \to \mathbb{R}^m$ which to any price vector assigns an asset as defined earlier. Price-dependent assets serve only a technical purpose here, but they might have some interest in themselves. Clearly, the notion of a pseudoequilibrium can be extended straightforwardly to allow for price-dependent asset structures.

We shall need the standard concept of a Walras equilibrium in the underlying economy $\mathcal{E}$, which is what was considered in Chapter 13, Section 2. Again, such equilibria are not achievable given

Economies with Incomplete Markets 385

the incompleteness of markets, and they are used only as a technical device. We may then state the following existence results for pseudoequilibria.

Theorem 15.2. *Let $\mathcal{E}$ be an economy and $(a_1(\cdot),\ldots,a_k(\cdot))$ a price-dependent asset structure. Assume that consumer preferences satisfy Assumption 0.2 and monotonicity and that the asset structure has the following property*

(∗) *For each Walras equilibrium $(z_1,\ldots,z_m,p)$ of $\mathcal{E}$,*

 (a) *the vectors in $B(p) = \{p \square a_1(p),\ldots,p \square a_k(p)\}$ are linearly independent,*
 (b) *the set $\{z \in \mathbb{R}^S \mid z_h = 0,\ h > S - k\}$ is not contained in span $B(p)$.*

Then the economy $\mathcal{E}$ has a pseudoequilibrium.

We prove Theorem 15.2 through a series of lemmas.

Lemma 15.2. *Let $(z_1,\ldots,z_m,p)$ be a competitive equilibrium. Then*

 (i) *$B(p)^{\perp}$, the orthogonal complement of span $B(p)$ in $\mathbb{R}^r$, has dimension $r - k$,*
 (ii) *$W_{(1,\ldots,r-k)}(u_1,\ldots,u_{S-k}) \neq 0$ for any basis $u_1,\ldots,u_{r-k}$ of $B(p)^{\perp}$.*

Proof. Part (i) of the lemma is an immediate consequence of property (∗) in the statement of Theorem 15.2. For $u_1,\ldots,u_{r-k}$, a basis of $B(p)^{\perp}$, $W_{(1,\ldots,r-k)} = 0$ implies that the $(r - k)$-vectors $(u_{\tau 1},\ldots,u_{\tau,S-k})$, $\tau = 1,\ldots,r-k$, are not linearly independent, so that $B(p)^{\perp}$ contains a vector $u \neq 0$ with $u_h = 0$ for $h = 1,\ldots,r-k$, and consequently, every z with $z_h = 0$ for $h > r - k$ must belong to span $B(p)$, contradicting the second part of the assumption. □

The next step is to introduce a suitable parametrization of $\mathbb{L}_k^r$. We use the orthogonal complements of the k-subspaces, so that the parameter space is a subset of $\mathbb{R}^{r^{-k}}$. Let

$$\mathbb{B}_+ = \left\{ w \in \mathbb{R}^{r^{-k}} \,\middle|\, \|w\| \leq 1,\ w_{1,\ldots,r-k} \geq 0 \right\},\ \mathbb{S}_+ = \left\{ w \in \mathbb{B}_+ \,\middle|\, \|w\| = 1 \right\}.$$

386   *Theory of General Economic Equilibrium*

We use $\mathbb{B}_+$ as a parameter space for subspace selection; clearly, $\mathbb{B}_+$ is convex and compact.

The following lemma uses only lower boundedness of the consumption sets, and it is stated without proof.

Lemma 15.3. *There is a constant $Q > 0$ such that for every pseudoequilibrium (z, p, L), $z_{ih} \leq Q$ for all i and h.*

In view of Lemma 15.3, one may define truncated choice sets $\hat{Z}_i = \{z_i \in Z_i \mid z_{ih} \leq Q\}$ for each i; let $\hat{Z} = \prod_{i=1}^{m} \hat{Z}_i$. Each $\hat{Z}_i$ is convex and compact.

Let $\mathbf{S}$ be the set of ordered $(r - k)$-tuples $\sigma = (i_1, \ldots, i_{r-k})$ from $\{1, \ldots, r\}$, and for each $w \in \mathbb{B}_+ \backslash \{0\}$, let $\mathbf{S}(w)$ be the subset of $\mathbf{S}$ consisting of all $\sigma = (i_1, \ldots, i_{r-k})$, such that $w_\sigma = w_{i_1, \ldots, i_{r-k}} \neq 0$. For each $\sigma \in \mathbf{S}(w)$, we define

$$U^\sigma(w) = u^\sigma(w_\sigma^{-1}w),$$

which is the basis for an $(r - k)$-dimensional subspace of $\mathbb{R}^r$ associated with w, and under suitable circumstances, its orthogonal complement will be the subspace L figuring in the definition of a pseudoequilibrium.

Choose a number $\varepsilon > 0$, used as slackness parameter in the following, arbitrarily. For consumer 1, the budget correspondence $\gamma_1^\varepsilon : \Delta \times \mathbb{B}_+ \rightrightarrows \hat{Z}_1$ is defined as

$$\gamma_1^\varepsilon(p, w) = \{z_1' \in \hat{Z}_1 \mid p \cdot z_1' < 0\}; \tag{4}$$

thus, γ_1^ε does not depend on w and ε, and it has open graph and non-empty convex values.

For consumers $i \neq 1$, budget correspondences are slightly more complex. Define $\gamma_i^\varepsilon : \Delta \times \mathbb{B}_+ \rightrightarrows \hat{Z}_i$ as follows: For $w \in \mathbb{B}_+ \backslash \mathbf{S}_+$, let

$$\gamma_i^\varepsilon(p, w) = \left\{ z_i' \in \hat{Z}_i \mid -\varepsilon < p \cdot z_i' < 0 \right\} \tag{5}$$

for $w \in \mathbf{S}_+$ and $\sigma \in \mathbf{S}(w)$, let

$$\gamma_i^{\sigma, \varepsilon}(p, w) = \left\{ z_i' \in \hat{Z}_i \mid -\varepsilon < p \cdot z_i' < 0, \, p \mathbin{\square} z_i' \in U^\sigma(w)^\perp + B_\varepsilon \right\},$$

Economies with Incomplete Markets
387

where B_ε is the open ball in $\mathbb{R}^k$ with center in 0 and radius ε, and define $\gamma_i^\varepsilon(p, w)$ as

$$\gamma_i^\varepsilon(p, w) = \mathrm{conv}\left(\{\gamma_i^{\sigma,\varepsilon}(p, w) \mid \sigma \in \mathbf{S}(w)\}\right).$$

The definition of $\gamma_i^{\sigma,\varepsilon}$ reflects the use of the subspace parameters w. Clearly, the set $\gamma_i^\varepsilon(p, w)$, defined as the convex hull of several possibly distinct budget sets, has no economic interpretation unless $\gamma_i^{\sigma,\varepsilon}$ turns out to be independent of σ, as indeed it does in equilibrium.

Lemma 15.4. *The correspondences γ_i^ε, $i = 2, \ldots, m$, have open graph and non-empty convex values.*

Proof. Let $(p, w) \in \Delta \times \mathbb{B}_+$ and $z_i' \in \gamma_i^\varepsilon(p, w)$ be arbitrary. If w belongs to the open set $\mathbb{B}_+ \backslash \mathbf{S}_+$, then γ_i^ε has the form (4) in an open set containing (p, w), so that $z_i' \in \gamma_i^\varepsilon(p', w')$ in a neighborhood of (p, w). If $w \in \mathbf{S}_+$, then there is an open set $G_{(p,w)}$ containing (p, w) such that $\sigma \in \mathbf{S}(w)$ entails $\sigma \in \mathbf{S}(w')$ for all $(p', w') \in G_{(p,w)}$, and

$$z_i' \in \mathrm{conv}\left(\{\gamma_i^{\sigma,\varepsilon}(p', w') \mid \sigma \in \mathbf{S}(w)\}\right) \subset \mathrm{conv}\left(\{\gamma_i^{\sigma,\varepsilon}(p', w') \mid \sigma \in \mathbf{S}(w')\}\right) \tag{6}$$

for all (p', w') in some open subset of $G_{(p,w)}$ containing (p, w). But the set on the right-hand side in (5) is a subset of $\gamma_i^\varepsilon(p', w')$ for (p', w') in a small enough neighborhood of (p, w). We conclude that γ_i^ε has an open graph. Convexity of $\gamma_i^\varepsilon(p, w)$ at each (p, w) is a straightforward consequence of the definition. $\qquad\square$

For $i = 1, \ldots, m$, define the generalized demand correspondence $\varphi_i^\varepsilon : \hat{Z} \times \Delta \times \mathbb{B}_+ \rightrightarrows \hat{Z}_i$ by

$$\varphi_i^\varepsilon(z, p, w) = \begin{cases} \gamma_i^\varepsilon(p, w) & \text{if } z_i \notin \mathrm{cl}\, \gamma_i^\varepsilon(p, w), \\ \gamma_i^\varepsilon(p, w) \cap P_i(z_i) & \text{otherwise,} \end{cases}$$

where $\mathrm{cl}\, A$ denotes the closure of the set A. Then φ_i^ε is irreflexive (in the sense that $z_i \notin \varphi_i^\varepsilon(z, p, w)$), and it has open graph and convex values. Also, $\varphi_i^\varepsilon(z, p, w) = \emptyset$ implies that $z_i \in \mathrm{cl}\, \gamma_i^\varepsilon(p, w)$ and $P_i(z_i) \cap \gamma_i^\varepsilon(p, w) = \emptyset$.

The price formation is taken care of by the correspondence ψ^ε : $\hat{Z} \rightrightarrows \Delta$ by

$$\psi^\varepsilon(z) = \left\{ p' \in \Delta \,\middle|\, p' \cdot \sum_{i=1}^m z_i > 0 \right\}.$$

Clearly, ψ^ε has convex (possibly empty) values and an open graph.

Finally, there is need for a correspondence regulating subspace selection. Define $\Omega : \Delta \rightrightarrows \mathbb{S}_+$ by

$$\Omega(p) = \left\{ w \in \mathbb{S}_+ \mid \exists v_1, \ldots, v_{r-k} \in B(p)^\perp, \ W(v_1, \ldots, v_{r-k}) = w \right\}.$$

Lemma 15.5. *The correspondence Ω has closed graph and non-empty values.*

Proof. Non-emptyness of $\Omega(p)$ for each p is obvious, as $B(p)$ consists of at most k linearly independent vectors.

To prove closedness of the graph of Ω, let $(p^v, w^v)_{v=1}^\infty$ be a sequence in the graph of Ω converging to some $(p^0, w^0) \in \Delta \times \mathbb{S}_+$. For each v, choose an orthonormal basis $(v_1^v, \ldots, v_{r-k}^v)$ for an $(r-k)$-dimensional subspace of $\mathbb{R}^r$ with

$$v_r^v \cdot (p^v \,\square\, a_j(p^v)) = 0, \quad r = 1, \ldots, r-k, \ j = 1, \ldots, k,$$

and $\|(W(v_1^v, \ldots, v_{r-k}^v)\| = w^v$. Taking subsequences if necessary, it may be assumed that the sequence $((v_1^v, \ldots, v_{r-k}^v))_{v=1}^\infty$ converges to some $(v_1^0, \ldots, v_{r-k}^0) \in (\mathbb{R}^r)^{r-k}$. By continuity of W, one has that $\|(W(v_1^0, \ldots, v_{r-k}^0)\| = w^0$, so that the family $\{v_1^0, \ldots, v_{r-k}^0\}$ spans an $(r-k)$-dimensional subspace of $\mathbb{R}^r$. Continuity of $p \,\square\, a$ for each a implies that

$$v_\tau^0 \cdot (p^0 \,\square\, a_j(p^0)) = 0, \quad \tau = 1, \ldots, r-k, \quad, j = 1, \ldots, k,$$

and it can be concluded that $w^0 \in \Omega(p^0)$. $\qquad\square$

Now a fixed point theorem can be applied.

Lemma 15.6. *There exists $(z^\varepsilon, p^\varepsilon, w^\varepsilon) \in \hat{Z} \times \Delta \times \mathbb{B}_+$, such that*

(i) *for $i = 1, \ldots, m$, $\varphi_i^\varepsilon(z^\varepsilon, p^\varepsilon, w^\varepsilon) = \emptyset$,*
(ii) *$\psi^\varepsilon(z^\varepsilon) = \emptyset$,*
(iii) *$w^\varepsilon \in \Omega(p^\varepsilon)$, in particular, $w^\varepsilon \in \mathbb{S}_+$.*

Economies with Incomplete Markets 389

Proof. The existence of z^ε, p^ε, and w^ε) satisfying (i) and (ii) and such that $w^\varepsilon \in \text{conv} \ (\Omega(p^\varepsilon))$ follows from Theorem 1.7, since the correspondence conv (Ω) is uhc with closed convex and non-empty values. To show that the second property in (iii) holds, assume to the contrary that w^ε does not belong to $\Omega(p^\varepsilon)$, so that w^ε is a proper convex combination of points in $\Omega(p^\varepsilon)$; in particular, p^ε belongs to $\mathbb{B}_+ \backslash \$_+$. Then $\gamma_i^\varepsilon(p^\varepsilon, w^\varepsilon) = \{z_i \mid p^\varepsilon \cdot z_i < 0\}$ for all i, and from (i) it follows that $p \cdot z_i^\varepsilon \leq 0$ and $P_i(z_i^\varepsilon) \cap \{z_i \mid p^\varepsilon \cdot z_i < 0\} = \emptyset$ for each i, which together with (ii) and monotonicity of preferences gives that $(z^\varepsilon, p^\varepsilon)$ satisfies all the conditions of Definition 15.1 and consequently is a competitive equilibrium. Application of Lemma 15.1 then gives that $\Omega(p^\varepsilon)$ is a unique point in $\$_+$ and its $(1, \dots, r - k)$-coordinate is > 0, so that conv $\Omega(p^\varepsilon) = \Omega(p^\varepsilon)$, a contradiction which proves (iii). $\qquad \square$

The following lemma is a restatement of the previous one in terms which facilitate the concluding limit argument in the slack parameter ε.

Lemma 15.7. *For each $\varepsilon > 0$, there is $(z^\varepsilon, p^\varepsilon, L^\varepsilon) \in \hat{Z} \times \Delta \times \mathbf{L}$ such that*

(i) $\sum_{i=1}^m z^\varepsilon \leq 0$,
(ii) (a) $p^\varepsilon \cdot z_1^\varepsilon = 0, -\varepsilon < p^\varepsilon \cdot z_i^\varepsilon \leq 0$ for $i = 2, \dots, m$,
  (b) $p^\varepsilon \ \Box \ z_1^\varepsilon \in L^\varepsilon + 2(m-1)B_\varepsilon, p^\varepsilon \ \Box \ z_i^\varepsilon \in L^\varepsilon + B_\varepsilon$ for $i = 2, \dots, m$.
  (c) if $z_1' \in P_1(z_1^\varepsilon)$, then $p^\varepsilon \cdot z_1' > 0$,
  (d) for $i = 2, \dots, m$, if $z_i' \in P_i(z_i^\varepsilon)$, then $p^\varepsilon \cdot z_i' \notin [-\varepsilon, 0]$ or $p^\varepsilon \ \Box \ \cdot z_i' \notin L^\varepsilon + B_\varepsilon$.
(iii) $p^\varepsilon \ \Box \ a_j(p^\varepsilon) \in L^\varepsilon, j = 1, \dots, k$.

Proof. Statement (i) is an immediate consequence of Lemma 15.6(ii) and the definition of ψ^ε. By Lemma 15.6(iii), $w^\varepsilon \in \$_+$, so w^ε is uniquely determined, and we may define L^ε as $U^\sigma(w^\varepsilon)^\perp$, where the definition is independent of $\sigma \in S(w^\varepsilon)$. By the definition of Ω, we have that $p^\varepsilon \ \Box \ a_j(\varepsilon) \in L^\varepsilon$ for each j; this proves statement (iii).

(ii): For each i, it follows from Lemma 15.6(i) that $z_i^\varepsilon \in \text{cl} \ \gamma_i^\varepsilon(p^\varepsilon, w^\varepsilon)$ and that $P_i(z^\varepsilon) \cap \gamma_i^\varepsilon(p^\varepsilon, w^\varepsilon) = \emptyset$. For $i = 1$, this together with the monotonicity of preferences implies that

$$p^\varepsilon \cdot z_1^\varepsilon = 0 \text{ and } p^\varepsilon \cdot z_1' > 0 \text{ whenever } z_1' \in P_1(z_1^\varepsilon). \qquad (7)$$

390   *Theory of General Economic Equilibrium*

From Lemma 15.6(iii), we get that the subspace $L^\varepsilon = U^\sigma(w^\varepsilon)^\perp$ is independent of σ, and reasoning as above we get for $i \geq 2$ that

$$-\varepsilon < p^\varepsilon \cdot z_i^\varepsilon \leq 0, \; p^\varepsilon \;\square\; z_i^\varepsilon \in L^\varepsilon + B_\varepsilon \tag{8}$$

and that $z_i' \in P_i(z_i^\varepsilon)$ implies $p^\varepsilon \cdot z_i' \notin \;]-\varepsilon, 0]$ or $p^\varepsilon \;\square\; z_i' \notin L^\varepsilon + B_\varepsilon$. Thus, we have proved statements (ii.a)–(ii.d) of the lemma for $i \neq 1$.

It remains only to show that $p^\varepsilon \;\square\; z_1^\varepsilon \in L^\varepsilon + m B_\varepsilon$. For this, note that

$$p^\varepsilon \;\square\; \textstyle\sum_{i=2}^m z_i^\varepsilon \in L^\varepsilon + (m-1) B_\varepsilon$$

by (8), and also $p^\varepsilon \;\square\; \sum_{i=1}^m z_i^\varepsilon \in L^\varepsilon + B_\varepsilon$ follows by part (i) together with the fact that $p^\varepsilon \cdot \sum_{i=1}^m z_i^\varepsilon \geq -(m-1)\varepsilon$, and from this it may be concluded that $p^\varepsilon \;\square\; z_1^\varepsilon \in L^\varepsilon + 2(m-1) B_\varepsilon$ as desired. $\square$

Proof of Theorem 15.2. Choose a sequence $(\varepsilon_\nu)_{\nu=1}^\infty$ of values of ε converging to 0 and for each ε_ν in the sequence a triple $(z^{\varepsilon_\nu}, p^{\varepsilon_\nu}, L^{\varepsilon_\nu}) \in \hat{Z} \times \Delta \times \mathbf{L}$ satisfying (i)–(iii) of Lemma 15.7. Taking subsequences if necessary, we may assume that the sequence converges to $(z^0, p^0, L^0) \in \hat{Z} \times \Delta \times \mathbb{L}_k^r$ (where $\mathbb{L}_k^r$ has the Fell topology (also known as the topology of closed convergence, see, e.g. Hildenbrand (1974)), on closed subsets of $\mathbb{R}^r$.

We must check that (i)–(iii) of Definition 15.1 are satisfied. Here, (i) follows easily from Lemma 15.7(i) and (iia). To prove (iii) of Definition 15.1, we note that each of the vectors $p^{\varepsilon_\nu} \;\square\; a_j \in \mathbb{R}^r$ belong to L^{ε_ν}, and using the continuity of $\cdot \;\square\; a_j$, we get that the sequence $(p^{\varepsilon_\nu} \;\square\; a_j)_{\nu=1}^\infty$ converges to $p^0 \;\square\; a_j$, which therefore must belong to L^0. It remains only to check condition (ii).

From Lemma 15.7(ii), we get that $p^0 \cdot z_i^0 = 0$ and $p^0 \;\square\; z_i^0 \in L^0$ for all i. For $i = 1$, suppose that there is $z_1' \in P_1(z_1^0)$ with $p^0 \cdot z_1' = 0$. By the open graph property of P_1 and the fact that $0 \in \mathrm{int} Z_1$, we get that there is some $z_1'' \in Z_1$ with $p \cdot z_1'' < 0$, such that $z_1'' \in P_i(z_1^{\varepsilon_\nu})$ for large enough ν. But then $p^{\varepsilon_\nu} \cdot z_1'' > 0$ by Lemma 15.7(ii.c), and we have a contradiction, showing that (ii) of Definition 5.3 is satisfied for $i = 1$.

For $i > 1$, suppose again that there is $z_i' \in Z_i$ with $z_i' \in P_i(z_i^0)$ and $p^0 \cdot z_i' = 0$, $p^0 \;\square\; z_i' \in L^0$. By the properties of P_i, we may then find $z_i'' \in Z_i$ such that $p^0 \cdot z_i'' < 0$, $p^0 \;\square\; z_i'' \in L^0$. Since the

Economies with Incomplete Markets 391

correspondence $(p \ \square \ \cdot \ )^{-1} : \mathbb{R}^\ell \rightrightarrows \mathbb{R}^\ell \times \mathbb{R}^r$ to $\{z \in \mathbb{R}^\ell \mid p \ \square \ z = u\}$ is lower hemicontinuous, we can find a sequence $(\tilde{z}_i^{\varepsilon_\nu})_{\nu=1}^\infty$ such that $p^{\varepsilon_\nu} \ \square \ \tilde{z}_i^{\varepsilon_\nu} \in L^{\varepsilon_\nu} + B_{\varepsilon_\nu}$ and $\tilde{z}_i^{\varepsilon_\nu} \to z_i''$. We have that $\tilde{z}_i^{\varepsilon_\nu} \in P_i(z_i^{\varepsilon_\nu})$ and $p^{\varepsilon_\nu} \cdot \tilde{z}_i^{\varepsilon_\nu} < 0$ for large enough ν, contradicting Lemma 15.7(ii.d). This shows that (ii) of Definition 15.1 holds for all i, so that (z^0, p^0, L^0) is a pseudoequilibrium. $\qquad \square$

2.3 Existence of pseudoequilibria II

The results in the preceding section establish the existence in a model which was rather more general than what is usually seen, since assets might depend on prices, but which on the other hand satisfies the specific property $(*)$ in Theorem 15.2, stating that at any Walras equilibrium (x, p) of the economy, the vectors $p \ \square \ a_j(p)$ for $j = 1, \ldots, k$ are linearly independent and are not all 0 in the last $r - k$ coordinates. Since our main interest is in assets that do not depend on prices, we may obtain a somewhat sharper result than that of Theorem 15.2, replacing the specific assumptions in $(*)$ by a simpler one.

Theorem 15.3. *Let $\mathcal{E}$ be an economy and $(a_1, \ldots, a_k)$ an (ordinary) asset structure. Assume that consumer preferences satisfy Assumption 0.2 and monotonicity.*

If $\mathcal{E}$ has finitely many Walras equilibria, then $\mathcal{E}$ has a pseudoequilibrium.

Proof. Let $\nu \in \mathbb{N}$ be chosen arbitrarily, and define an economy $\mathcal{E}^\nu$ with a price-dependent asset structure as follows.

Let $\{p_1, \ldots, p_T\}$ be the set of prices p such that (z, p) is a competitive equilibrium in $\mathcal{E}$ for some allocation z, and let U_t with $p_t \in U_t$, $t = 1, \ldots, T$ be mutually disjoint open subsets of Δ. For $t = 1, \ldots, T$, it is possible to select vectors $a_j^{(t,\nu)} \in \mathbb{R}^\ell$ with $\|a_j^{(t,\nu)} - a_j\| < \dfrac{1}{\nu}, j = 1, \ldots, k$, such that

(a) $p_t \ \square \ a_j^{(t,\nu)}$ for $j = 1, \ldots, k$ are linearly independent,

(b) $(p_t \ \square \ a_j^{(t,\nu)})_h \neq 0$ for all $h > r - k$, each j.

Let $\psi_t : \Delta \to [0,1]$ be a continuous function with $\psi_t(p_t) = 1, \psi_t(p') = 0$ for $p' \notin U_t, t = 1, \ldots, T$, and define $\psi_0 : \Delta \to [0,1]$ by $\psi_0(p) =$

$1 - \sum_{t=1}^{T} \psi_t(p)$. The price-dependent asset structure α_j^v, $j = 1, \ldots, k$ is defined by

$$\alpha_j^v(p) = \sum_{t=0}^{T} \psi_t(p) a_j^{(t,v)}.$$

Now, the economy $\mathcal{E}^v$ with a price-dependent asset structure α_j^v, $j = 1, \ldots, k$ has property $(*)$, and by Theorem 15.2, it has a pseudoequilibrium (z^v, p^v, L^v).

Without loss of generality, it may be assumed that the sequence $(z^v, p^v, L^v)_{v \in \mathbb{N}}$ in the space $\prod_{i=1}^{m} \hat{Z}_i \times \Delta \times \mathbb{L}_k^r$ converges to some (z^0, p^0, L^0) (using that $\mathbb{L}_k^r$ is compact in the Fell topology). It is easily checked that (z^0, p^0, L^0) satisfies the conditions of a pseudoequilibrium in the economy $\mathcal{E}$. $\qquad\square$

The result of Theorem 15.3 has not been done away completely with the restrictions posed by property $(*)$, since it is assumed that the economy has only finitely many equilibria. Some extensions are indicated in the exercises.

2.4 Proof of Lemma 15.1

The proof of Lemma 15.1 which is given below is adapted from Merslyakov (1987). For the proof of the proposition, we need two more lemmas.

Lemma 15.8. *The map W is continuous, and if $z \in \mathrm{Im} W$, then*

$$\sum_{t=1}^{k+1} (-1)^t z_{i_1, \ldots, i_{k-1}, j_t} z_{j_1, \ldots, j_{t-1}, j_{t+1}, \ldots, j_{k+1}} = 0$$

for all sets of indices $\{i_1, \ldots, i_{k-1}, j_1, \ldots, j_{k+1}\}$.

Proof. Continuity is obvious from the definition of W,

$$W_{(i_1, \ldots, i_k)}(v_1, \ldots, v_k) = \det \begin{pmatrix} v_{1 i_1} & \cdots & v_{1 i_k} \\ \vdots & & \vdots \\ v_{k i_1} & \cdots & v_{k i_k} \end{pmatrix}.$$

Economies with Incomplete Markets 393

To prove the second statement, for each t we expand the determinant

$$z_{i_1,\ldots,i_{k-1},j_t} = \sum_{s=1}^{k}(-1)^{k+s}v_{sj_t}D_{i_1,\ldots,i_{k-1},j_t}^{s,j_t},$$

where

$$D_{i_1,\ldots,i_{k-1},j_t}^{s,j_t} = \det \begin{pmatrix} v_{1i_1} & \cdots & v_{1i_{k-1}} \\ \vdots & & \vdots \\ v_{t-1i_1} & \cdots & v_{t-1i_{k-1}} \\ v_{t+1i_1} & \cdots & v_{t+1i_{k-1}} \\ \vdots & & \vdots \\ v_{ki_1} & \cdots & v_{ki_{k-1}} \end{pmatrix}$$

is the subdeterminant obtained from the columns $i_1,\ldots,i_{k-1},j_t$ by deleting column j_t and row s. Then

$$\sum_{t=1}^{k+1}(-1)^t z_{i_1,\ldots,i_{k-1},j_t} z_{j_1,\ldots,j_{t-1},j_{t+1},\ldots,j_{k+1}}$$

$$= \sum_{t=1}^{k+1}(-1)^t \sum_{s=1}^{k}(-1)^{k+s}v_{sj_t}D_{i_1,\ldots,i_{k-1},j_t}^{s,j_t} z_{j_1,\ldots,j_{t-1},j_{t+1},\ldots,j_{k+1}}$$

$$= \sum_{s=1}^{k}(-1)^{k+s-1}D_{i_1,\ldots,i_{k-1},j_t}^{s,j_t} \sum_{t=1}^{k+1}(-1)^{t+1}v_{sj_t}z_{j_1,\ldots,j_{t-1},j_{t+1},\ldots,j_{k+1}}.$$

Here, the last sum can be identified as an expansion along the first row of the determinant

$$\det \begin{pmatrix} v_{sj_1} & \cdots & v_{sj_{k+1}} \\ v_{1j_1} & \cdots & v_{1j_{k+1}} \\ \vdots & & \vdots \\ v_{kj_1} & \cdots & v_{sj_{k+1}} \end{pmatrix},$$

394   *Theory of General Economic Equilibrium*

and this determinant is zero as the matrix contains two identical rows.

Lemma 15.9. *If $z \in \mathrm{Im}\, W$, then*

$$z_{i_1,\ldots,i_k} z_{j_1,\ldots,j_k} = \sum_{t=1}^{k} z_{i_1,\ldots,i_{s-1},j_t,i_{s+1},\ldots,i_m} z_{j_1,\ldots,j_{t-1},i_s,j_{t+1},\ldots,j_k},$$

where s is a fixed number with $1 \le s \le k$.

Proof. Apply Lemma 15.8 replacing $i_1,\ldots,i_{s-1},i_s,\ldots,i_{m-1}$, $j_1,\ldots,j_m,j_{m+1}$ by $i_1,\ldots,i_{s-1},i_{s+1},\ldots,i_m,j_1,\ldots,,j_m,i_s$ and use the skewsymmetry of $z_{i_1,\ldots,i_k}$ in the indices. $\qquad\qquad\square$

Proof of Lemma 15.1. Linearity follows immediately from the definition. For property (i), we note that the matrix

$$\begin{pmatrix} u_{1i_1} & \cdots & u_{1i_k} \\ \vdots & & \vdots \\ u_{ki_1} & \cdots & u_{ki_k} \end{pmatrix}$$

with u_{ji_h} defined in (3), is a unit matrix when $z_{i_1,\ldots,i_k} = 1$.

It remains only to show (ii), that is for each subset $\{j_1,\ldots,j_k\}$ of $\{1,\ldots,d\}$, we have that

$$\det \begin{pmatrix} u_{1j_1} & \cdots & u_{1j_k} \\ \vdots & & \vdots \\ u_{kj_1} & \cdots & u_{kj_k} \end{pmatrix} = z_{j_1,\ldots,j_k}. \tag{9}$$

We use induction on the number τ of indices in $\{j_1,\ldots,j_k\}$ which are not in the set $\{i_1,\ldots,i_k\}$. For $\tau = 0$, the result follows from (i). Assume that (9) holds for some $0 \le \tau < k$, and consider any subset $\{j_1,\ldots,j_k\}$ with $\tau + 1$ indices not in $\{i_1,\ldots,i_k\}$.

Expanding the determinant $\mathbf{D}$ on the left-hand side in (9) after column j_s, we obtain

$$\mathbf{D} = \sum_{t=1}^{k} (-1)^{s+t} u_{tj_s} D^{t,j_s}_{j_1,\ldots,j_{s-1},j_{s+1},\ldots,j_k}.$$

Now,

$$D^{s,j_s}_{j_1,\ldots,j_{s-1},j_{s+1},\ldots,j_k} = z_{j_1,\ldots,j_{s-1},i_s,j_{s+1},\ldots,j_k}$$

by the induction hypothesis. For $t \neq s$,

$$D^{t,j_s}_{j_1,\ldots,j_{s-1},j_{s+1},\ldots,j_k} = (-1)^{s+t} z_{j_1,\ldots,j_{s-1},i_t,j_{s+1},\ldots,j_k}$$

(replace column j_s by column i_t and apply the induction hypothesis). Inserting the expansion and using the definition of u_{tj_s}, one gets

$$\mathbf{D} = \sum_{t=1}^{k} z_{i_1,\ldots,i_{t-1},j_s,i_{t+1},\ldots,i_k} z_{j_1,\ldots,j_{s-1},i_t,j_{s+1},\ldots,j_k} = z_{i_1,\ldots,i_k} z_{j_1,\ldots,j_k},$$

where the last equality follows from Lemma 15.9. Since $z_{i_1,\ldots,i_k} = 1$, the desired result is obtained. $\quad\square$

3. Exercises

(1) Let $\mathcal{E}$ be an economy with two periods $t = 0, 1$, and with two goods in each period, none of which can be stored. At $t = 1$, there are two possible states of nature s_1 and s_2. There are two consumers with utility functions

$$u_1(x^0, x^1(s_1), x^1(s_2)) = x_1^0 x_2^0 + \frac{1}{2} x_1^1(s_1) x_2^1(s_1) + \frac{1}{2} x_1^1(s_2) x_2^1(s_2),$$

$$u_1(x^0, x^1(s_1), x^1(s_2)) = x_1^0 (x_2^0)^2 + \frac{1}{2} x_1^1(s_1)(x_2^1(s_1))^2 + \frac{1}{2} x_1^1(s_2)(x_2^1(s_2))^2.$$

Consumer 1 has the endowment $(1, 1, 1)$ in period 0 and $(0, 0, 3)$ in period 1 independent of state, and consumer 2 has $(2, 1, 0)$ in each period and state.

There are two financial assets, namely A paying 1 unit of account in each period, and B paying 2 units of account in s_1 and 1 unit in s_2. Find a Radner equilibrium.

(2) Check the welfare properties of the equilibrium found in Exercise 1. Is the allocation Pareto-optimal? Restricted Pareto optimal?

(3) Consider an economy over two periods $t = 0, 1$, with three goods in each period, none of which can be stored. At $t = 1$, there are two possible states of nature s_1 and s_2. There are two consumers with utility functions

$$u_1(x^0, x^1(s_1), x^1(s_2)) = x_1^0 x_2^0 x_3^0 + \frac{1}{2}x_1^1(s_1)x_2^1(s_1)x_3^1(s_1)$$

$$+ \frac{1}{2}x_1^1(s_2)x_2^1(s_2)x_3^1(s_2),$$

$$u_1(x^0, x^1(s_1), x^1(s_2)) = x_1^0(x_2^0)^2 x_3^0 + \frac{1}{2}x_1^1(s_1)(x_2^1(s_1))^2 x_3^1(s_1)$$

$$+ \frac{1}{2}x_1^1(s_2)(x_2^1(s_2))^2 x_3^1(s_2).$$

Consumer 1 has the endowment $(1, 1, 1)$ in period 0 and $(0, 0, 3)$ in period 1 independent of state, and consumer 2 has $(2, 1, 0)$ in each period and state.

There are two real assets, namely

Security	State s_1	State s_2
A	$(1, 0, 0)$	$(3, 1, 0)$
B	$(0, 1, 1)$	$(0, 2, 1)$

Find a pseudoequilibrium.

(4) Let (z, p, L) be a pseudoequilibrium in the economy $\mathcal{E}$ with real assets $a_1, \ldots, a_k$, such that $p \square a_1, \ldots, p \square a_k$ are linearly independent. Show that (z, p) is a (Walras) equilbrium with incomplete markets.

(5) A *flag* in $\mathbb{R}^n$ is a finite sequence $(L_1, L_2, \ldots, L_s)$ of subspaces of $\mathbb{R}^r$ with $L_1 \subset \cdots \subset L_k$. Suppose that the financial markets are

Economies with Incomplete Markets 397

hierarchical in the sense that consumers in the subset M_1 have access to all securities in $A_1 = \{a_1, \ldots, a_k\}$, those of a larger subset M_2 have access to a subset A_2 of A_1, etc., and all consumers have access to $A_s \subset \cdots \subset A_1$.

Define a pseudoequilibrium in this setup, where the subspaces in Definition 15.1 are replaced by flags. Consider how the existence proof should be adapted to this situation.

Chapter 16

General Equilibrium and Money

1. Money and Decentralized Exchange

We have already discussed (pairwise) exchanges in many contexts, and it is now high time to give them further consideration. For a given exchange economy $\mathcal{E} = (X_i, P_i, \omega_i)_{i=1}^m$, a system of exchanges is an array $(x_{ij})_{i=1\ j=1}^{m\ \ m}$ of vectors $x_{ij} \in \mathbb{R}^l$, whereby x_{ij} is interpreted as specifying the amount (positive or negative) of commodities $h = 1, \ldots, l$ that consumer i received from consumer j. In accordance with the interpretation, the system of exchanges is assumed to satisfy the condition of skewsymmetry,

$$x_{ij} = -x_{ji}, \quad i, j = 1, \ldots, m, \tag{1}$$

so that in particular, $x_{ii} = 0$ for all i.

The system of exchanges $(x_{ij})_{i=1\ j=1}^{m\ \ m}$ defines an allocation in the obvious way, namely by

$$x_i = \omega_i + \sum_{j=1}^m x_{ij}, \quad i = 1, \ldots, m, \tag{2}$$

and it follows from (1) and (2) that

$$\sum_{i=1}^m x_i = \sum_{i=1}^m \omega_i + \sum_{i=1}^m \sum_{j=1}^m x_{ij} = \sum_{i=1}^m \omega_i,$$

where we used the skewsymmetry from (1), so that the allocation satisfies the condition of aggregate feasibility. However, a given allocation can in principle emerge from many different systems of exchanges, and the specific choice may not be obvious, even in situations, where the allocation is an equilibrium (in some sense which we do not need to specify here) given a price vector p.

This lack of a specific procedure for determining the relevant pairwise exchanges may be a problem if we want to consider the market mechanism as an absolutely decentralized device for choosing allocations in society. Given that an equilibrium of supply and demand has been found, with associated equilibrium price, for the full decentralization of the allocation process, we need a set of rules prescribing what consumer i has to do, in the sense of meeting the other consumers and exchanging goods with them.

1.1 Realizing net trades through pairwise trades

Following the approach of Ostroy and Starr (1974), we are looking for trading rules which can be applied to allocation–price pairs (x, p), or, more conveniently, to the set $\mathcal{Z}$ of all triples (ω, z, p) with $p \in \mathbb{R}^l_{++}$, where $z = (z_1, \ldots, z_m) \in \mathbb{R}^{lm}$ are arrays of net trades satisfying

$$\sum_{i=1}^{m} z_i = 0,$$

(3)

$$p \cdot z_i = 0, z_i \geq -\omega_i, \quad i = 1, \ldots, m.$$

The net trades associated with a Walras equilibrium satisfy (3), but the other arrays of net trades may be used as well as point of departure.

The process of trade is assumed to take place sequentially over several periods, where in each period a trader can meet only one other trader. There is a specified ordering in which consumers meet each other, formalized as a sequence σ^t of permutations of $\{1, \ldots, m\}$ such that $\sigma^t(i) = j$ implies $\sigma^t(j) = i$ for all i, j and t, and $\sigma^t(i) = i$ for at most one t. It is possible to define a sequence of T permutations, where $T = m - 1$ for m even and $T = m$ for m odd so that each consumer meets each other exactly once. The collection of T periods is called a *round* of the trading process.

At the beginning of period t, consumer i has a modified endowment ω_i^t and meets $\sigma^t(i) = j$ with modified endowment ω_j^t, trading $z_i^t = x_{i\sigma^t(i)}^t$. As a result of the trading, the holdings of consumer i are modified to ω_i^{t+1}, by the pairwise trade $z_i^t = \omega_i^{t+1} - \omega_i^t$, and similarly, endowment of j is modified further to ω_j^{t+1} with $z_j^t = \omega_j^{t+1} - \omega_j^t$. The residual net trade v_i^t for consumer i at date t is given by $v_i^1 = z_i$ for $t = 1$ and

$$v_i^t = v^{t-1} - \sum_{\tau=1}^{t-1} z_i^\tau$$

for $t > 1$.

The following constraints on z_i^t for $t = 1, \ldots, T$ and $i = 1, \ldots, m$ are reasonable,

(A1) $\omega_i^t + z_i^t \geq 0$ (non-negativity of holdings at any time),
(A2) $z_i^t = -z_{\sigma^t(i)}^t$ (skewsymmetry or conservation of commodities),
(A3) $p \cdot z_i^t = 0 = p \cdot z_{\sigma(j)}^t$ (exchange of equal values, *quid-pro-quo*).

If the three conditions are satisfied, the sequence of trades is said to be admissible.

We would like the process of pairwise exchanges to be completed in the course of a single round, giving the condition (B) defined by

$$\sum_{\tau=1}^{T} z_i^\tau = z_i, \quad i = 1, \ldots, m. \tag{4}$$

So far nothing has been said about the meaning of decentralization in the present context. For the process of realizing net trades through pairwise exchanges to be independent of any authority or outside agent, the rules for determining the pairwise exchange between i and j to be effectuated when matched at date t should depend only on data available to i and j. Writing this rule as

$$\rho(\omega_i^\tau, \omega_j^\tau; \mathcal{D}^\tau) = (z_i^\tau, z_j^\tau),$$

402   *Theory of General Economic Equilibrium*

Box 1. Obtaining net trades through pairwise trades: The classical example discussed here, which can be traced back to Wicksell (1901), indicates the non-trivial nature of the problem of arranging allocation. Suppose that there are four consumers and four goods, whereby consumer i is interested only in good i but has endowment only of good $i + 1$ (modulo 4). More specifically, we assume that $\omega_i = e_{i+1}$ for $i = 1, 2, 3$ and $\omega_4 = e_1$, and we consider the array of net trades $(z_1, \ldots, z_4)$ with

$$
\begin{aligned}
z_1 &= (1, -1, 0, 0), \\
z_2 &= (0, 1, -1, 0), \\
z_3 &= (0, 0, 1, -1), \\
z_4 &= (-1, 0, 0, 1).
\end{aligned}
\tag{5}
$$

For simplicity, we assume that all prices are equal to 1.

It is seen that no consumer can find another one with whom to trade and thereby obtain the desired commodity. The simplest way in which to effectuate the desired net trade would be to deliver the endowment to the neighbor to the left while getting the desired good from the other neighbor. This is, however, *not* a sequence of pairwise trades. The closest we can come is by letting one consumer, say consumer 1, act as a middleman, trading with each of the other consumers. After the trade between 1 and 2, their desired net trades are modified to

$$
v_1^2 = (1, 0, -1, 0), \quad v_2^2 = (0, 0, 0, 0),
$$

trade between 1 and 3 changes them to

$$
v_1^3 = (1, 0, 0, -1), \quad v_3^3 = (0, 0, 0, 0),
$$

and after the final trade between 1 and 4, all desired net trades are reduced to $(0, 0, 0, 0)$.

---

the degree of decentralization of the rule can be expressed as

(D1) $\mathcal{D}^\tau = \{(v_i^\tau, v_j^\tau)\}$ (only residual net trades matter)

(D2) $\mathcal{D}^\tau = \{(v_i^\tau, v_j^\tau), (i, j)\}$ (residual net trades *and* the labels of the individuals may matter),

(D3) $\mathcal{D}^\tau = \{(v_i^1, \ldots, v_i^\tau, v_j^1, \ldots, v_j^\tau), (i, j)\}$ (the rule may also depend on the trading history of the relevant individuals).

General Equilibrium and Money 403

The above forms of dependence of the trading rule on data are all compatible with the intuitive notion of a procedure which can be implemented without any outside control. If we move beyond D3, allowing for dependence on data which are not immediately related to i and j, we are implicitly assuming some form of outside monitoring, which means that we do not have a decentralized procedure any more. We formulate a specific version as the condition

(C) $\mathcal{D}^\tau = \left\{(v_1^1, \ldots, v_1^\tau, v_2^1, \ldots v_2^\tau, \ldots, v_m^1, \ldots, v_m^\tau), (i, j)\right\}.$

Here, the trading histories of *all* individuals may enter as an argument in the trading rule for i and j at date t.

If we accept centralization, then there is a trading rule which gives rise to an admissible sequence of trades achieving the given net trade array in a single round of pairwise trades.

Theorem 16.1. *Let $(\omega, z, p) \in Z$ be a triple consisting of an endowment, a net trade array and a price, such that* (3) *is satisfied. Then there is a trading rule which satisfies the conditions* (A), (B) *and* (C).

Proof. Choose a consumer i_1 with $z_{i_1 h_1} < 0$ (so that i_1 can deliver good h_1 in some pairwise exchange). Since $\sum_{i=1}^m z_{ih_1} = 0$, there is i_2 such that $z_{i_2 h_1} > 0$, and since $p \cdot z_{i_2} = 0$, there is some commodity h_2, such that $z_{i_2 h_2} < 0$. Proceeding in this way, we obtain a sequence of individuals and goods $i_1 h_1 i_2 h_2 i_3 \ldots$ such that for each $j \geq 1$, $z_{i_j h_j} < 0$ and $z_{i_{j+1} h_j} > 0$. For each such sequence, if $j \geq m + 1$, then some individual i_0 (which without loss of generality can be chosen as i_1) must occur more than one time, giving a sequence of the form

$$c = (i_1 \, h_1 \, i_2 \, h_2 \, i_3 \ldots h_k \, i_1),$$

to be referred to as a cycle (of length $k \leq m$).

Given the cycle c, we construct an associated system of pairwise trades between i_j and i_{j+1}, and between i_k and i_1. If the trade must satisfy (A3), then the value should be constant in each pairwise exchange, and if this value is $\delta > 0$, then associated exchanges y_{ij}

have the form

$$y_{i_jh} = \begin{cases} \delta/p_{i_j} & h = i_{j-1} \text{ (with } i_{j-1} = k \text{ for } k = 1\text{)}, \\ -\delta/p_{i_j} & h = i_j \\ 0 & \text{otherwise}, \end{cases}$$

for $j = 1, \ldots, k$. Choose $\delta > 0$ maximal with the property that $0 \le y_{i_jh} \le z_{i_j}$ for $z_{i_jh} > 0$ and $0 \ge y_{i_jh} \ge z_{i_j}$ for $z_{i_jh} < 0$. Then the modified system of net trades $(z^1, \ldots, z^m)$ with

$$z_i^1 = z_i - y_i, i \in \{i_1, \ldots, i_k\}$$

is such that $z_{ih}^1 \le z_{ih}$ for $z_{ih} \ge 0$, $z_{ih}^1 \ge z_{ih}$ for $z_{ih} \le 0$, and the number of pairs (i, h) such that the net trade of i in commodity h is non-zero has been reduced by at least one.

Repeating the above procedure on the reduced net trades, one eventually obtains a finite set of pairwise trades y_{ij}^r, the sum of which can be reformulated as net trades x_{ij} implementing the given net trades in a single round of at most t trading periods. We leave the details to the reader. $\qquad\square$

The theorem shows that the conditions for achieving net trades through the process of pairwise exchanges is not too restrictive to permit a solution, at least when we allow for the use of all information relevant for the process. As pointed out earlier, this is not compatible with the intuitive notion of decentralization, since the rule for exchanging goods between i and j, who shall meet only once, could be determined only involving all the previous residual net trades.

1.2 Decentralized pairwise trades

If less information is available, things change considerably. Indeed, it was shown in Ostroy and Starr (1974) that there is no rule satisfying (A), (B), (C) and even the weakest version (D3) of decentralization. We shall limit ourselves to showing that (A), (B), (C) and (D1) cannot be simultaneously satisfied. For this, we exhibit two simple

General Equilibrium and Money 405

examples (ω^1, z^1, p^1), (ω^2, z^2, p^2) such that if the rule works in the first one, it fails in the second (for a suitable ordering of the pairwise exchanges). We shall assume that $m = l = 5$ and that the net trades have a structure similar to that of Box 1, so that they can be represented by the two (5×5)-matrices

$$
\begin{pmatrix}
1 & -1 & 0 & 0 & 0 \\
0 & 1 & -1 & 0 & 0 \\
0 & 0 & 1 & -1 & 0 \\
0 & 0 & 0 & 1 & -1 \\
-1 & 0 & 0 & 0 & 1
\end{pmatrix}
\quad \text{and} \quad
\begin{pmatrix}
1 & -1 & 0 & 0 & 0 \\
0 & 1 & -1 & 0 & 0 \\
0 & 1 & 0 & 0 & -1 \\
0 & -1 & 1 & 0 & 0 \\
-1 & 0 & 0 & 0 & 1
\end{pmatrix}
$$

with rows representing the net trades of the five individuals in each of the two cases. The price is chosen as $p^1 = p^2 = (1, 1, \ldots, 1)$, and initial endowments are chosen to be minimal such that (3) is satisfied.

Suppose that there is a rule satisfying (A), (B), (C) and (D1) such that z^1 can be obtained. As we remarked in Box 1, the only way of establishing z^1 in a single row is by the choice of a middleman, subsequently, exchanging good h for good $h + 1$ with agent h until finally obtaining the desired good. Renumbering agents and goods if necessary, we may assume that individual 1 is the middleman, so the rule must specify that whenever two agents have exactly opposite residual net trades in one commodity, and for all the other commodities the residual net trade is 0 for at least one of the agents, then this commodity should be transferred from the agent with negative residual net trade to the agent with positive residual net trade.

Turning now to the net trade z^2, and assuming that the sequence of trades is such that agent 1 meets all the other agents before the remaining meetings occur, the rule will describe that agent 1 changes the residual net trade to $(1, 0, -1, 0, 0)$, and the meeting with agent 3 results in no trade. Then, meeting with agent 4 shifts the residual net trade back to $(1, -1, 0, 0, 0)$, and the meeting with agent 5 again results in no trade. Since the possibilities of trade for agent 1 have not

406   *Theory of General Economic Equilibrium*

been exhausted, the original net trade of agent 1 cannot be realized in one round of trades.

Summing up, we have established the following result.

Theorem 16.2. *There is no trading rule satisfying* (A), (B), (C) *and* (D1) *which realizes z for all* $(\omega, z, p) \in \mathcal{Z}$.

While a general result on decentralization of pairwise trades is beyond reach, there are important special cases where it is indeed possible. One such case — and indeed the one that we have been heading at throughout this section — is obtained when there is a particular good which acts as a *medium of exchange,* so that trade can be performed using this good as an intermediate.

Theorem 16.3. *Let* $(\omega, z, p) \in \mathcal{Z}$, *and suppose that there is* $h \in \{1, \ldots, k\}$ *such that*

$$\omega_{ih} - \max\{0, z_{ih}\} \geq \sum_{k \neq h : z_{ik} > 0} \frac{p_k z_{ik}}{p_h}$$

for each h. Then there is a trading rule realizing z and satisfying (A), (B), (D1) *and* (E).

Intuitively, the rule should be such that in the pairwise trade between agents i and j, agent i buys as much as possible from j of any good $k \neq h$ for which the residual net trades of i and j are, respectively, positive and negative, while j similarly buys from i whenever j has a positive and i a negative residual net trade.

Proof of Theorem 16.3. For ease of notation, after a change of units, if necessary, we may assume that $p = (1, 1, \ldots, 1)$. Define the rule $\rho^*(\omega_i^t, \omega_j^t, v_i^t, v_j^t)$ by

$$\rho_k^*(\omega_i^t, \omega_j^t, v_i^t, v_j^t) = \begin{cases} \min\left\{v_{ik}^t, -v_{jk}^t\right\} & v_{ik}^t > 0, v_{jk}^t < 0, \\ -\min\left\{-v_{ik}^t, v_{jk}^t\right\} & v_{ik}^t > 0, v_{jk}^t < 0, \\ 0 & \text{otherwise,} \end{cases} \quad (6)$$

for $k \neq h$, and

$$\rho_h^*(\omega_i^t, \omega_j^t, v_i^t, v_j^t) = -\sum_{k \neq h} \rho_k^*(\omega_i^t, \omega_j^t, v_i^t, v_j^t). \tag{7}$$

It is clear from the definition that pairwise exchanges according to the rule ρ^* satisfy (A) and that the residual net trades v_i^{t+1} satisfy

$$0 \leq v_{ik}^{t+1} \leq v_{ik}^t \text{ if } v_{ik}^t > 0, \ v_{ik}^t \leq v_{ik}^{t+1} \leq 0 \text{ if } v_{ik}^t < 0, \tag{8}$$

for $k \neq h$.

Consider now a round of T trades whereby all pairs of individuals have met once, and choose an agent i arbitrarily. Suppose that $v_{ik}^{T+1} \neq 0$ for some commodity $k \neq h$, say $v_{ik}^{T+1} > 0$. Since v^{T+1} satisfies (3), there must be at least one agent j such that $v_{jk}^{T+1} < 0$. However, after the trade t between i and j according to (6), either $v_{ik}^{t+1} = 0$ or $v_{jk}^t = 0$, and by the monotonicity property (8), we get a contradiction. We conclude that $v_{ik}^{T+1} = 0$ for all i and $k \neq h$. But then it follows from (3) that $v_{ih}^{T+1} = 0$ for all i, so that z is indeed realized after the round of T trades following the rule ρ^*. $\qquad\qquad\square$

It should be noted that in our discussion so far, we have been dealing only with the purely technical aspects of arranging pairwise exchanges, abstracting from questions of incentives. It may therefore happen that some of the pairwise exchanges may leave a consumer worse off in the sense that the endowment before the exchange is preferred to the endowment as revised by the exchange. Following a tradition introduced by Jevons (1875), such exchanges are known as *monetary* exchanges, and it was shown in Madden (1976) that if we use the above approach to achieve net trades associated with Walras equilibria, then there will be some economies for which some exchanges are monetary, at least when the numbers of goods and individuals are not too small. This means that one has to take a somewhat relaxed stand on the demand that each pairwise exchange should leave the parties no worse off.

What has emerged from our discussion is that money in the sense of a medium of exchange also matters in the context of general equilibrium, once we consider the process of implementing the

408 *Theory of General Economic Equilibrium*

equilibrium. We shall look closer into the role and nature of media of exchange in the following.

1.3 Pairwise exchanges of dated commodities: The problem of distrust

Even if the purely technical problems of finding suitable pairwise trades for the realization of a given net trade can be solved, there may be other problems which are related to the very nature of the commodities involved. As pointed out in Kiyotaki and Moore (2002), a pairwise trade in dated commodities, agreed upon at date 0, implies that at some later date, the agent delivers the commodity in question, and if there is a possibility that the agent may default on this obligation, the pairwise exchanges must reflect this possibility. To this phenomenon of possible distrust in agents should be added the additional complications arising if a commodity is acquired by an agent with the purpose of using it in another pairwise trade, since now it is not a question of distrust in the other party of the exchange but in the commodity exchanged.

This problem can be illustrated by the following example with three agents and one good for delivery at three different dates 1 (now), 2 and 3, whereby agent 1 wants to consume at date 1 but is endowed with the good only at $t = 3$, agent 2 wants the good at date 2 and has it at date 1 and agent 3 wants the good in date 3 and has the good at date 2, so that we get the following desired net trades in matrix form

$$\begin{pmatrix} 1 & 0 & -1 \\ 1 & 1 & 0 \\ 0 & -1 & 1 \end{pmatrix},$$

where as usual the net trades of the individuals are the rows of the matrix.

To achieve these net trades, one of the agents must act as a middleman, so that we could have agent 1 selling the physical unit of the good to agent 2 against a promise to deliver the good at date 2, possibly in the form of a paper, which then is sold by agent 1 to agent 3 against a promise to deliver the good at date 3. Technically, this works well, but it may be upset if some agent has doubts about the fulfillment of the promises, in particular, the last exchange, where agent 1 attempts to sell a promise of another agent to deliver the good, that is a paper issued by some other individual.

The possibility to resell a promise presupposes a particularly strong form of trust which does not come by itself. It could be established if the papers issued by the consumers were to be verified and guaranteed by some independent authority, but this solution runs counter to the idea of decentralization which has been the driving force of our investigations so far. Fortunately, public — and, consequently, decentralized — verification is a possibility, at least in theory, in the form of the blockchain technology, which we briefly survey in Section 2.

2. Cybercurrencies and Blockchain

In this section, we briefly consider what seems to be a digression of little relevance for general equilibrium, namely cybercurrencies and in particular the backbone of cybercurrencies, which is the blockchain technology. The blockchain was designed in connection with the launch of BitCoin (Nakamoto, 2008), the first electronic currency but it has eventually found many other applications. The survey of blockchain and its use discussed here follows Zheng *et al.* (2017).

We begin with the construction of electronic means of payment. The basic idea is that an agent should be able to transfer a unit of the currency to another agent digitally. In the BitCoin architecture, this is achieved by digitally signing a coded message, a *hash*, of the previous transaction together with the public address of the other agent and adding this to the coin, which in this way takes the form of a long list of transactions. The procedure is illustrated in Fig. 1.

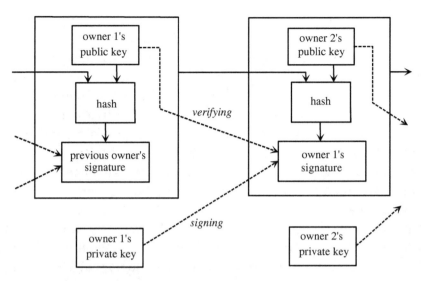

Fig. 1. Transactions with BitCoin: The owner uses the public key and the hash of previous transactions together with a private key to sign a transfer. The last transaction is then added to the list of previous transactions.

In order to make this trustworthy, it must be prevented that the same coin can be given out several times. One way of achieving this would be to have a central authority monitoring all transactions, but BitCoin instead uses a system of public announcement, so that all transactions are visible to everyone. For this to work, there must be an arrangement by which all eventually agree on a single history of transactions.

This is where the proof-of-work technique comes in: A number of new transactions are collected to form a *block*. Each block has a header containing (among other things) a hash of the previous block and a *nonce*, a 4-byte array which is 0 at the beginning. To verify the block, an agent in the network must perform repeated trials and errors, each time increasing the nonce by 1, so that the hash of the other content of the header together with the nonce either obtains a prescribed value or differs from it by an allowable distance. Here, the hash function is a given function transforming large data arrays to smaller ones in such a way that it is easy to perform a hash of a given array but very complicated to find an array which produces

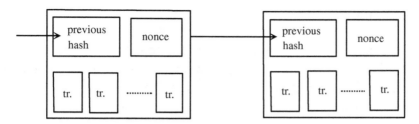

Fig. 2. Verification of transactions ("tr." in the figure) by proof-of-work: The header of the block, containing, among other things, a hash of the previous block, with the nonce appended, should result in a series of 0 when hashed. To obtain this, the nonce is changed step by step. When the result is obtained, the block is appended to the previous blocks as another block in the chain.

a given result. Once a solution has been found, the relevant agent publishes it, and everyone can verify that it is correct and that the transactions in the block are legitimate. From now, the block can be used as parent block for new proof-of-work verifications.

The procedure is illustrated in Fig. 2, where the block to the right is subjected to the proof-of-work verification. Once it has succeeded, the block is included in the chain and the transactions in the block can be considered as verified. This means that the acquirer of the BitCoin can use it in future transactions.

The main achievement of BitCoin, at least in our present context, is that it demonstrates the possibility of a totally decentral means of payment. The particular version of such a decentral electronic currency has some limitations (such as its small capacity, which has to do with the choice of block size, the particular proof-of-work selected, which is energy-consuming when carried out on a large scale), but this need not detract us here: What matters in our context is that (1) decentralized realization of Walras allocations is possible and (2) that this is achieved using the blockchain technology, which essentially can be reduced to keeping a publicly accessible ledger which shows all the transactions that have been carried out using the electronic currency involved. This presence of this form of *memory* is what allows BitCoin and the subsequent versions of cybercurrencies to fulfil the role of money. We shall have more to say about memory in Section 3.

3. Money and Memory

We consider an overlapping generations economy, where each generation consisting of m agents lives for two consecutive periods of time, consuming bundles of l goods in each period. Each individual i in generation t has endowments ω_i^t, ω_i^{t+1} of the l goods in each period as well as preferences defined by a preference correspondence P_i^t on $\mathbb{R}_+^{2l}$, which can be represented by a separable utility function of the form

$$u_i(x_i^t) + \beta u_i(x_i^{t+1}),$$

where $\beta \in [0, 1]$ is a discount factor assumed to be the same for all agents. For simplicity, we assume that there is no production and goods cannot be stored.

Allocation in the economy is performed in subsets of agents alive at date t. Following Kocherlakota (1998), we let the $2m$ agents at date t be partitioned into subsets, each of which is called a *match*, and all reallocation of endowments is performed within the subsets. We may think of a match as the outcome of some random process of matching individuals for the purpose of reallocation, but for simplicity, we assume the choice of partition at date t to be deterministic.

For each individual i, let $S^t(i)$ be the set of individuals who are in agent i's match at date t (if there are any) and define $Q^t(i)$ recursively by

$$Q^1(i) = S^1(i),$$

$$Q^t(i) = \bigcup_{j \in S^t(i)} Q^{t-1}(j).$$

This, $Q^1(i)$ contains the immediate contacts of i, the previous contact of those, etc. In our present overlapping generations framework, we have that

$$j \in S^t(i) \Rightarrow Q^{t-1}(j) \cap Qt - 1(i) = \emptyset. \tag{9}$$

General Equilibrium and Money 413

For every match S^t at date t, an allocation is an array $(x_i^t)_{i \in S^t}$ of consumption bundles $x_i^t \in \mathbb{R}^l$, and it is feasible if

$$x_t^t \in \mathbb{R}_+^l, \text{ each } i \in S^t \text{ (individual feasibility in the match),}$$

$$\sum_{i \in S^t} x_i^t \leq \sum_{i \in S^t} \omega_i^t \text{ (aggregate feasibility w.r.t. the match).} \tag{10}$$

At a later state we shall need the following assumption.

Assumption 16.1. For any match S^t, there is no feasible allocation $(x_i^t)_{i \in S^t}$ which Pareto dominates the endowment allocation $(\omega_i^t)_{i \in S^t}$ (in the sense that $u_i(x_i^t) \geq u_i(\omega_i^t)$ for all i and $u_{i^0}(x_{i^0}^t) > u_{i^0}(\omega_{i^0}^t)$ for some i^0).

Since we shall be comparing the role of money and of memory, we need to introduce *money* into the model. We assume that there is an initial total endowment of money to the size of M which is distributed among the initial m individuals as money holdings μ_i^0. For every match S^t at date t, a *money allocation* is an array $\mu^t = (\mu_i^t)_{i \in S^t}$ such that

$$\sum_{i \in S^t} \mu_i^0 \leq \sum_{i \in S^t} \mu_i^{t-1}.$$

Since utilities have already been defined and depend only on the consumption of the goods, money has no intrinsic value to the consumers.

In order to add *memory* to our model, we need to specify what should be remembered, and to do so, we need to introduce a *trading mechanism* to be used for constructing allocations in a given match. This mechanism is given by

(i) a set $\mathcal{A}$ consisting of sets $A(t, i, S^t, \mu^{t-1})$ of actions a_i open to individual i at date t, given the match S^t and the initial money allocation μ^{t-1}, for each i, t, S^t and μ^{t-1}, and

414 *Theory of General Economic Equilibrium*

(ii) an outcome function, which for given t, S^t, and μ^{t-1} assigns to each array $\mathbf{a}$ of actions $a_i \in A(t, i, S^t, \mu^{t-1})$ a feasible allocation

$$f(\mathbf{a}; t, S^t, \mu^{t-1})$$

(of goods and money) in S^t.

To describe the choice of action, we need to specify the *information* available to the agents. In our simple setting, information takes the form of particular variables on which the choice can be made dependent, so that a *strategy* is a mapping from the available information to the set of actions. The exact form of the information, in particular the part which will be designated as *public* will vary depending on the situation, and in each case, it will be assumed that the strategy can depend only on this part of the information. More specifically, we say that a strategy is a *perfect public equilibrium* (PPE) if for each value of the public information, the strategy selects an action which is optimal given the other members' choices. An allocation that is obtained in a PPE is said to be a PPE allocation.

Before proceeding, we must add a further restriction on the choices of strategies in the mechanism, since we are discussing the role of money or alternatively of memory in the context of allocation. If allocations in society can be obtained merely by pointing them out, then the formation of subsets trading with each other and the process of selecting the right changes could be sidestepped, and no money would be needed at all. Therefore, we introduce a limitation for the type of mechanisms to be considered, more specifically, we shall assume that the mechanism is voluntary in the sense that participating in trading using the mechanism will not make the agent worse off than staying outside.

Definition 16.1. A trading mechanism $(\mathcal{A}, f)$ is voluntary if for each date t and each match S^t, there is $a_i \in A(t, i, S^t, \mu^{t-1})$, such that

$$f((a_i, \mathbf{a}_{-i}), t, S^t, \mu^{t-1}) = (\omega_i^t, \mu_i^t)$$

for some strategy array $\mathbf{a}_{-i}$ of the other agents in S^t.

General Equilibrium and Money   415

An example of a voluntary trading mechanism is given by letting $\mathcal{A}^0$ be given by

$$A^0(t, i, S^t, \mu^{t-1}) = \left\{ (x^t, \mu^t) \,\middle|\, x^t \in \mathcal{F}(S^t), \mu^t \in \times[0, M]^{S^t} \right\}, \qquad (11)$$

and the outcome function is

$$f^0(\mathbf{a}, t, S^t, \mu^{t-1}) = \begin{cases} y_1^t & a_i = y_1^t, \text{ all } i \in S^t, \\ (\omega_i^t, \mu_i^{t-1})_{i \in S^t} & \text{otherwise.} \end{cases} \qquad (12)$$

Thus, in $(\mathcal{A}^0, f^0)$, the agents submit the desired allocations, which are implemented if all agree, and in the case of disagreement, everyone in the match is left with the initial endowments and money holdings. This means that $(\mathcal{A}^0, f^0)$ is a voluntary mechanism, since for any specification of the desired allocation in goods and money, each individual can obtain the endowment and initial money holding by stating another allocation.

The mechanism $(\mathcal{A}^0, f^0)$ is a *direct mechanism* in the sense that the individuals submit information about the desired result of trading. The following is a version of the *revelation principle* from mechanism theory adapted to PPEs of voluntary mechanisms.

Theorem 16.4. *Let* $(\mathcal{A}, f)$ *be a voluntary trading mechanism. Then every PPE allocation of* $(\mathcal{A}, f)$ *can be obtained as a PPE allocation of the direct voluntary trading mechanism* $(\mathcal{A}^0, f^0)$.

Proof. There are three different cases to be considered which are as follows:

(i) **No money or memory is involved in the trading process:** Suppose that the PPE of $(\mathcal{A}, f)$ results in the feasible allocation $(x_i^t)_{i \in S^t}$ in each match. Then the strategy array in $(\mathcal{A}^0, f^0)$ which for each match S^t lets all individuals in S^t choose the allocation $(x_i^t)_{i \in S^t}$ will result in this allocation.

Suppose that some individual $i \in S^t$ selects another allocation. Then all individuals in S^t remain with their individual endowment, and in all other matches, present as well as future, nothing is changed. However, since $(\mathcal{A}, f)$ is voluntary, there is an action for individual i in S^t which will also result in the initial endowment to everyone without any changes for other matches. It follows that i cannot improve by changing action, so that $(x_i^t)_{i \in S^t}$ is also a PPE allocation in $(\mathcal{A}^0, f^0)$.

(ii) **With money:** The reasoning is as above. If $(x_i^t, \mu_i^t)_{i \in S^t}$ is the allocation of commodities and money at date t and in the match S^t, then the strategy of individual i in $(\mathcal{A}^0, f^0)$ will select (x_i^t, μ_i^t) in the match S^t. If an individual i deviates at the match S^t, then every individual $j \in S^t$ gets the endowment ω_j^t and the initial money holding μ_j^{t-1}, with possible consequences for the allocation and thereby for the endowment of the individuals in future periods. But again, by the assumption of voluntariness, the same could be obtained by a suitable action in $(\mathcal{A}, f)$, so that $(x_i^t, \mu_i^t)_{i \in S^t}$ is a PPE allocation in $(\mathcal{A}^0, f^0)$.

(iii) **With memory:** Here, we specify the strategies in $(\mathcal{A}^0, f^0)$ as follows: For any t, S^t, and $i \in S^t$, if all individuals in $Q^t(i)$ have chosen the PPE allocation of $(\mathcal{A}, f)$ in all previous matches, then i should choose $(x_j^t)_{j \in S^t}$; otherwise, the choice is the endowment allocation $(\omega_j^t)_{j \in S^t}$. A deviation from this strategy by an individual i will result in the endowment allocation in the current period and in all future periods, and this could be obtained in $(\mathcal{A}, f)$ also, so the specified strategy is a PPE in $(\mathcal{A}^0, f^0)$. $\square$

Since by Theorem 16.4, allocations obtained in a PPE in some voluntary mechanism can also be obtained as PPE allocations in the direct mechanism, we can omit the reference to the mechanism when dealing with such allocations, in the sequel called *incentive-feasible* allocations. Since we have several different scenarios (neither money nor memory, with money but without memory or with memory but without money), it is to be understood that comparing incentive-feasible allocations with or without money, the comparison refers only to the allocation of commodities.

We are now ready for the main result of this section.

Theorem 16.5. *An allocation which is incentive feasible with money is incentive feasible with memory.*

General Equilibrium and Money 417

Proof. Let $(x_i^t, \mu_i^t)_{i \in S^t, S^t \in S}$ be an incentive-feasible allocation with money. Then, there is a PPE in some voluntary mechanism $(\mathcal{A}, f)$ which results in this allocation. Here, the choice of action a_i of any individual in any match is a function of the identities of agents in the match and of their money holdings.

For $t = 1$ and S^1, a match at $t = 1$, the equilibrium strategies in $(\mathcal{A}, f)$ select an action $\mathbf{a}^1$ such that

$$f(\mathbf{a}^1, S^1, \mu^0) = (x_i^1, \mu_i^1)_{i \in S^1}, \tag{13}$$

or, equivalently, for the given goods allocation, there is a final money holding μ_i^1 and an action array for the individuals in the match S^1 such that (13) is satisfied. Similarly, for each t and match S^t and date t, there is μ^{t-1} and $\mathbf{a}^t$ such that

$$f(\mathbf{a}^t, S^t, \mu^{t-1}) = (x_i^t, \mu_i^t)_{i \in S^t}, \tag{14}$$

meaning that final money holdings can be reconstructed from the initial holdings and the goods allocation so as to satisfy (14).

From Theorem 16.4, we know that the goods-and-money allocation can be obtained as a PPE allocation of the direct mechanism with money. Since the money holdings can be reconstructed from the commodity allocation given the knowledge of all commodity trades in $Q^t(i)$, the PPE equilibrium, the strategy which chooses the relevant commodity allocation whenever the initial money holding is compatible with the trading history, and the endowment allocation otherwise, is a PPE equilibrium in the direct mechanism without money. $\qquad \square$

The result shows that whatever can be achieved using money can be obtained using memory instead, so that in this sense, memory is as good as money. Actually, there are contexts, as explored in the exercises, where memory performs better than money, sustaining efficient allocations which could not be reached if only money can be used. Thus, it may be conjectured that the blockchain technology, as a real-world version of memory, has something to contribute when choosing allocations in society.

418   *Theory of General Economic Equilibrium*

4. Exercises

(1) (Bilateral trades and transaction cost) Assume that there is cost connected with the transfer of a commodity from one agent to another, expressed as a share $\lambda(h, i, j) \in [0, 1]$ of commodity h used up when this commodity is transferred from agent i to agent j.

    Show that under suitable assumptions on the structure of $\lambda(h, i, j) \in [0, 1]$, this may give rise to the identification of a particular commodity as money in the sense of medium of exchange.

(2) (Feldman, 1973). Consider an exchange economy with two commodities and three consumers, all having consumption set $\mathbb{R}_+^2$ and preferences described by the utility function $u(x_1, x_2) = x_1 x_2$. The initial endowments are $\omega_1 = (1, 9)$, $\omega_2 = (5, 5)$, $\omega_3 = (5, 5)$. Show that the exchange $x_{13} = (2, -6)$ transforms the endowment allocation to a Pareto-optimal allocation.

    Consider next the economy with three commodities and consumers, given by

Consumer	Utility function	Endowment
1	$u_1(x_1) = 3x_{11} + 2x_{12} + x_{13}$	$(0, 1, 0)$
2	$u_2(x_2) = 2x_{21} + x_{22} + 3x_{23}$	$(1, 0, 0)$
3	$u_3(x_3) = x_{31} + 3x_{32} + 2x_{33}$	$(0, 0, 1)$

Show that (1) the endowment allocation is not Pareto-optimal, and (2) there is no way of achieving a Pareto-optimal allocation by voluntary exchanges.

(3) Consider the following economy over time: The set of agents is $[0, 1]$, and there are three types of infinitely lived agents, each with measure $1/3$, and there are three consumption goods, which are non-storable and indivisible. Agents of type i can produce commodity i using a fourth (input) commodity as input, and the technology is (Kocherlakota, 1998).

$$Y_i = \{(y_1, y_4) \in \mathbb{R} \times \mathbb{R}_- \mid y_1 \leq -y_4\}, \quad i = 1, 2, 3.$$

At date t, agents of type i have utility $u(c_{i+1})-y_i$, where u is strictly increasing, strictly concave and differentiable, and $i + 1$ should be considered modulo 3. The consumers discount future utility by a factor $\beta > 0$. All agents have the endowment $(0,0,0,\omega)$ in each period.

Suppose that matching is such that an agent is paired with an agent of each type with equal probability.

Consider the allocation where agents of type i give agents of type $i - 1$ (modulo 3) y^* units of output (of commodity i) whenever paired, whereby y^* satisfies $u'(y^*) = 1$. Show that this allocation is Pareto-optimal.

(4) Show that the allocation considered in Exercise 3 is incentive feasible with memory if the discount factor β is large enough (one may use an argument known from folk theorems, whereby the right behavior is sustained by a threat of punishment in future periods).

(5) Show that the allocation in Exercise 3 is *not* incentive feasible with money (agents must be offered an increase in future utility to undertake production). Conclude that memory may be superior to money in some environments.

Bibliography

R. Abraham and J. Robbin. *Transversal Mappings and Flows*. WA Benjamin, New York, 1967.

F. Ackerman. Still dead after all these years: Interpreting the failure of general equilibrium theory. *Journal of Economic Methodology*, 9:119–139, 2002.

C.D. Aliprantis and D.J. Brown. Equilibria in markets with a Riesz space of commodities. *Journal of Mathematical Economics*, 11:189–207, 1983.

C.D. Aliprantis, D.J. Brown, and O. Burkinshaw. *Existence and Optimality of Economic Equilibrium*. Springer-Verlag, New York, 1990.

C.D. Aliprantis, M. Florenzano, and R. Tourky. General equilibrium analysis in ordered topological general equilibrium analysis in ordered topological vector spaces. *Journal of Mathematical Economics*, 40:247–269, 2004a.

C.D. Aliprantis, P.K. Monteiro, and R. Tourky. Non-marketed options, non-existence of equilibria, and non-linear prices. *Journal of Economic Theory*, 114:345–357, 2004b.

A. Araujo and P.K. Monteiro. Equilibrium without uniform conditions. *Journal of Economic Theory*, 48:416–427, 1989.

A.P. Araujo and M.R. Páscoa. Bankruptcy in a model of unsecured claims. *Economic Theory*, 20:455–481, 2002.

K.J. Arrow. The organization of economic activity: issues pertinent to the choice of market versus non-market allocation. In *The Analysis and Evaluation of Public Expenditures: The PPB System*, pp. 47–64. Joint economic committee of the Congress of the United States, Washington D.C., 1969.

K.J. Arrow and G. Debreu. Existence of an equilibrium for a competitive economy. *Econometrica*, 22:265–290, 1954.

K.J. Arrow and F.H. Hahn. *General Competitive Analysis*. Holden-Day, San Francisco, 1971.

J.-P. Aubin. *Mathematical Methods of Game and Economic Theory*. North Holland, Amsterdam, 1979.

J.-P. Aubin. Cooperative fuzzy games. *Mathematics of Operations Research*, 6:1–13, 1981.

R.J. Aumann. Markets with a continuum of traders. *Econometrica*, 32:39–50, 1964.

R.J. Aumann and M. Maschler. The bargaining set for cooperative games. In M. Dresher, L.S. Shapley, and A.W. Tucker, editors, *Advances in Game Theory*, pp. 443–476. Princeton University Press, Princeton, 1964.

R.J. Aumann and L.S. Shapley. *Values of Non-atomic Games*. Princeton University Press, Princeton, 1974.

Y. Balasko. Some results on uniqueness and on stability of equilibrium in general equilibrium theory. *Journal of Mathematical Economics*, 2: 95–118, 1975.

T. Bergstrom. When was Coase right? Technical report, University of California, Santa Barbara, 2017.

T.F. Bewley. Existence of equilibrium in economies with infinitely many commodities. *Journal of Economic Theory*, 4:514–540, 1972.

G. Bonanno. General equilibrium theory with imperfect competition. *Journal of Economic Surveys*, pages 297–328, 1990.

J.-M. Bonnisseau. Regular economies with non-ordered preferences. *Journal of Mathematical Economics*, 39:153–174, 2003.

J.-M. Bonnisseau and B. Cornet. General equilibrium theory with increasing returns: the existence problem. In *Equilibrium Theory and Applications. Proceedings of the Sixth International Symposium in Economic Theory and Econometrics*, pp. 65–82. Cambridge University Press, Cambridge, 1991.

A. Borglin and H. Keiding. Existence of equilibrium actions and of equilibrium. *Journal of Mathematical Economics*, 3:313–316, 1976.

A. Borglin and H. Keiding. *Optimality in Infinite Horizon Economies*. Springer-Verlag, Berlin, Heidelberg, 1986.

N. Bourbaki. *Theory of Sets*. Springer, Berlin, Heidelberg, 2004.

X. Calsamiglia. On the size of the message space under non-convexities. *Journal of Mathematical Economics*, 10:197–203, 1982.

D. Cass and K. Shell. Do sunspots matter? *Journal of Political Economy*, 91: 193–227, 1983.

J.S. Chipman. Homothetic preferences and aggregation. *Journal of Economic Theory*, 8:26–38, 1974.

R.H. Coase. The problem of social cost. *Journal of Law and Economics*, 3: 1–44, 1960.

W.A. Coppel. *Foundations of Convex Geometry*. Cambridge University Press, Cambridge, 1998.

B. Cornet. General equilibrium theory and increasing returns: Presentation. *Journal of Mathematical Economics*, 17:103–118, 1988.

V.I. Danilov and A.I. Sotskov. A generalized economic equilibrium. *Journal of Mathematical Economics*, 19:341–356, 1990.

C. d'Aspremont and L. Gerard-Varet. Incentives and incomplete information. *Journal of Public Economics*, 11:25–45, 1979.

A.V. Deardorff. The possibility of factor price equalization, revisited. *Journal of International Economics*, 36(1–2):167–175, 1994.

A.V. Deardorff. Does growth encourage factor price equalization? *Review of Development Economics*, 5(2):169–181, 2001.

G. Debreu. The coefficient of resource utilization. *Econometrica*, 19:273–292, 1951.

G. Debreu. *Theory of Value*. Wiley, New York, 1959.

G. Debreu. Economies with a finite set of equilibria. *Econometrica*, 38: 387–392, 1970.

G. Debreu. Excess demand functions. *Journal of Mathematical Economics*, 1: 15–21, 1974.

G. Debreu and H. Scarf. A limit theorem on the core of an economy. *International Economic Review*, 4:235–246, 1963.

D.W. Diamond and P.H. Dybvig. Bank runs, deposit insurance, and liquidity. *Journal of Political Economy*, 91:401–419, 1983.

E. Dierker. Gains and losses at core allocations. *Journal of Mathematical Economics*, 2(2):119–128, 1975.

E. Dierker and B. Grodal. The price normalization problem in imperfect competition and the objective of the firm. *Economic Theory*, 14:257–284, 1999.

A.K. Dixit and V. Norman. *Theory of International Trade*. Cambridge University Press, London, 1980.

J.H. Drèze and D. de la Vallee Poussin. A tâtonnement process for public goods. *Review of Economic Studies*, 38:133–150, 1971.

P. Dubey, J. Geanakoplos, and M. Shubik. Default and punishment in general equilibrium. *Econometrica*, 73:1–37, 2005.

N. Dunford and J.T. Schwartz. *Linear Operators*, volume I. Wiley, New York, 1957.

A. Edelman and E. Kostlan. How many zeros of a random polynomial are real? *Bulletin of the American Mathematical Society*, 32:1–37, 1995.

J. Eichberger, K. Rheinberger, and M. Summer. Credit risk in general equilibrium. *Economic Theory*, 57:407–435, 2014.

Ky Fan. Extensions of two fixed point theorems of F.E. Browder. *Mathematische Zeitschrift*, 112:234–240, 1969.

M.J. Farrell. The measurement of productive efficiency. *Journal of the Royal Statistical Society*, 120(3):253–290, 1957.

A.M. Feldman. Bilateral trading processes, pairwise optimality, and Pareto optimality. *Review of Economic Studies*, 40:463–473, 1973.

M. Florenzano. On the existence of equilibria in economies with an infinite dimensional commodity space. *Journal of Mathematical Economics*, 12: 207–219, 1983.

M. Florenzano. *General Equilibrium Analysis: Existence and Optimality Properties of Equilibria*. Kluwer Academic Publishers, Boston, Dordrecht and London, 2003.

M. Florig. Arbitrary small indivisibilities. *Economic Theory*, 22:831–843, 2002.

H. Föllmer and A. Schied. *Stochastic Finance: An Introduction in Discrete Time*. De Gruyter, Berlin/Boston, 4th edition, 2016.

D. Gale. *The Theory of Linear Economic Models*. McGraw-Hill, New York, 1960.

D. Gale and A. Mas-Colell. An equilibrium existence theorem for a general model without ordered preferences. *Journal of Mathematical Economics*, 2:9–15, 1975.

J. Geanakoplos. Overlapping generations models of general equilibrium. In J. Eatwell, M. Milgate, and P. Newman, editors, *General Equilibrium*. Palgrave Macmillan, London, 1989.

J. Geanakoplos. An introduction to general equilibrium with incomplete asset markets. *Journal of Mathematical Economics*, 19:1–38, 1990.

G. Giraud. Strategic market games: An introduction. *Journal of Mathematical Economics*, 39:355–375, 2003.

J.-M. Grandmont. Temporary general equilibrium theory. *Econometrica*, 45:535–572, 1977.

J.-M. Grandmont and Guy Laroque. On temporary Keynesian equilibrium. In *The Microeconomic Foundations of Macroeconomics*, pp. 41–61. Springer, 1977.

P. Griffiths and J. Harris. *Principles of Algebraic Geometry*. Wiley, New York, 1978.

T. Groves and J. Ledyard. Optimal allocation of public goods: A solution to the "free rider" problem. *Econometrica*, 45:783–810, 1977.

T. Groves and M. Loeb. Incentives and public inputs. *Journal of Public Economics*, 4:211–226, 1975.

H. Haller. Market Power, Objectives of the Firm, and Objectives of Shareholders. *Journal of Institutional and Theoretical Economics (JITE)/ Zeitschrift für die gesamte Staatswissenschaft*, 142:716–726, 1986.

P. Halmos. The range of a vector measure. *Bulletin of American Mathematical Society*, 54:418–421, 1948.

P.J. Hammond, M. Kaneko, and M.H. Wooders. Continuum economies with finite coalitions: Core, equilibria, and widespread externalities. *Journal of Economic Theory*, 49:113–134, 1989.

O.D. Hart. On the optimality of equilibrium when the market structure is incomplete. *Journal of Economic Theory*, 11:418–443, 1975.

J.R. Hicks. *Value and Capital*, 2nd edition. Clarendon Press, 1946.

W. Hildenbrand. *Core and Equilibrium of a Large Economy*. Princeton University Press, Princeton, 1974.

W. Hildenbrand and A.P. Kirman. *Introduction to Equilibrium Analysis*. North Holland, Amsterdam, 1976. ISBN 978-1-4832-7526-0. GoogleBooks-ID: 8zijBQAAQBAJ.

M.W. Hirsch. *Differential Topology*, volume 33. Springer Science & Business Media, 2012.

G. Huberman. A simple approach to arbitrage pricing theory. *Journal of Economic Theory*, 28:183–191, 1982.

L. Hurwicz. Outcome functions yielding Walrasian and Lindahl allocations at Nash equilibrium points. *Review of Economic Studies*, 46:217–225, 1979.

L. Hurwicz. What is the Coase theorem. *Japan and the World Economy*, 7: 60–74, 1995.

W.S. Jevons. *Money and the Mechanism of Exchange*. D. Appleton and Co., London, 1875.

W.S. Jevons. *Commercial Crises and Sun-Spots*. Nature Publishing Group, 1878.

A. Kajii. A general equilibrium model with fuzzy preferences. *Fuzzy Sets and Systems*, 26:131–133, 1988.

S. Kakutani. A generalization of Brouwer's fixed point theorem. *Duke Mathematical Journal*, 8:457–459, 1941.

E. Kalai, A. Postlewaite, and J. Roberts. A group incentive compatible mechanism yielding core allocations. Discussion Paper 329, Northwestern University, May 1978.

H. Keiding. Cores and equilibria in an infinite economy. In J. Los and M.W. Los, editors, *Computing Equilibria: How and why*. North Holland, Amsterdam, pp. 65–73, 1976.

J.L. Kelley. *General topology*. Springer-Verlag, New York, 1975.

M.C. Kemp and M. Okawa. Market structure and factor price equalization. *The Japanese Economic Review*, 49(3):335–339, 1998.

N. Kiyotaki and J. Moore. Credit cycles. *Journal of Political Economy*, 105: 211–248, 1997.

N. Kiyotaki and J. Moore. Evil is the root of all money. *The American Economic Review*, 92:62–66, 2002.

N.R. Kocherlakota. Money is memory. *Journal of Economic Theory*, 81: 232–251, 1998.

J. Lindenstrauss. A short proof of Lyapunov's convexity theorem. *Journal of Mathematics and Mechanics*, 15:971–972, 1966.

426 Bibliography

L. Ljungqvist and T.J. Sargent. *Recursive Macroeconomic Theory*. MIT Press, Cambridge, Massachusetts, 2018.

A.A. Lyapunov. On completely additive vector-functions. *Izv.Akad.Nauk SSSR*, 4:465–478, 1940.

R.P. MacLean and A. Postlewaite. Excess functions and nucleolus allocations of pure exchange economies. *Games and Economic Behavior*, 1: 131–143, 1989.

P.J. Madden. A theorem in decentralized exchange. *Econometrica*, 44:787–791, 1976.

L. Makowski and J.M. Ostroy. Vickrey-Clarke-Groves mechanisms and perfect competition. *Journal of Economic Theory*, 42:244–261, 1987.

R.R. Mantel. Homothetic preferences and community excess demand functions. *Journal of Economic Theory*, 12:197–201, 1976.

K. Marx. *Capital, Vol. 1: A Critique of Political Economy*. Vintage: New York, 1867.

A. Mas-Colell. The price equilibrium existence problem in topological vector lattices. *Econometrica*, 54:1039–1053, 1986.

A. Mas-Colell. An equivalence theorem for a bargaining set. *Journal of Mathematical Economics*, 18:129–139, 1989.

L.W. McKenzie. On the existence of general equilibrium for a competitive economy. *Econometrica*, 27:54–71, 1959.

L.W. McKenzie. The classical theorem on existence of competitive equilibrium. *Econometrica*, 49:819–842, 1981.

J. Meade. External economies and diseconomies in a competitive situation. *Economic Journal*, 62:54–67, 1952.

Yu.I. Merslyakov. *Rational Groups (in Russian)*. Nauka, Moscow, 1987.

J.W. Milnor and D.W. Weaver. *Topology from the Differentiable Viewpoint*. Princeton University Press, Princeton, 1997.

K. Mount and S. Reiter. The informational size of message spaces. *Journal of Economic Theory*, 8:161–192, 1974.

S. Nakamoto. Bitcoin: A peer-to-peer electronic cash system, 2008. URL https://bitcoin.org/bitcoin.pdf.

T. Negishi. Welfare economics and the existence of an equilibrium for a competitive economy. *Metroeconomica*, 12:92–97, 1960.

H. Nikaido. *Monopolistic Competition and Effective Demand*. Princeton University Press, Princeton, 1975.

H. Osana. On the informational size of message spaces for resource allocation processes. *Journal of Economic Theory*, 17:66–78, 1978.

J.M. Ostroy. On the existence of Walrasian equilibrium in large-square economies. *Journal of Mathematical Economics*, 13:143–163, 1984.

J.M. Ostroy and R.M. Starr. Money and the decentralization of exchange. *Econometrica*, 42:1093–1113, 1974.

E.A. Pazner and D. Schmeidler. Egalitarian equivalent allocations: A new concept of economic equity. *The Quarterly Journal of Economics*, 92: 671–687, 1978.

A.C. Pigou. *The Economics of Welfare*, 3rd edition. Macmillan, London, 1929.

L. Qi. Conditions for Factor Price Equalization in the Integrated World Economy Model. *Review of International Economics*, 11:899–908, 2003.

R. Radner. Existence of equilibrium of plans, prices, and price expectations in a sequence of markets. *Econometrica*, 40:289–303, 1972.

D. Ricardo. *On the Principles of Political Economy and Taxation*. John Murray, London, 1817.

R.T. Rockafellar. *Convex Analysis*. Princeton University Press, Princeton, 1970.

S.A. Ross. The arbitrage theory of capital asset pricing. *Journal of Economic Theory*, 13:341–360, 1976.

W. Rudin. *Functional Analysis*, 2nd edition. McGraw-Hill, Boston, 1991.

P.A. Samuelson. International trade and the equalisation of factor prices. *The Economic Journal*, 58:163–184, 1948.

A. Sard. The measure of the critical values of differentiable maps. *Bulletin of the American Mathematical Society*, 48(12):883–890, 1942.

H. Scarf. The core of an N person game. *Econometrica*, 35:50–69, 1967.

H.H. Schaefer. *Topological Vector Spaces*. Macmillan, New York, 1966.

S. Schecter. On the structure of the equilibrium manifold. *Journal of Mathematical Economics*, 6:1–5, 1979.

D. Schmeidler. The nucleolus of a characteristic function game. *SIAM Journal of Applied Mathematics*, 17:1163–1170, 1969.

D. Schmeidler. Walrasian analysis via strategic outcome functions. *Econometrica: Journal of the Econometric Society*, pages 1585–1593, 1980.

D. Schmeidler and K. Vind. Fair Net Trades. *Econometrica*, 40:637–642, 1972.

W.J. Shafer. The nontransitive consumer. *Econometrica*, 42:913–919, 1974.

L.S. Shapley. A value for n-person games. In H.W. Kuhn and A.W. Tucker, editors, *Contributions to the Theory of Games*, volume II. Princeton University Press, Princeton, 1953.

H. Sonnenschein. Market excess demand functions. *Econometrica*, 40:549–563, 1972.

E. Sperner. Neuer Beweis für die Invarianz der Dimensionszahl und des Gebietes. *Abh.Math.Sem.Univ.Hamburg*, 6:265–272, 1928.

R.M. Starr. Quasi equilibria in markets with non-convex preferences. *Econometrica*, 37:25–38, 1969.

R.M. Starr. *General Equilibrium Theory: An Introduction*. 2nd edition. Cambridge University Press, Cambridge, Massachusetts, 2011.

D.A. Starrett. Fundamental nonconvexities in the theory of externalities. *Journal of Economic Theory*, 4:180–199, 1972.

M. Tvede. *Overlapping Generations Economies.* Palgrave Macmillan, John Murray, Albemarle Street, London, 2010.

H. Uzawa. Walras' existence theorem and Brouwer's fixed point theorem. *Economic Studies Quarterly*, 8:59–62, 1962.

H R. Varian. A remark on boundary restrictions in the Global Newton method. *Journal of Mathematical Economics*, 4:127–130, 1977.

K. Vind. Edgeworth-allocations in an exchange economy with many traders. *International Economic Review*, 5:165–177, 1964.

J. von Neumann and O. Morgenstern. *Theory of Games and Economic Behavior.* Princeton University Press, Princeton, 1944.

M. Walker. A simple incentive compatible scheme for attaining Lindahl allocations. *Econometrica*, 49:65–71, 1981.

L. Walras. *Eléments d'économie politique pure.* L. Corbaz, Lausanne, 1874.

J. Werner. Equilibrium in economies with incomplete financial markets. *Journal of Economic Theory*, 36:110–119, 1985.

K. Wicksell. *Lectures on Political Economy*, volume 1. Routledge and Kegan Paul, London, 1901.

S. Willard. *General Topology.* Addison-Wesley, Reading, Massachusetts, 1970.

S. Wong and K.K. Yun. The lens condition with two factors. *Review of International Economics*, 11:692–696, 2003.

N.C. Yannelis and W.R. Zame. Equilibrium in Banach lattices without ordered preferences. *Journal of Mathematical Economics*, 15, 1986.

L.A. Zadeh. Fuzzy sets. *Information and Control*, 8:338–353, 1965.

W.R. Zame. Efficiency and the role of default when security markets are incomplete. *The American Economic Review*, 83:1142–1164, 1993.

Z. Zheng, S. Xie, H. Dai, X. Chen, and H. Wang. An Overview of Blockchain Technology: Architecture, Consensus, and Future Trends. In *2017 IEEE International Congress on Big Data (BigData Congress)*, pages 557–564, Honolulu, HI, USA, June 2017. IEEE. ISBN 978-1-5386-1996-4. doi: 10.1109/BigDataCongress.2017.85. URL http://ieeexplore.ieee.org/document/8029379/.

Index

A

admissible (commodity space), 162
admissible preference, 41
aggregate feasibility, 22
allocation, 22
arbitrage, 362
arbitrage opportunity, 365
arbitrage pricing theory (APT), 358
arbitrage-free market, 366
atom, 156
atom (for measure), 150

B

balanced family of coalitions, 117
balanced NTU game, 117
bankruptcy (financial markets), 358
BitCoin, 409
blockchain, 409
bundle (of commodities), 2

C

capital asset pricing model (CAPM), 355
cardinal balancedness, 121
characteristic function (of game), 127
coalition production economy, 120
Coase independence, 285
Coase theorem, 285
Cobb–Douglas function, 31
coefficient of resource utilization, 105

commodities, 1
commodity space, 2
comparative advantage, 74
competitive resource allocation process, 260
consistency of coalitions, 126
consumer, 3
consumption plan, 3
consumption set, 4
continuity (correspondence), 17
continuum of traders, 146
convexity (of consumption set), 4
convexity (of preference correspondence), 6
core of an economy, 116
core of NTU game, 117
cost–benefit analysis, 100
Cournot–Walras equilibrium, 242
critical equilibrium, 205
cybercurrencies, 409
cycles in OLG model, 328

D

dated commodities, 2
decisive (allocation process), 260
decomposable (matric), 64
degree of dissatisfaction, 134
demand, 9

E

Edgeworth box, 23
efficiency (of allocation), 90

429

egalitarian allocation, 88
egalitarian-equivalent allocation, 89
envy-free allocation, 89
equal-treatment allocation, 123
equilbrium with price-setting firms, 293
equilibrium manifold, 199
equilibrium of plans, prices and price expectations, 337
exceptional set (of consumers), 143
excess (of imputation), 133
exchange economy, 22
exchanges (system of), 15
expected number of equilibria, 210
expected utility, 336
external effects, 276

F

factor price equalization (FPE), 77, 216
factor proportion, 75
fair allocation, 89
Farrell measure, 106
feasible, 23
feasible production, 11
financial assets, 377
FPE equilibrium, 220
full appropriation mechanism, 269
fundamental non-convexities, 280
fuzzy cooperative game, 135
fuzzy core of economy, 137
fuzzy sets, 135

G

game, 252
game form, 125, 251
GDMN lemma, 32
graph (of correspondence), 17
Groves–Ledyard mechanism, 289

H

hash function, 410
Heckscher–Ohlin structure theorem, 79

Heckscher–Ohlin–Samuelson model, 73
hierarchic equilibrium, 39
homothetic consumers, 211

I

imperfect competition, 233
improvement (of allocation), 116
indecomposable (matrix), 64
indeterminacy of Walras equilibrium, 325
individual feasibility, 22
infinite-dimensional commodity spaces, 158
input, 10
input efficiency, 106
irreducibility, 37
irreflexivity, 6

K

Kakutani fixed-point theorem, 29
Kakutani space, 170
Koopmans diagram, 87

L

large economy, 142
large-square economy, 171
law of one price, 369
lens condition, 218
lhc, *see* lower hemicontinuity
Lindahl equilibrium, 286
local non-satiation, 9
lower hemicontinuity, 18
Lyapunov's theorem, 154

M

marginal product mechanism, 269
marginal rate of substitution, 75
market agent, 16
market equilibrium, 92
market failure, 275
martingale, 365
Marx, 65
match, 412
measure space of agents, 147

Index

measures of efficiency, 106
mechanism, 253
medium of exchange, 406
minimum-wealth, 36
monopolist, 234
multimap, 17

N

national accounting, 101
natural monopolies, 291
natural projection, 204
net trade, 10
von Neumann model, 67
von Neumann–Morgenstern utility, 336
no-arbitrage condition, 363
no-trade equilibrium, 206
non-atomic, 156
non-satiation, 6
non-wasteful (allocation process), 260
nonatomic (measure), 150
nonatomic measure space of agents, 148
nonce, 410
NTU game, 117
nucleolus allocations, 133

O

objective demand, 239
output, 10
output efficiency, 106
overlapping generations (OLG), 313

P

pairwise trade, 400
Pareto optimal (allocation), 84
Pareto-frontier approach, 34
path-connectedness (of set of equilbria), 207
perceived demand, 233
performance correspondence, 251
Perron–Frobenius theorem, 62
Pigou tax, 281
Plücker coordinates, 383
preference correspondence, 5

preference relation, 5
price normalization, 245
price-dependent asset, 384
privacy preserving (allocation process), 260
production, 10
production function, 13
production set, 11
productive (matrix), 60
profit maximization, 13–14
profit maximization problem, 244
proof-of-work, 411
pseudoequilibirum, 382
public goods, 284
purely competitive sequence, 154

Q

quantity-setting firms, 240
quasi-equilibrium, 27

R

Radner equilibrium, 341
Radon–Nikodym derivative, 366
real assets, 381
real wealth maximization, 247
reduced model, 318
regular equilibrium, 205
replica economies, 122
resource relatedness, 38
Ricardo, 74
risk aversion, 338

S

second-best allocation, 102
self-financing trading strategy, 370
separation of convex sets, 108
separation theorem, 110
sequences of finite economies, 151
set-valued mapping, 17
Shapley value, 128
Shapley–Folkman theorem, 155
social optimum, 98
social utility function, 98
standard (preference correspondence), 6

Index

states of nature, 334
strategic market games, 253
strategies, 125
strong Nash equilibrium, 126
subjective demand, 233
sunspot equilibrium, 348

T

trading mechanism, 413
trading strategy, 370
transformation curve, 11
TU (transferable utility) game, 127
two-fund separation, 358

U

uhc, *see* upper hemicontinuity
uncertainty, 333
uniqueness property (of allocation process), 259
upper comprehensive, 4

upper hemicontinuity, 18
utility function, 6
utility representation of preferences, 7

V

value allocation, 129
vector lattice, 170
vector measure, 149
Vickrey–Clarke–Groves (VCG) mechanism, 268

W

Walras equilibrium, 24
Walras equilibrium (with production), 55
Walras' law, 26
Walrasian mechanism, 270
welfare theorem I, 93
welfare theorem II, 95
well-behaved (economy), 161

CPSIA information can be obtained
at www.ICGtesting.com
Printed in the USA
BVHW041034200620
581572BV00006B/10